Climate Justice, Policy Reform, and the Future of Environmental Law

Enas Mohammad Alqodsi
https://orcid.org/0000-0003-4643-9660
United Arab Emirates University, UAE

Iyad Mohammad Jadalhaq
University of Sharjah, UAE

Mohammed El Hadi El Maknouzi
University of Al Dhaid, UAE

IGI Global
Scientific Publishing
Publishing Tomorrow's Research Today

Vice President of Editorial Melissa Wagner
Director of Acquisitions Mikaela Felty
Director of Book Development Jocelynn Hessler
Production Manager Mike Brehm
Cover Design Jose Rosado

Published in the United States of America by
IGI Global Scientific Publishing
701 East Chocolate Avenue
Hershey, PA, 17033, USA
Tel: 717-533-8845 | Fax: 717-533-7115
Website: https://www.igi-global.com E-mail: cust@igi-global.com

Library of Congress Cataloging-in-Publication Data

Names: Alqodsi, Enas Mohammed, 1978- editor | Jadalhaq, Iyad Mohammad, 1976- editor | El Maknouzi, Mohammed El Hadi, 1977- editor
Title: Climate justice, policy reform, and the future of environmental law / Enas Mohammed Alqodsi, Iyad Mohammad Jadalhaq, Mohammed El Hadi El Maknouzi.
Description: Hershey : IGI Global Scientific Publishing, 2026. | Includes bibliographical references and index. | Summary: "This book aims to explore and critically assess the interface between climate justice and the legal responses that underpin environmental governance at local, national, regional, and international levels"-- Provided by publisher.
Identifiers: LCCN 2025046628 (print) | LCCN 2025046629 (ebook) | ISBN 9798337366357 hardcover | ISBN 9798337366364 paperback | ISBN 9798337366371 ebook
Subjects: LCSH: Climatic changes--Law and legislation | Climate justice | Indigenous peoples | Environmental policy | Environmental law | Environmental protection
Classification: LCC K3585 .C55 2026 (print) | LCC K3585 (ebook) | DDC 340/.115--dc23/eng/20251119
LC record available at https://lccn.loc.gov/2025046628
LC ebook record available at https://lccn.loc.gov/2025046629

British Cataloguing in Publication Data
A Cataloguing in Publication record for this book is available from the British Library.

Table of Contents

Detailed Table of Contents

This chapter discusses climate vulnerability and justice of the law for the Indigenous people, climate migrants, and other disenfranchised people, such as women, children, and disabled individuals. Through the intersectional lens, it looks at the exacerbating effect of confluent social, cultural, and economic inequalities in increasing climate risks and analyzes deficiencies of existing legal systems. Community-based resilience and legal advocacy through case studies are complemented by pools of emerging responses in policy and best practices that help define the pathways to inclusive governance. The recommendations call on the need to enhance human rights-based climate policies, institutionalize protection of climate migrants, integrate intersectionality into climate changes, and improve the involvement of the community to promote just and sustainable climate mitigation measures.

Research shows that the Philippines shows massive indications of extreme climate vulnerability (ND-GAIN Index, 2024; IPCC, 2022). Recent research shows (IMF, 2020) that the climate impact is exacerbated by resource-extracting environmental crimes like unregulated logging, illegal mining, and pillaging of natural resources. These crimes produce a collapse of ecosystems, weaken our natural defenses against the climate, and impact the climate-vulnerable more than the rest of the world. These include the Indigenous, farming, and coastal communities and peoples (Global Witness, 2023). Corporations are still under-implementing climate adaptation actions, even though there is strategic alignment with Environmental, Social, and Governance (ESG) frameworks. These climate actions still neglect the systemic causes of vulnerability, stemming from environmental degradation and supply chain exploitation (UNEP, 2021). Effective adaptation research shows the need for policy on the inequities causing precarity.

This chapter explores the changing role of the environmental judiciary in resolving environmental disputes. It reviews foundational concepts, various institutional frameworks, jurisdictional setups, and enforcement tools, emphasizing judicial remedies, preventative strategies, and restorative methods. The analysis covers challenges like standing, evidence complexities, and compliance, while comparing different jurisdictions' experiences. Overall, the chapter shows how courts strengthen environmental governance, ensure accountability, and promote sustainable development, underscoring the judiciary's vital role in safeguarding ecosystems and public rights to a healthy environment.

Many natural and human factors impact the environment. Since the right to live in a healthy environment is one of the fundamental human rights, Achieving this sustainability requires the involvement of the legislative, judicial and executive authorities. judiciary, is considered one of the most important tools for enforcing legal provisions. Therefore, some countries have established specialized environmental courts, often referred to as "green courts." Others have opted to create specialized circles for environmental disputes within existing civil or criminal courts, some countries have chosen to train specific judges to specialize in such disputes. Beyond the judiciary, or alongside it. In all cases, This highlights the need to develop an advanced legal and judicial framework as an effective tool for environmental protection. Therefore, this article seeks to compare the rules of environmental litigation across legal systems in different geographic regions to identify the best proposals for achieving environmental sustainability.

This chapter analyses a comparative examination of legal measures in India and the Maldives, addressing the link between human rights and climate change. In the face of climate catastrophes like global warming, cloudbursts, and dry winters, it explores judicial pronouncements and legal systems that recognise climate rights as fundamental human rights in these vulnerable nations. It studies how both judiciaries interpret and enforce constitutional and international laws to shield communities from climate impacts, highlighting India's incorporation of human rights into climate law and urging the Maldives to enhance its legal framework through specialised climate courts and global cooperation. The chapter also contributes to academic discourse and offers policy suggestions to strengthen judicial capacity in addressing complex climate-related challenges.

Climate change poses a severe legal and ethical challenge, disproportionately impacting Indigenous peoples who contribute least to global emissions. This chapter offers a comparative analysis of Indigenous rights and climate justice under international human rights and environmental law. It scrutinizes key instruments like UNDRIP, ILO Convention 169, the UNFCCC, and the Paris Agreement, alongside pivotal cases such as Saramaka v. Suriname and Teitiota v. New Zealand. The findings highlight a critical gap: human rights law protects Indigenous land and culture but lacks climate-specific mandates, while environmental law promotes climate action without recognizing Indigenous peoples as binding right-holders. To reconcile these regimes, the chapter proposes integrating Indigenous rights and judicial accountability into climate governance and formally institutionalizing Indigenous ecological knowledge for a just and effective global response.

The concept of intergenerational justice, and the right of future generations to a clean, healthy, and sustainable environment, involves the duty of the current generation to pass on the environment to future generations, based on the principles of comparable options, quality, and access. If this right is legally recognized, it must be protected through various legal means, especially litigation. This chapter confirms that intergenerational justice is a legal matter, as evidenced by its study of a number of judicial precedents in several jurisdictions, constitutional texts, international treaties, and declarations, in order to understand the challenges facing litigation in this field, such as challenges related to legal standing, justiciability, and causation. This chapter argues for expanding the concept of legal standing, adopting a precautionary and cumulative approach to causation, and for the explicit legal recognition of intergenerational justice in national laws.

International climate law has experienced rapid normative consolidation, as evidenced by the recognition of binding obligations in recent international jurisprudence for prevention, due diligence, cooperation, and reparation. Despite this progress, institutional mechanisms for systematic enforcement remain inadequate. Existing courts are limited by narrow jurisdiction, restricted standing, and procedural frameworks that are not well-suited for complex, science-intensive climate disputes. This chapter contends that an International Court for the Environment represents the institutional realization of an already mature normative framework, rather than a mere political innovation. It presents a doctrinal and institutional blueprint for climate adjudication, addressing jurisdiction, procedure, evidentiary standards, remedies, compliance mechanisms, and participatory guarantees. By integrating scientific expertise, expanding access to justice, and facilitating structural remedies, such a court could bridge the gap between normative ambition and practical accountability.

Policy reform and legal innovation are interdependent for a climate-resilient future. ESG frameworks align corporate duties with societal goals, but require robust environmental law to avoid greenwashing. A just climate economy demands accessible, participatory rulemaking, transparent metrics, and enforceable standards that scale from local to levels. Innovations include performance-based regulations, climate risk disclosure, and dynamic permitting that update with science. Legal tools such as citizen lawsuits, right-to-information, and climate justice considerations channel accountability toward vulnerable communities. Integrating ESG with environmental justice requires harmonizing fiduciary duties with public interest, ensuring minority and frontline voices input into policy design. Cross-border cooperation, standardized reporting, and adaptive governance enable timely responses to climate shocks. State capacity, funding for transition, and clear decarbonization pathways reduce uncertainty for investors and workers alike, fostering a just, regenerative economy.

Climate change increasingly affects the built environment, placing greater pressure on cities and communities, particularly in regions experiencing rapid urbanization and environmental stress. This chapter examines the role of sustainable architecture in addressing climate challenges through design strategies that promote environmental responsibility, resilience, and social equity. Focusing on architectural practice in the Egyptian context, the chapter explores climate-responsive design approaches such as passive cooling, energy-efficient building systems, sustainable materials, and context-sensitive urban design. It highlights how sustainable architectural solutions can reduce environmental impact, improve thermal comfort, and enhance the quality of life for diverse communities. The chapter emphasizes that architecture plays a crucial role in shaping resilient and sustainable built environments capable of responding to climate change while supporting more balanced and inclusive urban development.

In India, environmental crime is challenging climate justice, often impacting the marginalised communities and ecologically sensitive classes that depend on the environment for their lives and livelihoods. This chapter focuses on the increasingly common application of remote sensing, satellite imagery, artificial intelligence, and digital forensics in the detection, investigation, and prosecution of environmental crimes. By contextualising technology within a climate justice and rights-based framework, this study examines the nature of data generated through technology and its compatibility with standards of evidence in India, specifically under the Bharatiya Sakshya Adhiniyam (BSA), 2023, and jurisprudence through the National Green Tribunal (NGT). Using doctrinal analysis, along with a review of Indian case studies and comparative insights from the EU and the US, the chapter identifies evidentiary uncertainties, algorithmic opacity, privacy risks, and institutional gaps.

This chapter examines how international climate commitments are incorporated into domestic law through constitutional environmental rights and how climate justice reshapes state accountability. It addresses the gap between non-binding agreements, such as the Paris Agreement, and the enforcement of constitutional provisions in rights-based climate litigation. Drawing on a comparative analysis of cases from Germany, Colombia, South Africa, and the Netherlands, it shows that courts increasingly treat environmental degradation as a violation of fundamental rights, including life, health, and intergenerational equity.The chapter argues that constitutional environmental rights are becoming enforceable legal norms and that judges rely on international principles to require stronger climate action. It also

Preface

INTRODUCTION

Climate change stands as one of the greatest challenges of our time—an urgent and complex global phenomenon that intersects deeply with human rights, governance, economic justice, and environmental sustainability. Its impacts transcend borders, disproportionately affecting vulnerable populations and exposing deep-rooted inequalities in legal and institutional frameworks. In this context, climate justice has emerged as a vital paradigm that calls for equitable solutions rooted in fairness, inclusivity, and accountability.

As editors, we present this volume, *Climate Justice, Policy Reform, and the Future of Environmental Law*, with the conviction that law must play a transformative role in addressing the climate crisis. This book aims to critically explore how legal systems—both domestic and international—can respond to the mounting challenges posed by climate change while ensuring justice and protecting the rights of the most affected communities. By interrogating the intersections between environmental law, social justice, and public policy, this volume seeks to provide a rigorous scholarly foundation for legal reforms and innovations that support sustainable, fair, and enforceable climate action.

The primary aim of this edited collection is to explore and critically assess the interface between climate justice and the legal responses that underpin environmental governance at local, national, regional, and international levels. It bridges theoretical, doctrinal, and empirical perspectives on legal frameworks, highlighting how law and legal reform can contribute to equitable and sustainable climate outcomes. In doing so, the book fills a significant gap in the academic literature by advancing multidisciplinary legal approaches to climate action, resilience, and sustainability. The contributors examine key areas including public international law, comparative environmental law, constitutional environmental rights, procedural and substantive environmental justice, and the evolving role of courts in shaping environmental policy and governance.

This volume is intended for a diverse and engaged readership, including faculty members at universities and higher education institutions, particularly in law, political science, environmental studies, and public policy; legal scholars and researchers specializing in environmental, constitutional, international, and administrative law; policy analysts and advisors working on environmental reform; judges, legal practitioners, and regulatory authorities; government agencies and legislators; international organizations and NGOs; postgraduate students in law, environmental governance, and human rights; as well as climate activists and advocacy networks.

We are deeply grateful to the distinguished contributors whose expertise and insights have enriched this collection. Their collective scholarship not only advances academic understanding but also offers practical pathways toward more just and effective climate law and policy. It is our hope that this book will serve as a valuable resource for scholars, practitioners, and policymakers committed to building a more equitable and sustainable future in the face of the climate emergency.

ORGANIZATION OF THE BOOK

Chapter 1 introduces the United Arab Emirates as an emerging leader in climate mitigation through its strategic pivot from oil dependency toward green hydrogen. We highlight how the chapter systematically evaluates the country's legal and policy frameworks, revealing both notable progress and critical gaps. While the UAE demonstrates strong alignment with international climate goals and a commitment to resilience, the discussion underscores the need for improved enforcement, stakeholder inclusion, and regulatory coherence to ensure that green hydrogen development advances a genuinely sustainable and equitable low-carbon future.

Chapter 2 situates climate change within the broader framework of international environmental law, emphasizing its deeply embedded ethical, legal, and social dimensions. We present this chapter as a conceptual cornerstone of the volume, advancing climate justice as an essential lens for rethinking responsibility, equity, and accountability in global governance. The chapter challenges structural inequalities in the international system and calls for transformative legal reform grounded in fairness and legitimacy.

Chapter 3 turns to the global textile industry, exposing its significant yet often overlooked role in environmental degradation across borders. We underscore how the chapter bridges constitutional law and international legal frameworks to examine judicial interventions in advancing environmental justice. By focusing on the Global South, it reveals disparities in both environmental harm and regulatory burdens, while proposing pathways for more equitable and sustainable governance of transnational industries.

Chapter 4 expands the discussion to natural resource governance, where climate change intensifies existing inequalities in access and control. We highlight the chapter's integration of climate justice principles—distributive fairness, procedural equity, and human rights—into legal frameworks governing land, water, and minerals. Its critical examination of historical injustices and forward-looking legal strategies offers a compelling roadmap for achieving sustainable and inclusive resource management.

Chapter 5 centers on climate vulnerability among marginalized populations, including Indigenous peoples, climate migrants, and other at-risk groups. We emphasize the chapter's intersectional approach, which reveals how overlapping inequalities heighten climate risks and expose gaps in existing legal protections. Through case studies and policy analysis, the chapter advances practical recommendations for inclusive governance and rights-based climate action.

Chapter 6 examines the Philippine context, where environmental crimes such as illegal mining and deforestation exacerbate climate vulnerability. We present this chapter as a critical inquiry into the limitations of corporate climate adaptation strategies, particularly in addressing systemic inequalities. It calls for stronger regulatory frameworks and accountability mechanisms to align corporate practices with human rights and environmental sustainability.

Chapter 7 explores the evolving role of the judiciary in environmental governance. We emphasize how courts serve not only as arbiters of disputes but also as key actors in promoting accountability, preventive action, and ecological restoration. By comparing institutional models and addressing procedural challenges, the chapter underscores the judiciary's central role in advancing sustainable development.

Chapter 8 complements this discussion by focusing specifically on environmental litigation as a tool for achieving sustainability. We highlight its comparative analysis of legal systems and judicial mechanisms, including specialized environmental courts. The chapter makes a strong case for strengthening legal infrastructures and judicial capacity as essential components of effective environmental protection.

Chapter 9 offers a comparative perspective on the intersection of human rights and climate change in India and the Maldives. We underscore how judicial innovation in these jurisdictions is shaping climate jurisprudence, particularly through the recognition of environmental rights as fundamental human rights. The chapter also identifies opportunities for strengthening legal frameworks and judicial capacity in vulnerable regions.

Chapter 10 provides a critical examination of Indigenous rights within international climate and environmental law. We highlight its identification of a key disconnect between human rights protections and climate governance mechanisms. By advocating for the integration of Indigenous knowledge and stronger legal recognition, the chapter advances a more inclusive and effective approach to global climate justice.

Chapter 11 addresses the emerging field of intergenerational justice, focusing on the legal challenges of representing future generations in climate litigation. We emphasize the chapter's analysis of doctrinal barriers such as standing, causation, and justiciability, alongside its proposals for expanding legal recognition and adopting precautionary approaches. It positions intergenerational equity as a vital component of climate law.

Chapter 12 presents a bold institutional proposal for an international court dedicated to climate issues. We frame this chapter as both visionary and pragmatic, offering a detailed blueprint for addressing enforcement gaps in international climate law. By integrating scientific expertise and expanding access to justice, the proposed court represents a significant step toward bridging the divide between legal norms and practical accountability.

Chapter 13 explores the intersection of environmental law, policy reform, and ESG frameworks in shaping a just climate economy. We highlight the chapter's focus on legal innovation, transparency, and participatory governance as tools to prevent greenwashing and ensure accountability. Its emphasis on harmonizing economic and environmental priorities offers a comprehensive vision for sustainable development.

Chapter 14 shifts attention to the built environment, examining how sustainable architecture can contribute to climate resilience and social equity. We underscore the chapter's focus on design strategies—such as passive cooling and energy efficiency—within the context of rapid urbanization. It demonstrates how architecture can play a transformative role in shaping inclusive and environmentally responsible communities.

Chapter 15 introduces technological innovation into the realm of environmental enforcement, focusing on the use of remote sensing, artificial intelligence, and digital forensics in India. We highlight its critical analysis of evidentiary challenges, privacy concerns, and institutional gaps, while emphasizing the potential of technology to enhance accountability and support climate justice.

Chapter 16 examines transboundary climate litigation and the growing role of constitutional law in enforcing international climate commitments. We emphasize the chapter's comparative analysis of judicial practices across multiple jurisdictions, illustrating how courts are increasingly recognizing environmental harm as a violation of fundamental rights. It highlights the emergence of new forms of accountability and calls for stronger constitutional integration of climate justice principles.

Chapter 17 concludes the volume by exploring innovative climate finance mechanisms through the Tropical Forest Forever Facility. We present this chapter as a forward-looking contribution that links legal design, financial governance, and environmental justice. By emphasizing enforceability, institutional coherence, and

Indigenous participation, it offers a compelling model for equitable and sustainable climate finance.

CONCLUSION

In bringing this volume to a close, we are reminded that climate change is not merely an environmental challenge, but a profound legal, ethical, and institutional test of our time. The chapters assembled in this collection collectively demonstrate that climate justice is no longer an abstract aspiration; it is an emerging legal paradigm reshaping how we understand responsibility, rights, and governance in a rapidly changing world. Across jurisdictions, sectors, and disciplines, the contributors illuminate both the progress achieved and the significant gaps that remain in aligning law with the urgency and complexity of the climate crisis.

As editors, we observe a number of unifying themes that run through this work. First, there is a clear shift toward rights-based approaches, where courts, policymakers, and communities increasingly frame climate harm as a violation of fundamental human rights. Second, the growing prominence of equity—whether intergenerational, transnational, or intersectional—signals a necessary rebalancing of legal frameworks to address historical injustices and present-day vulnerabilities. Third, the role of institutions, from domestic courts to proposed international tribunals, emerges as pivotal in translating normative commitments into enforceable obligations.

At the same time, this volume underscores that legal innovation must be accompanied by institutional capacity, political will, and inclusive participation. Whether in the governance of natural resources, the regulation of global industries, or the deployment of emerging technologies, effective climate action depends on systems that are transparent, accountable, and responsive to those most affected. The recurring emphasis on Indigenous knowledge, community engagement, and participatory governance reflects an important recognition: durable solutions cannot be imposed from above but must be co-created with those on the frontlines of climate change.

We also note the expanding interface between law and other domains—technology, finance, architecture, and corporate governance—highlighting the increasingly interdisciplinary nature of climate solutions. From digital forensics in environmental enforcement to ESG-driven economic transformation and climate-responsive design in the built environment, the law is both shaping and being shaped by broader societal shifts. This interplay presents opportunities for innovation, but also demands careful calibration to ensure that efficiency does not come at the expense of justice.

Looking ahead, the contributions in this volume collectively call for a more coherent, integrated, and forward-looking legal order—one capable of bridging the gap between global commitments and local realities. Strengthening enforcement mechanisms, enhancing access to justice, and embedding climate considerations into constitutional and institutional frameworks are no longer optional; they are essential to securing a sustainable future.

Ultimately, this book affirms that the pursuit of climate justice is inseparable from the pursuit of human dignity, ecological integrity, and intergenerational equity. It is our hope that this volume will serve not only as a reference for scholars and practitioners, but also as a catalyst for continued dialogue, critical reflection, and meaningful reform. The path forward is complex, but as the chapters herein demonstrate, the law—when thoughtfully designed and courageously applied—remains one of our most powerful tools for shaping a just and resilient climate future.

Enas Mohammad Alqodsi
United Arab Emirates University, UAE

Iyad Mohammad Jadalhaq
University of Sharjah, UAE

Mohammed El Hadi El Maknouzi
University of Al Dhaid, UAE

Acknowledgment

The editors would like to express their profound gratitude to the United Arab Emirates University for its unwavering and generous support, which has been fundamental to the successful development of this volume. The University's strong commitment to research excellence, innovation, and international academic collaboration has provided an enabling environment that significantly contributed to the realization of this work.

We also extend our sincere appreciation to our respective institutions for fostering a dynamic and intellectually stimulating academic atmosphere that encourages interdisciplinary scholarship and critical engagement with contemporary global challenges.

The editors are particularly grateful to IGI Global Scientific Publishing for offering us the valuable opportunity to contribute to this publication. We deeply appreciate their continued guidance, professionalism, and commitment to high-quality scholarly publishing throughout all stages of the book development process.

Finally, we would like to acknowledge and thank all contributing authors, reviewers, and academic colleagues whose expertise, dedication, and insightful contributions have greatly enriched this volume and enhanced its scholarly impact.

Chapter 1
From Oil to Green Hydrogen:
The UAE's Legal Strategies for Climate Mitigation and Adaptation

Enas Mohammed Alqodsi
http://orcid.org/0000-0003-4643-9660
United Arab Emirates University, UAE

Ibrahim Al Shawabkeh
http://orcid.org/0000-0001-6548-258X
United Arab Emirates University, UAE

Omar Farahat
http://orcid.org/0000-0002-5862-1809
Faculty of Law, McGill University, Canada

Habiba Al Shamisi
http://orcid.org/0009-0006-9861-2161
United Arab Emirates University, UAE

Iyad Mohammad Jadalhaq
University of Sharjah, UAE

Mohammed ElMaknouzi
http://orcid.org/0000-0003-3396-8985
University of Al Dhaid, UAE

ABSTRACT

Amidst the global urgency to tackle climate change, the United Arab Emirates (UAE) is emerging as a regional pioneer by integrating innovative solutions such as green hydrogen into its energy strategy. This chapter examines the legal frameworks supporting climate change mitigation and adaptation in the UAE, analyzing statutes, regulations, and policy documents to evaluate strengths, weaknesses, and alignment with international best practices. Findings reveal a robust system promoting mitigation, adaptation, and resilience, yet gaps remain in enforcement, stakeholder engagement, and harmonization with evolving global agreements. Addressing these gaps is essential to ensure responsible green hydrogen development, protect consumer interests, and advance a low-carbon, climate-resilient future.

DOI: 10.4018/979-8-3373-6635-7.ch001

1. INTRODUCTION

Climate change is one of the most pressing and complex global challenges of the 21st century, affecting ecosystems, economies, and communities across the globe. The scientific consensus on the need for immediate and coordinated action is overwhelming, as rising temperatures, extreme weather events, and sea-level rise continue to disrupt the natural balance. Nations worldwide are grappling with the challenge of reducing carbon emissions while maintaining economic growth and securing energy supplies(Alqodsi, 2024). Among the countries taking significant strides in this domain is the United Arab Emirates (UAE), a nation traditionally known for its oil wealth but now recognized as a key player in the global sustainability movement.

The UAE is uniquely positioned within this context. As one of the world's major energy producers, the country faces the dual challenge of maintaining its economic reliance on hydrocarbons while simultaneously transitioning towards a low-carbon, sustainable future. This balancing act is reflected in the nation's proactive environmental policies and its commitment to international climate agreements, such as the Paris Agreement(Morganti et al., 2024). Over the past decade, the UAE has embarked on an ambitious journey to diversify its energy sources(Nobanee, El Hadi El Maknouzi, Sadok, & Alodat, 2024), investing heavily in renewable energy projects such as solar and wind, as well as in emerging technologies like green hydrogen.

Central to this transition is the development of a comprehensive legal framework designed to regulate and guide the country's climate change mitigation and adaptation efforts. The UAE has taken a leadership role in the region(Alqodsi, Jadalhaq, & El Hadi El Maknouzi, 2023), not only through technological innovation but also through forward-thinking legislation aimed at fostering a sustainable economy. These laws and policies are vital tools for driving the country's ambitious environmental goals, which include achieving net-zero emissions by 2050 and positioning itself as a global leader in clean energy production.

However, the path toward sustainability is not without its challenges(Dahiyat, 2025). The rapid pace of technological advancements, coupled with evolving global climate policies, demands a legal framework that is both flexible and robust. The effectiveness of the UAE's legal approach to climate change depends on its ability to address these challenges while ensuring that economic growth and environmental protection go hand in hand(Diaz et al., 2024). Key areas of concern include the need for stronger enforcement mechanisms, the integration of international best practices, and the development of regulations that encourage innovation while protecting the public interest.

This study seeks to explore and evaluate the UAE's legal framework in the context of its climate change initiatives, specifically focusing on its efforts to reduce carbon emissions and build resilience to climate impacts. The research delves into the various laws and regulations that govern the UAE's environmental strategy, examining their strengths, limitations, and alignment with global climate goals. Through a detailed analysis of legal instruments, policy documents, and regulatory frameworks, this research aims to provide a comprehensive understanding of the UAE's approach to climate governance.

Furthermore, the study takes into account the broader geopolitical and economic landscape in which the UAE operates. As a key player in both regional and global energy markets, the UAE's climate policies have significant implications not only for its domestic agenda but also for international efforts to combat climate change(Dewan & Godina, 2024). The country's commitment to reducing its carbon footprint, while simultaneously maintaining its position as a leader in energy production, presents a unique case study for how nations can navigate the complexities of climate governance in an increasingly interconnected world(Morganti et al., 2024).

Ultimately, this research endeavors to assess the effectiveness of the UAE's legal framework in promoting sustainable development, environmental stewardship, and climate resilience. By identifying potential gaps and areas for improvement, the study offers actionable recommendations for enhancing the country's legal instruments and aligning them with evolving international climate standards. In doing so, it contributes to the ongoing discourse on how nations can develop legal strategies that are not only responsive to the challenges of climate change but also capable of driving transformative, long-term solutions.

2. LEGAL FRAMEWORK FOR CLIMATE CHANGE MITIGATION

The United Arab Emirates (UAE), a nation known for its rapid economic development and thriving energy industry(Al-Sarihi & Mason, 2020), has embarked on a significant journey towards sustainable practices in the face of global climate challenges (Al Naqbi, Tsai, & Mezher, 2019) As the impacts of climate change become more pronounced, the UAE recognizes the importance of addressing both mitigation and adaptation strategies to ensure a low-carbon and climate-resilient future (Al-Sarihi, 2018) This introduction outlines key areas of focus within the UAE's comprehensive approach:

2.1 Mitigation Initiatives

With a commitment to reducing its carbon footprint, the UAE has established a robust legal framework designed to mitigate the effects of climate change. Through emission reduction initiatives, the nation seeks to play its part in global efforts to limit temperature rise. This includes specific targets for reducing greenhouse gas emissions, as well as policies to promote the adoption of renewable energy sources(-Buchner et al., 2014).

The United Arab Emirates (UAE), cognizant of the pressing need to address climate change, has put in place a comprehensive suite of legislations and policies aimed at mitigating its carbon footprint. These measures were introduced to align with international climate goals, such as those outlined in the Paris Agreement. The UAE's legislative journey in climate mitigation commenced with significant legal developments, beginning around the mid-2000s (Luomi, 2014).

2.2 One Pivotal Legislation is the "Renewable Energy Law," Enacted in 2006

This law laid the foundation for the promotion of renewable energy sources within the country(Mezher, Dawelbait, & Abbas, 2012). It mandated the establishment of the "Masdar Initiative", a pioneering project aiming to develop sustainable and renewable energy technologies (Siddiqui, Shohan, & Afzal). The law also paved the way for the creation of the "Abu Dhabi Future Energy Company (Masdar)(Reiche, 2010)"(City), which focuses on driving the advancement and implementation of renewable energy projects and technologies, this pioneering legislation marked a crucial milestone in the country's commitment to promoting renewable energy and reducing its carbon footprint(Tsai et al., 2017).

Foundation for Promoting Renewable Energy Sources:

The Renewable Energy Law served as a fundamental framework for the UAE's efforts to diversify its energy mix by integrating renewable energy sources into its overall energy strategy. By focusing on renewable energy, the law aimed to decrease the nation's reliance on traditional fossil fuels and contribute to global efforts to combat climate change(Al-Amir & Abu-Hijleh, 2013).

Case Study: Masdar—A Qualitative Analysis of the UAE's Green Hydrogen Strategy

Masdar, established in 2008, exemplifies the UAE's commitment to renewable energy and its pioneering role in green hydrogen technology(Pandita, Pradeep, Anjukandan, & Vignesh, 2024). This case study examines Masdar's contributions, focusing on its qualitative impact on the UAE's low-carbon transition(Tahir, Tzes, & Yousaf, 2024).

Masdar's Green Hydrogen Initiatives: Masdar's strategic focus on green hydrogen is evident in its numerous projects globally(Kováčová, 2024), encompassing various stages of the hydrogen value chain, from production and storage to transportation and end-use applications(Dsilva, Shaikh, & Usmani, 2024).

Qualitative Impact:

Renewable Energy Capacity: Masdar's portfolio of operational renewable energy projects (solar and wind) significantly contributes to the UAE's clean energy targets. These projects demonstrably reduce carbon emissions(Tahir et al., 2024).

Green Hydrogen Production: Masdar actively works toward becoming a leading green hydrogen producer. Its ongoing projects are steadily increasing green hydrogen production, with plans for significant future expansion. This green hydrogen is intended for use in various sectors, including heavy industry and transportation, further reducing reliance on carbon-intensive energy sources(Abo-Khalil, 2024).

Economic Impact: Masdar's investments have spurred economic growth, creating jobs and stimulating related industries in the UAE(Koch, 2024). Its activities have attracted considerable foreign direct investment (FDI), further boosting the national economy(Tahir et al., 2024).

Technological Advancement: Masdar's commitment to research and development (R&D) in advanced green hydrogen technologies has driven innovation and the development of improved solutions(Issac, 2024). These advancements contribute to enhanced efficiency and cost-effectiveness within the green hydrogen sector(El-Refae & Belarbi, 2024).

Challenges and Future Outlook: While Masdar’s initiatives represent considerable progress, challenges persist. These include the need for significant infrastructure investments, securing sustainable water resources for hydrogen production, and developing robust regulatory frameworks to support this nascent industry(Jaber, 2025). Continued R&D, government support, and collaboration between public and private sectors are critical for overcoming these obstacles and realizing the full potential of green hydrogen in the UAE(Ibrahim & Ottman, 2025).

Conclusion: Masdar's success underscores the UAE's dedication to a sustainable energy future, highlighting green hydrogen's potential. This case study emphasizes Masdar's significant qualitative impact while acknowledging remaining challenges. The analysis demonstrates the substantial contributions of Masdar to the UAE's clean energy transition, providing valuable insights into the multifaceted dimensions of this strategic initiative.

Through the Renewable Energy Law, the UAE not only laid the groundwork for the development of renewable energy projects but also set the stage for fostering research, innovation, and collaboration in the clean energy sector. The legislation was a crucial stepping stone towards the nation's ambitious goals of achieving a significant share of renewable energy in its energy mix, as outlined in subsequent strategic plans such as the Energy Strategy 2050 and the UAE Energy Strategy for 2030.

In essence, the Renewable Energy Law of 2006 played a pivotal role in initiating the UAE's journey towards a more sustainable energy future, fostering technological innovation, and establishing the UAE as a player on the global stage in the realm of renewable energy and climate action.

2.3 In 2015, the UAE Further Demonstrated its Dedication to Mitigation by Submitting its Intended Nationally Determined Contribution (INDC) under the Paris Agreement

The Paris Agreement: Overview and UAE's Involvement (Herold, Siemons, & Herrmann, 2018)

The Paris Agreement (Pauw, Castro, Pickering, & Bhasin, 2020)is a landmark international treaty within the United Nations Framework Convention on Climate Change (UNFCCC) that was adopted in December 2015 and entered into force on November 4, 2016 (King & Van Den Bergh, 2019) The agreement represents a global effort to combat climate change by limiting global warming to well below 2 degrees Celsius above pre-industrial levels, while striving to limit the increase to 1.5 degrees Celsius. It aims to achieve this through a combination of mitigation (reducing greenhouse gas emissions) and adaptation (building resilience to the impacts of climate change) measures.

Key Features of the Paris Agreement:

Nationally Determined Contributions (NDCs): Each participating country, known as a Party, submits its Intended Nationally Determined Contribution (INDC) outlining its climate action plans, targets, and strategies. These contributions represent the country's commitment to reducing its greenhouse gas emissions and enhancing its climate resilience (Michaelowa, Shishlov, & Brescia, 2019).

Global Stocktake:

The Paris Agreement includes a mechanism for a global stocktake every five years, where Parties collectively assess progress toward meeting the agreement's goals and consider the need for increased ambition in their climate actions (Estrada & Botzen, 2021).

Transparency and Accountability: The agreement emphasizes transparency in reporting emissions and progress, facilitating the comparison of efforts between countries (Oberthür & Groen, 2017). Parties are also accountable for their contributions and are encouraged to enhance their efforts over time (Dimitrov, 2016).

Adaptation and Finance: The agreement recognizes the importance of adaptation efforts to address the impacts of climate change (Bodansky, 2016). Developed countries are also committed to providing financial resources to support developing countries in their climate actions and adaptation efforts (Klein et al., 2008).

UAE's Involvement and Intended Nationally Determined Contribution (INDC):

In line with its commitment to address climate change, the United Arab Emirates (UAE) joined the Paris Agreement and submitted its Intended Nationally Determined

Contribution (INDC) in 2015(Rahman & Miah, 2016). The UAE's INDC outlines its specific climate action plans and contributions to the global effort to limit global warming (Bran, 2016).

Key points from the UAE's INDC include:

The UAE committed to a 50% reduction in its carbon emissions by 2030 compared to business-as-usual scenarios (Gandhi, Apostoleris, & Sgouridis, 2022). This ambitious target reflects the UAE's dedication to a sustainable low-carbon development pathway (Dargin, 2021).

To achieve this target, the UAE planned to increase the share of clean energy in its energy mix, specifically aiming for 24% of its energy to be generated from clean sources by 2021(Elrahmani, Hannun, Eljack, & Kazi, 2021).

The INDC also emphasized the importance of diversifying the UAE's economy away from fossil fuels and promoting sustainable development while ensuring energy security (El Anshasy & Khalid, 2023).

The UAE highlighted its commitment to investing in renewable energy projects, improving energy efficiency, and implementing carbon capture, utilization, and storage (CCUS) technologies.

By submitting its INDC, the UAE showcased its commitment to global climate efforts and demonstrated its role as a responsible participant in addressing climate change. The UAE's climate actions, as outlined in its INDC, align with its broader national strategies for sustainable development and the transition to a more resilient and low-carbon future.

In conclusion, the Paris Agreement is a pivotal international accord that shapes global climate action. The UAE's participation in the agreement and submission of its INDC underscore its dedication to mitigating climate change and contributing to the global effort to limit global warming. Through its INDC commitments, the UAE aims to promote sustainable development and transition towards cleaner and more sustainable energy sources.

2.4 UAE's Energy Strategy 2050: Advancing Towards a Sustainable Energy Future

The Energy Strategy 2050 is a visionary and ambitious plan unveiled by the United Arab Emirates (UAE) in 2017(Matz, 2018). It outlines the nation's comprehensive roadmap for its energy sector, with a primary focus on transitioning to a sustainable and diversified energy mix, reducing carbon emissions, and ensuring energy security for future generations (Almheiri, 2015). This long-term strategy reflects the UAE's commitment to addressing climate change, promoting economic growth, and securing its energy needs.

Key Goals of the Energy Strategy 2050:

Clean Energy Contribution: A central goal of the Energy Strategy 2050 is to increase the share of clean energy in the UAE's total energy mix to 50% by 2050(Zaiter, Ramadan, Bouabid, El-Fadel, & Mezher, 2023). This is a significant shift from conventional fossil fuel-based energy sources to cleaner and more sustainable alternatives (Lim, Foo, & Tan, 2018).

Diversification: The strategy emphasizes the importance of diversifying the energy mix to reduce dependence on fossil fuels (Alsalman, Ahmed, & Saboor), enhance energy security, and safeguard against fluctuations in global energy markets.

Carbon Emission Reduction: By embracing cleaner energy sources, the UAE aims to substantially reduce its carbon footprint and contribute to global efforts to mitigate climate change. The reduction in greenhouse gas emissions aligns with international climate commitments (Sbia, Shahbaz, & Hamdi, 2014).

Energy Efficiency: The strategy also underscores the importance of improving energy efficiency across sectors (Xie, Sheng, & Guo, 2015). This involves implementing measures to optimize energy consumption and reduce waste, leading to overall sustainability gains.

Research and Innovation: The Energy Strategy 2050 recognizes the significance of research, innovation, and technology development in achieving its goals (Chaube, Chapman, Shigetomi, Huff, & Stubbins, 2020). It promotes investment in advanced energy technologies, innovative solutions, and knowledge-sharing to drive progress.

Elements of the Energy Strategy 2050: Renewable Energy Expansion: To achieve the 50% clean energy target, the UAE aims to significantly expand its renewable energy capacity. This includes the development of large-scale solar and wind power projects, harnessing the abundant sunlight and wind resources available in the region.

Nuclear Energy: The strategy acknowledges the importance of nuclear energy as a low-carbon and reliable energy source (Chenic et al., 2022). The UAE plans to deploy nuclear power plants to contribute to its clean energy goals while ensuring energy security.

Carbon Capture and Utilization: The Energy Strategy 2050 recognizes the potential of carbon capture, utilization, and storage (CCUS) technologies in mitigating emissions from industrial processes and power generation (Santos, Sgouridis, & Alhajaj, 2021).

Energy Efficiency Measures: The strategy outlines plans to enhance energy efficiency across sectors (Dubey & Krarti, 2017), including buildings, transportation, and industries. This involves implementing energy-efficient technologies, regulations, and practices to reduce energy wastage (Ahmed & Alipour, 2019).

Demand-Side Management: The strategy also emphasizes the importance of demand-side management, encouraging consumers and industries to use energy more efficiently and responsibly (Afshari, Nikolopoulou, & Martin, 2014).

Significance and Impact: The Energy Strategy 2050 represents a comprehensive approach to transforming the UAE's energy landscape and aligning it with global sustainability goals. By increasing the contribution of clean energy sources and diversifying its energy mix, the UAE aims to mitigate climate change, reduce environmental impact, and secure its energy future. The strategy's emphasis on innovation, efficiency, and technology underscores the UAE's commitment to staying at the forefront of global energy developments while contributing positively to the environment and society.

In conclusion, the Energy Strategy 2050 is a forward-looking plan that places the UAE on a trajectory toward a more sustainable, low-carbon, and resilient energy future. By embracing renewables, nuclear energy, and energy efficiency, the UAE aspires to lead by example in the global transition to cleaner and more sustainable energy systems.

2.5 Renewable Energy Promotion in the UAE: Advancing Sustainability through Solar and Wind Power

The United Arab Emirates (UAE) has demonstrated a strong commitment to sustainable development and combating climate change through the promotion of renewable energy sources, particularly solar and wind power. This drive for sustainability is not only a response to global environmental concerns but also aligns with the UAE's strategic goals of diversifying its energy mix, reducing carbon emissions, and securing its energy future.

Key Objectives of Renewable Energy Promotion:

1: Reducing Carbon Footprint: By transitioning from fossil fuel-based energy sources to renewable energy, the UAE aims to substantially reduce its carbon footprint and mitigate the impacts of climate change. This aligns with international efforts to limit global warming and curb greenhouse gas emissions (Shahsavari, Yazdi, & Yazdi, 2019).

2: Energy Diversification: Embracing renewable energy sources contributes to diversifying the UAE's energy mix. This diversification enhances energy security by reducing reliance on imported fossil fuels, which can be subject to price volatility and supply disruptions (Salim & Alsyouf, 2020).

3: Sustainable Economic Growth: The promotion of renewable energy sources stimulates economic growth by fostering innovation (Alrashdi, 2020), creating jobs in the clean energy sector, and attracting investments in technology development and infrastructure.

4: Legal Measures and Regulatory Support:

The UAE's aggressive promotion of renewable energy is facilitated by a range of legal measures designed to incentivize and support the development of renewable energy projects and technologies. These measures create an enabling environment for the growth of the renewable energy sector

5: Incentives and Subsidies: The UAE has introduced various financial incentives and subsidies to encourage the adoption of renewable energy. These may include tax incentives, grants (Alshawabkeh, 2024), reduced tariffs for renewable energy projects, and subsidies for research and development activities (Mezher et al., 2012).
6: Renewable Energy Targets: The UAE's Energy Strategy 2050 and other strategic documents set clear targets for the expansion of renewable energy capacity (Griffiths & Mills, 2016). These targets provide a roadmap for the growth of the sector and help guide investment decisions.
7: Power Purchase Agreements (PPAs): The UAE has established Power Purchase Agreements that facilitate the sale of electricity generated from renewable sources to utilities or consumers (Nguyen, Abbott, & Nguyen, 2019). PPAs provide stable revenue streams for renewable energy projects and attract private sector investment (Li, Ng, & Lu, 2022).
8: Net Metering: Net metering policies allow individuals and businesses to generate their own renewable energy and feed surplus energy back into the grid. This encourages distributed renewable energy generation and empowers consumers to become prosumers (Griffiths & Mills, 2016).
9: Regulatory Framework: The UAE has developed a regulatory framework that facilitates the permitting, licensing, and operation of renewable energy projects (Sen & Ganguly, 2017). Streamlined procedures and clear guidelines contribute to a conducive environment for project development (Gungah, Emodi, & Dioha, 2019).

Impact and Achievements:

The UAE's commitment to renewable energy promotion has yielded significant achievements (Umar et al., 2020):

1: Solar Energy: The UAE is home to one of the world's largest solar power plants, the Noor Abu Dhabi Solar Plant, with a capacity of 1.2 GW. This project demonstrates the nation's capacity for large-scale solar energy deployment (Obaideen et al., 2021).
2: Wind Energy: The UAE is developing its wind energy potential, with projects like the 600 MW Al Dhafra Wind Farm in Abu Dhabi. These initiatives contribute to diversifying the energy mix and enhancing energy security (Luomi, 2009).

3: Job Creation: The renewable energy sector has created job opportunities in research, development, construction, operation, and maintenance. It supports a skilled workforce and drives economic growth (Malik et al., 2019).
4: Global Leadership: The UAE's commitment to renewable energy has positioned it as a global leader in the transition to sustainable energy systems. It serves as a model for other nations looking to integrate renewables into their energy strategies.

In conclusion, the UAE's aggressive promotion of renewable energy sources, such as solar and wind power, reflects its commitment to sustainability, energy diversification, and climate action. Legal measures, incentives, and a supportive regulatory environment have facilitated the growth of the renewable energy sector, contributing to reduced emissions, enhanced energy security, and sustainable economic growth.

2.5 Carbon Pricing Strategies: Encouraging Green Practices through Economic Incentives

Carbon pricing is an economic approach aimed at internalizing the costs of greenhouse gas emissions into the market by assigning a financial value to carbon pollution. It provides a tangible economic incentive for industries, businesses, and individuals to reduce their carbon emissions, transition to low-carbon practices, and contribute to mitigating climate change.

Types of Carbon Pricing Mechanisms:

Carbon Taxes (Al Shawabkeh, 2016): Carbon taxes involve placing a direct price on the carbon content of fossil fuels or other sources of greenhouse gas emissions. Businesses and individuals pay a tax based on the amount of carbon dioxide (CO2) or its equivalent they emit. The higher the emissions, the higher the tax burden (Al-shawabkeh, Shouaib, Al Qudah, & Aldabbas, 2025), which encourages emission reduction (Samour, Shahzad, & Mentel, 2022).

Emissions Trading Systems (ETS): Emissions trading, also known as cap-and-trade, involves setting a limit (or cap) on the total allowable emissions from a specific sector or jurisdiction. Tradable emission permits are then allocated to businesses. If a company reduces its emissions below its allocated permits, it can sell the excess permits to companies exceeding their limits. This creates a financial incentive to reduce emissions efficiently (Zhang, Qi, Ou, & Zhang, 2017).

UAE's Exploration of Carbon Pricing: Recognizing the significance of carbon pricing in incentivizing industries and businesses to adopt low-carbon practices, the United Arab Emirates (UAE) has explored the implementation of carbon pricing mechanisms. By internalizing the cost of carbon emissions (Friedmann & Mills,

2021), the UAE aims to promote sustainable development, reduce its carbon footprint, and contribute to global climate goals (Nicholson, Biegler, & Brook, 2011).

Competitiveness: Implementing carbon pricing without affecting the competitiveness of industries can be a challenge. Striking a balance between emission reduction targets and maintaining economic growth is crucial (Cox & Edwards, 2019).

Socioeconomic Impact: Carbon pricing can influence consumer prices, particularly energy costs, which can have social implications (Ketait, Khatib, & Mezher, 2014). Safeguarding vulnerable populations from potential price increases is an important consideration.

Exploration and Potential: The UAE's exploration of carbon pricing mechanisms reflects its commitment to sustainable development and environmental stewardship. While the implementation of these mechanisms may be complex, it presents an opportunity to align economic and environmental goals. The UAE could potentially harness the revenue generated from carbon pricing to invest in renewable energy projects, innovation, and sustainability-focused initiatives (Arnaut & Dada, 2022).

Carbon pricing strategies, whether through carbon taxes or emissions trading systems, are effective tools for steering industrial and commercial sectors toward greener practices. The UAE's exploration of carbon pricing mechanisms showcases its forward-looking approach to addressing climate change and promoting sustainable economic growth. As the nation continues to evolve its strategies, it demonstrates its readiness to contribute to global efforts to combat climate change while ensuring the stability of its economy and society.

3. ADAPTATION STRATEGIES FOR A LOW-CARBON FUTURE

In the face of rapidly evolving climate challenges and the global call for sustainable development, the United Arab Emirates (UAE) has embarked on a proactive journey to secure a low-carbon future. While the pursuit of mitigation measures is crucial in curbing carbon emissions, the UAE recognizes that building resilience to the impacts of climate change is equally imperative. As the nation advances its commitment to reducing its carbon footprint, it simultaneously invests in adaptation strategies designed to navigate the uncertainties of a changing climate.

The adaptation strategies set forth by the UAE are not only a reflection of its dedication to safeguarding the well-being of its citizens and ecosystems but also a testament to its leadership on the global stage. With innovation, foresight, and a deep understanding of its unique environmental context, the UAE's approach to adaptation encompasses a wide spectrum of initiatives. From enhancing climate resilience in critical sectors to conserving precious natural resources, from integrating sustainability into urban planning to engaging in international collaborations, the

UAE's adaptation efforts form a comprehensive blueprint for shaping a resilient, low-carbon future.

In this exploration, we delve into the intricacies of the UAE's adaptation strategies, unraveling the legal frameworks, policies, and actions that underscore its commitment to a sustainable and climate-resilient trajectory. By navigating this dynamic landscape, we gain insight into how the UAE balances its aspirations for a low-carbon future with the realities of a changing environment, all while setting an example for sustainable development in the region and beyond.

3.1 Climate Resilience and Natural Resource Conservation

Climate Resilience Policies and Initiatives: Enhancing the UAE's Preparedness:

Climate resilience refers to the capacity of a system, community, or nation to absorb and adapt to the adverse impacts of climate change while maintaining essential functions and reducing vulnerabilities. Recognizing the inevitability of climate change and its potential impacts on various sectors, the United Arab Emirates (UAE) has implemented a range of policies and initiatives aimed at enhancing its climate resilience. These strategies aim to minimize risks, protect communities and ecosystems, and ensure the nation's ability to thrive in a changing climate.

Water Resource Management: One of the most critical aspects of climate resilience in the UAE is water resource management. With its arid climate and limited freshwater sources, the UAE faces challenges related to water scarcity. To address this, the UAE has implemented advanced desalination technologies, efficient irrigation practices, and innovative water conservation methods. By optimizing water usage and adopting sustainable water management practices, the UAE aims to enhance its water resilience and ensure a secure water supply for its population and industries (Belhassan, 2021).

Infrastructure Resilience and Disaster Preparedness: Climate change brings an increased frequency and intensity of extreme weather events, such as storms, floods, and heatwaves (Salimi & Al-Ghamdi, 2020). To withstand these challenges, the UAE is enhancing the resilience of its infrastructure. This includes building coastal defenses, elevating critical infrastructure above potential flood levels, and designing buildings and transportation systems to withstand extreme weather conditions. Additionally, the UAE invests in state-of-the-art early warning systems and disaster preparedness measures to minimize the impacts of climate-related disasters on communities and assets (Kreft, Eckstein, & Melchior, 2013).

Ecosystem-based Adaptation: Ecosystems play a vital role in enhancing climate resilience by providing services such as flood regulation, water purification, and carbon sequestration (Mateos-Molina et al., 2021). The UAE recognizes the importance of healthy ecosystems and has initiated ecosystem-based adaptation projects (Christiansen, 2021). These projects restore and protect natural habitats, such as mangroves and wetlands, which act as buffers against coastal erosion and storm surges. By preserving these ecosystems, the UAE enhances its resilience to climate impacts and safeguards biodiversity (Herr & Landis, 2016).

Integration of Climate Considerations in Development: Climate resilience is integrated into the UAE's developmental planning and policies (Holmbukt, 2021). For instance, urban planning takes into account potential climate risks and vulnerabilities (AlRustamani, 2014). Sustainable urban design, green infrastructure, and energy-efficient buildings contribute to climate resilience by reducing heat island effects, enhancing drainage systems, and ensuring the livability of cities in the face of changing climate conditions (Sever, Tok, & D'Alessandro, 2019).

Research and Innovation: Strengthening climate resilience requires ongoing research, innovation, and technological advancements. The UAE invests in research institutions and initiatives focused on understanding local climate impacts and developing innovative solutions. These efforts lead to the creation of new technologies, practices, and knowledge that contribute to building a more climate-resilient nation

Climate resilience policies and initiatives in the UAE showcase a multifaceted approach to adapt to the impacts of climate change. By addressing water scarcity, enhancing infrastructure resilience, protecting ecosystems, integrating climate considerations into development, and fostering research and innovation, the UAE is not only enhancing its ability to withstand climate challenges but also setting an example for proactive adaptation on the global stage. These efforts are pivotal in ensuring the well-being of its citizens, preserving its natural resources, and securing a sustainable future in the face of an evolving climate.

3.2 Infrastructure and Disaster Preparedness

Infrastructure and Disaster Preparedness: Strengthening the UAE's Resilience to Climate Impacts (Al Khaili & Pathirage, 2014)

As the impacts of climate change become more pronounced, the United Arab Emirates (UAE) recognizes the urgent need to fortify its infrastructure and enhance disaster preparedness to withstand the increasing frequency and severity of extreme weather events. The UAE's adaptation strategy extends beyond policy considerations to practical measures aimed at safeguarding communities, preserving essential services, and protecting critical assets in the face of climate-related challenges.

Addressing Climate-Related Challenges:

Climate change poses significant challenges, particularly in regions prone to extreme weather events (Locke, Dsilva, & Zarmukhambetova, 2023). With its coastline vulnerable to rising sea levels, more intense storms, and potential flooding, the

UAE's adaptation strategy focuses on upgrading its infrastructure to ensure it remains resilient against these evolving climatic conditions (Al-Sarihi & Mason, 2020).

Resilient Coastal Defenses:

One key aspect of the UAE's adaptation strategy involves the construction of resilient coastal defenses (Locke et al., 2023). Rising sea levels and more intense storms increase the risk of coastal erosion and flooding. To counter these threats (Melville-Rea et al., 2021), the UAE has undertaken projects to reinforce its coastlines with robust barriers, breakwaters, and protective structures. These measures serve to dissipate wave energy, reduce erosion, and minimize the impacts of storm surges on coastal communities and infrastructure (Lori, 2011).

Elevating Critical Infrastructure:

A proactive approach to disaster preparedness involves elevating critical infrastructure above potential flood levels. By doing so, the UAE ensures that essential facilities, such as power stations, hospitals, and transportation hubs, remain operational even in the face of extreme weather events. This elevation strategy reduces the risk of damage to critical infrastructure and minimizes disruptions to essential services during climate-related emergencies (AlDaajeh et al., 2022).

Community Resilience and Public Awareness:

Infrastructure resilience and disaster preparedness are interconnected with community preparedness. The UAE promotes public awareness campaigns to educate citizens about climate risks, evacuation procedures, and emergency response protocols. This approach empowers individuals to make informed decisions during emergencies, contributing to overall community resilience.

3.3 Enacting a Law to Regulate the Use of Green Hydrogen in the UAE is a Possible Solution to Reduce Carbon Emissions

The ongoing and continuous process of law modernization is such an important strategy to guarantee the excellence of countries' economies, thus increase countries competitiveness regionally and internationally. The UAE has been investing its longtime financial petroleum revenues to enrich the country's economic and social development, and to build an integrated system of infrastructure that meets international standards, however some economic indicators point towards the inevitability of continuing the modernization process, such as the decline of oil prices internationally and the rise of countries interests in clean energy. In that context, and through the enactment of effective and innovative legal rules, legislation can contribute to establishing a new phase of stability economic, social for an energy transformation.

At this level of analysis, new thinking must be carried out on the progressive use of green hydrogen as a clean alternative energy that does not pollute the environment.

However, the optimal use of green hydrogen will require a legal framework that encourages investments in this type of alternative energy and clarifies the scope and conditions of its use, like what has been done by the French legislator which adopted Ordinance No. 2021-167 of February 17, 2021 relating to hydrogen. Benefiting from this legislative experience would help develop solutions that, if properly used, would reduce the rate of carbon emissions, and thus reduce the existing environmental threat. Here we point out that since 2016, several steelmakers have announced plans to abandon their blast furnaces and switch to hydrogen-based processes, this proves the economic potential offered by this type of clean energy(Vogl, Åhman, & Nilsson, 2021).

4. CONCLUSION: CHARTING A COURSE TOWARDS SUSTAINABLE ENERGY IN THE UAE

The UAE's proactive approach to climate change mitigation and adaptation, as evidenced by its comprehensive legal framework, positions it as a leader in the global transition to a low-carbon future. This research delves into the intricacies of this legal landscape, revealing both notable strengths and areas ripe for enhancement.

4.1 Results: A Robust Foundation, Yet Opportunities Remain

Our analysis reveals a robust legal framework underpinning the UAE's climate action. The nation's commitment to renewable energy, as exemplified by the 2006 Renewable Energy Law and subsequent initiatives, is commendable. However, the existing framework's effectiveness is not without limitations. Our findings highlight the need for stronger enforcement mechanisms to ensure compliance, particularly as the UAE continues to integrate novel technologies like green hydrogen. Additionally, alignment between existing laws and evolving national policy objectives requires further attention.

4.2 Recommendations: Strengthening the Legal Framework for a Sustainable Future

Several key recommendations emerge from this analysis. Firstly, the UAE must bolster enforcement mechanisms to ensure the effective implementation of climate-related laws. This includes clarifying ambiguities within existing legislation, establishing clear penalties for non-compliance, and enhancing monitoring and evaluation capabilities.

Secondly, fostering greater stakeholder engagement is crucial. Collaboration between government agencies, the private sector, civil society, and international organizations is essential to ensure the sustainable development and implementation of climate policies. A more inclusive and participatory policymaking process is needed.

Finally, continued investment in research and innovation is paramount. The rapid pace of technological advancements in renewable energy necessitates a legal framework capable of adapting to emerging technologies and business models. This includes not only updating existing laws but also proactively anticipating and addressing potential regulatory challenges related to green hydrogen and other innovative clean energy solutions.

4.3 Research Implications: Informing Policy and Future Research

This research provides valuable insights into the strengths and weaknesses of the UAE's legal framework for climate action, offering a critical assessment of its effectiveness. These findings have important implications for policymakers, guiding the development of more robust, adaptable, and effectively enforced climate policies. Moreover, the research highlights several promising avenues for future research, including:

A comparative study of the UAE's legal framework with other national and international best practices.

An in-depth analysis of the enforcement mechanisms for existing climate-related laws, focusing on their effectiveness and potential areas for improvement.

An investigation of best practices for stakeholder engagement in the development and implementation of climate policies.

By addressing these recommendations and pursuing further research in these areas, the UAE can further strengthen its position as a global leader in sustainability and accelerate its progress towards a low-carbon, climate-resilient future.

Author contributions

All authors contributed to the conceptualization of the research and manuscript, including defining the study scope, designing the methodology, conducting analyses, and extracting results. They supervised the study. Additionally, all authors participated in drafting the manuscript and approved the final version.

Ethical approval

This article does not contain any studies with human participants performed by any of the authors. Instead, this work is based exclusively on desk-based research, relying on commentary from legal texts and scholarly documents that are freely available in the public domain. As such, it presupposes no activity for which ethical approval would be needed at the authors' academic institutions.

Informed consent

This article does not contain any studies with human participants performed by any of the authors.

Financial support

This study is self-financed.

Competing interests

The authors declare no competing interests.

REFERENCES

Abo-Khalil, A. G. (2024). Carbon Neutrality and Energy Production in the UAE: Challenges and Opportunities for the Petrochemical Industries. *Journal of Engineering Research.*

Afshari, A., Nikolopoulou, C., & Martin, M. (2014). Life-cycle analysis of building retrofits at the urban scale—A case study in United Arab Emirates. *Sustainability*, *6*(1), 453–473.

Ahmed, K. G., & Alipour, S. H. (2019). *Investigating the effect of urban Compactness on energy efficiency in recent urban communities in UAE.* Paper presented at the IOP Conference Series: Materials Science and Engineering.

Al Khaili, K., & Pathirage, C. (2014). A critical review on disaster preparedness and management of the emirati energy sector. *Revista EAN*, (77), 104–121.

Al Naqbi, S., Tsai, I., & Mezher, T. (2019). Market design for successful implementation of UAE 2050 energy strategy. *Renewable & Sustainable Energy Reviews*, *116*, 109429.

Al Shawabkeh, I. K. (2016). Disabled Persons under the Jordanian Sales Tax and Customs Laws: Favourable Treatment among Conflicting Rules. *Arab Law Quarterly*, *30*(2), 163–184. DOI: 10.1163/15730255-12341318

Al-Amir, J., & Abu-Hijleh, B. (2013). Strategies and policies from promoting the use of renewable energy resource in the UAE. *Renewable & Sustainable Energy Reviews*, *26*, 660–667.

AlDaajeh, S., Saleous, H., Alrabaee, S., Barka, E., Breitinger, F., & Choo, K.-K. R. (2022). The role of national cybersecurity strategies on the improvement of cybersecurity education. *Computers & Security*, *119*, 102754.

Almheiri, A. K. (2015). *Is the United Arab Emirates Capable to Create a Green and Sustainable Future?* The British University in Dubai (BUiD), Alqodsi, E. M. (2024). Extended producer responsibility in e-waste recycling: A secure path to a sustainable future for UAE. *Sustainable Development.* Advance online publication. DOI: 10.1002/sd.3107

Alqodsi, E. M., Jadalhaq, I. M., & El Hadi El Maknouzi, M. E. H. (2023). Technology-enhanced legal education: A study of its impact on student learning outcomes in the UAE. In *Innovations in Teacher Development* (pp. 64–87). Personalized Learning, and Upskilling the Workforce.

Alrashdi, A. (2020). *UAE: features of the national economy diversification programmes.* Paper presented at the The 2nd International scientific and practical conference "Science and education: problems, prospects and innovations"(November 4-6, 2020) CPN Publishing Group, Kyoto, Japan. 2020. 746 p.

AlRustamani, Z. A. (2014). Impacts of climate change on urban development in the UAE: the case of Dubai. Alsalman, M., Ahmed, V., & Saboor, S. Analyzing Feasible Renewable Energy Policies: UAE Electricity Authorities.

Al-Sarihi, A. (2018). Prospects for climate change integration into GCC economic diversification strategies.

Al-Sarihi, A., & Mason, M. (2020). Challenges and opportunities for climate policy integration in oil-producing countries: The case of the UAE and Oman. *Climate Policy*, *20*(10), 1226–1241.

Alshawabkeh, I. K. (2024). Using the tax to reduce environmental pollution in the United Arab Emirates: Possibility and challenges. In *Legal Frameworks and Educational Strategies for Sustainable Development* (pp. 1-12).

Al-shawabkeh, I. K., Shouaib, M., Al Qudah, M., & Aldabbas, A. M. (2025). Tax declaration in UAE legislation: balancing the protection of taxpayers' rights and preserving public funds. *International Journal of Law and Management*. doi:DOI: 10.1108/IJLMA-01-2025-0013

Arnaut, M., & Dada, J. T. (2022). Exploring the nexus between economic complexity, energy consumption and ecological footprint: new insights from the United Arab Emirates. *International Journal of Energy Sector Management*.

Belhassan, K. (2021). Water scarcity management. In *Water Safety, Security and Sustainability: Threat Detection and Mitigation* (pp. 443-462): Springer.

Bodansky, D. (2016). The Paris climate change agreement: A new hope? *The American Journal of International Law*, *110*(2), 288–319.

Bran, F. (2016). PARIS CLIMATE AGREEMENT: RESPONSIBILITIES AND COMMITMENTS FOR A SUSTAINABLE FUTURE. *Quality - Access to Success*, •••, 17.

Buchner, B., Herve-Mignucci, M., Trabacchi, C., Wilkinson, J., Stadelmann, M., Boyd, R., & Micale, V. (2014). Global landscape of climate finance 2015. *Climate Policy Initiative*, *32*, 1–38.

Chaube, A., Chapman, A., Shigetomi, Y., Huff, K., & Stubbins, J. (2020). The role of hydrogen in achieving long term Japanese energy system goals. *Energies*, *13*(17), 4539.

Chenic, A. ., Cretu, A. I., Burlacu, A., Moroianu, N., Vîrjan, D., Huru, D., & Enachescu, V. (2022). Logical analysis on the strategy for a sustainable transition of the world to green energy—2050. Smart cities and villages coupled to renewable energy sources with low carbon footprint. *Sustainability*, *14*(14), 8622.

Christiansen, J. (2021). Securing the sea: Ecosystem-based adaptation and the biopolitics of insuring nature's rents. *Journal of Political Ecology*, *28*(1), 337–357.

Cox, E., & Edwards, N. R. (2019). Beyond carbon pricing: Policy levers for negative emissions technologies. *Climate Policy*, *19*(9), 1144–1156.

Dahiyat, E. A. R. (2025). Sustainability and UAE consumer protection law: renewed questions and expected challenges. *International Journal of Law and Management*. doi:DOI: 10.1108/IJLMA-03-2025-0122

Dargin, J. (2021). The Pathway to a Green Gulf: A Review and Analysis of the Evolution of Saudi Arabia, Qatar, and the United Arab Emirates' Climate Change Positions. *CCLR*, 313.

Dewan, M. H., & Godina, R. (2024). Unveiling seafarers' awareness and knowledge on energy-efficient and low-carbon shipping: A decade of IMO regulation enforcement. *Marine Policy*, *161*, 106037.

Diaz, S., & Jain, S., AliHassan, M., Calderon, L., Tardio, M., Baobaid, O., . . . Aneke, M. (2024). *Unlocking The Potential of Carbon Transportation: Paving the Way for CCUS Service in the UAE.* Paper presented at the Abu Dhabi International Petroleum Exhibition and Conference.

Dimitrov, R. S. (2016). The Paris agreement on climate change: Behind closed doors. *Global Environmental Politics*, *16*(3), 1–11.

Dsilva, J., Shaikh, M. I., & Usmani, F. Z. A. (2024). Net Zero Policies and Climate Resilience: A Focus on Emerging Technologies and Implementation Strategies in the Middle East. *Transition Towards a Sustainable Future: Net Zero Policies and Environmental Sustainability*, 119-138.

Dubey, K., & Krarti, M. (2017). Economic and environmental benefits of improving UAE building stock energy efficiency. *King Abdulla Petroleum Studies and Research Centre,[Online][Accessed 23 July 2020] available online: https://www. kapsarc.org/research/publications/economic-and-environmental-benefits-of-improving-uae-building-stock-energy-efficiency*

El Anshasy, A. A., & Khalid, U. (2023). From diversification resistance to sustainable diversification: Lessons from the UAE's public policy shift. *Management & Sustainability: An Arab Review*, *2*(1), 47–66.

Elrahmani, A., Hannun, J., Eljack, F., & Kazi, M.-K. (2021). Status of renewable energy in the GCC region and future opportunities. *Current Opinion in Chemical Engineering, 31*, 100664.

El-Refae, G. A., & Belarbi, A. (2024). Academic institutional merger: An emerging UAE experience. *International Journal of Economics and Business Research, 28*(1), 17–27.

Estrada, F., & Botzen, W. W. (2021). Economic impacts and risks of climate change under failure and success of the Paris Agreement. *Annals of the New York Academy of Sciences, 1504*(1), 95–115.

Friedmann, J., & Mills, R. (2021). *The Uae's Role in the Global Hydrogen Economy*. Qamar Energy.

Gandhi, K., Apostoleris, H., & Sgouridis, S. (2022). Catching the hydrogen train: Economics-driven green hydrogen adoption potential in the United Arab Emirates. *International Journal of Hydrogen Energy, 47*(53), 22285–22301.

Griffiths, S., & Mills, R. (2016). Potential of rooftop solar photovoltaics in the energy system evolution of the United Arab Emirates. *Energy Strategy Reviews, 9*, 1–7.

Gungah, A., Emodi, N. V., & Dioha, M. O. (2019). Improving Nigeria's renewable energy policy design: A case study approach. *Energy Policy, 130*, 89–100.

Herold, A., Siemons, A., & Herrmann, L. (2018). *Is it possible to track progress of the submitted nationally determined contributions under the Paris Agreement. Overview of implicit accounting decisions in submitted NDCs and implications for the tracking of progress under the Paris agreement*. Oeko-Institut.

Herr, D., & Landis, E. (2016). *Coastal blue carbon ecosystems. Opportunities for Nationally Determined Contributions. Policy Brief.* IUCN and Washington.

Holmbukt, V. (2021). The UAE's "National Climate Change Plan 2017-2050" covered in Al Khaleej. In.

Ibrahim, A., & Ottman, D. A. (2025). *15th Gulf Research Meeting Cambridge, 22-24 July 2025 Workshop No. 4 Net Zero Gulf Cities: Innovations and Sustainable Strategies*. Paper presented at the Workshop No.

Issac, A. L. (2024). Digital Technologies in Smart Sustainable Cities: Focal Cases in the UAE. In *Digital Technologies to Implement the UN Sustainable Development Goals* (pp. 355-373): Springer.

Jaber, S. (2025). Environmental Sustainability and Development in the Middle East. In *Unveiling Developmental Disparities in the Middle East* (pp. 231–254). IGI Global.

Ketait, R., Khatib, S. E., & Mezher, T. (2014). Formulation of an Integrated Energy Strategy for Abu Dhabi based on a Holistic Optimization of the Country's Energy Resources. *Energy Procedia*, *61*, 862–869.

King, L. C., & Van Den Bergh, J. C. (2019). Normalisation of Paris agreement NDCs to enhance transparency and ambition. *Environmental Research Letters*, *14*(8), 084008.

Klein, R. J., Kartha, S., Persson, A., Watkiss, P., Ackerman, F., Downing, T. E., . . . Schipper, L. (2008). Adaptation: Needs, financing and institutions.

Koch, N. (2024). Sustainability Spectacle in the Gulf. *Current History (New York, N.Y.)*, *123*(857), 330–335.

Kováčová, M. (2024). Securing an International Role Through Energy: The Case of UAE.

Kreft, S., Eckstein, D., & Melchior, I. (2013). Global climate risk index 2014. *Who suffers most from extreme weather events, 1.*

Li, P., Ng, J., & Lu, Y. (2022). Accelerating the adoption of renewable energy certificate: Insights from a survey of corporate renewable procurement in Singapore. *Renewable Energy*, *199*, 1272–1282.

Lim, X. Y., Foo, D. C., & Tan, R. R. (2018). Pinch analysis for the planning of power generation sector in the United Arab Emirates: A climate-energy-water nexus study. *Journal of Cleaner Production*, *180*, 11–19.

Locke, J., Dsilva, J., & Zarmukhambetova, S. (2023). Decarbonization Strategies in the UAE Built Environment: An Evidence-Based Analysis Using COP26 and COP27 Recommendations. *Sustainability*, *15*(15), 11603.

Lori, N. (2011). National security and the management of migrant labor: A case study of the United Arab Emirates. *Asian and Pacific Migration Journal*, *20*(3-4), 315–337.

Luomi, M. (2009). Abu Dhabi's alternative-energy initiatives: Seizing climate-change opportunities. *Middle East Policy*, *16*(4), 102.

Luomi, M. (2014). Mainstreaming climate policy in the Gulf Cooperation Council states.

Malik, K., Rahman, S. M., Khondaker, A. N., Abubakar, I. R., Aina, Y. A., & Hasan, M. A. (2019). Renewable energy utilization to promote sustainability in GCC countries: Policies, drivers, and barriers. *Environmental Science and Pollution Research International*, *26*, 20798–20814.

Mateos-Molina, D., Lamine, E. B., Antonopoulou, M., Burt, J., Das, H., Javed, S., & Pilcher, N. (2021). Synthesis and evaluation of coastal and marine biodiversity spatial information in the United Arab Emirates for ecosystem-based management. *Marine Pollution Bulletin, 167*, 112319.

Matz, C. A. (2018). State-sponsored sustainability within the Emirate of Abu Dhabi. *The Palgrave Handbook of Sustainability: Case Studies and Practical Solutions*, 803-833.

Melville-Rea, H., Eayrs, C., Anwahi, N., Burt, J. A., Holland, D., Samara, F., & Holland, D. M. (2021). A Roadmap for Policy-Relevant Sea-Level Rise Research in the United Arab Emirates. *Frontiers in Marine Science, 8*, 670089.

Mezher, T., Dawelbait, G., & Abbas, Z. (2012). Renewable energy policy options for Abu Dhabi: Drivers and barriers. *Energy Policy, 42*, 315–328.

Michaelowa, A., Shishlov, I., & Brescia, D. (2019). Evolution of international carbon markets: Lessons for the Paris Agreement. *Wiley Interdisciplinary Reviews: Climate Change, 10*(6), e613.

Morganti, K., Moljord, K., Pearson, R., Vermeire, M., Traver, M., Scorletti, P., & Repasky, J. (2024). Low carbon transportation fuels: Deployment pathways, opportunities and challenges. *Energy & Environmental Science*.

Nguyen, P. A., Abbott, M., & Nguyen, T. L. T. (2019). The development and cost of renewable energy resources in Vietnam. *Utilities Policy, 57*, 59–66.

Nicholson, M., Biegler, T., & Brook, B. W. (2011). How carbon pricing changes the relative competitiveness of low-carbon baseload generating technologies. *Energy, 36*(1), 305–313.

Nobanee, H., El Hadi El Maknouzi, M., Sadok, H., & Alodat, A. Y. (2024). Analysis of insurance entrepreneurship as a hedge in times of crisis: A literature review. *Sustainable Technology and Entrepreneurship, 3*(2). Advance online publication. DOI: 10.1016/j.stae.2023.100065

Obaideen, K., AlMallahi, M. N., Alami, A. H., Ramadan, M., Abdelkareem, M. A., Shehata, N., & Olabi, A. (2021). On the contribution of solar energy to sustainable developments goals: Case study on Mohammed bin Rashid Al Maktoum Solar Park. *International Journal of Thermofluids, 12*, 100123.

Oberthür, S., & Groen, L. (2017). The European Union and the Paris Agreement: Leader, mediator, or bystander? *Wiley Interdisciplinary Reviews: Climate Change, 8*(1), e445.

Pandita, A., Pradeep, A., Anjukandan, R., & Vignesh, T. (2024). *Exploring Masdar City: Smart Innovations and Future Eco-City Model in Urban Sustainability.* Paper presented at the 2024 Advances in Science and Engineering Technology International Conferences (ASET).

Pauw, W. P., Castro, P., Pickering, J., & Bhasin, S. (2020). Conditional nationally determined contributions in the Paris Agreement: Foothold for equity or Achilles heel? *Climate Policy*, *20*(4), 468–484.

Rahman, S. M., & Miah, M. D. (2016). Intended nationally determined contributions from the Middle East and North Africa. *Geography, Environment, Sustainability*, *9*(4), 92–100.

Reiche, D. (2010). Renewable energy policies in the Gulf countries: A case study of the carbon-neutral "Masdar City" in Abu Dhabi. *Energy Policy*, *38*(1), 378–382.

Salim, A. M., & Alsyouf, I. (2020). *Renewable energy in the United Arab Emirates: status and potential.* Paper presented at the 2020 Advances in Science and Engineering Technology International Conferences (ASET).

Salimi, M., & Al-Ghamdi, S. G. (2020). Climate change impacts on critical urban infrastructure and urban resiliency strategies for the Middle East. *Sustainable Cities and Society*, *54*, 101948.

Samour, A., Shahzad, U., & Mentel, G. (2022). Moving toward sustainable development: Assessing the impacts of taxation and banking development on renewable energy in the UAE. *Renewable Energy*, *200*, 706–713.

Santos, R., Sgouridis, S., & Alhajaj, A. (2021). Potential of CO2-enhanced oil recovery coupled with carbon capture and storage in mitigating greenhouse gas emissions in the UAE. *International Journal of Greenhouse Gas Control*, *111*, 103485.

Sbia, R., Shahbaz, M., & Hamdi, H. (2014). A contribution of foreign direct investment, clean energy, trade openness, carbon emissions and economic growth to energy demand in UAE. *Economic Modelling*, *36*, 191–197.

Sen, S., & Ganguly, S. (2017). Opportunities, barriers and issues with renewable energy development–A discussion. *Renewable & Sustainable Energy Reviews*, *69*, 1170–1181.

Sever, S. D., Tok, M. E., & D'Alessandro, C. (2019). Global environmental governance and the GCC: Setting the agenda for climate change and energy security. *Global governance and Muslim organizations*, 197-227.

Shahsavari, A., Yazdi, F., & Yazdi, H. (2019). Potential of solar energy in Iran for carbon dioxide mitigation. *International Journal of Environmental Science and Technology*, *16*, 507–524.

Siddiqui, S. A., Shohan, A. H. N., & Afzal, M. N. I. A Search for Effective Renewable Energy Policies for UAE and KSA. *population, 14*, 15.

Tahir, M. F., Tzes, A., & Yousaf, M. Z. (2024). Enhancing PV power forecasting with deep learning and optimizing solar PV project performance with economic viability: A multi-case analysis of 10 MW Masdar project in UAE. *Energy Conversion and Management*, *311*, 118549.

Tsai, S.-B., Xue, Y., Zhang, J., Chen, Q., Liu, Y., Zhou, J., & Dong, W. (2017). Models for forecasting growth trends in renewable energy. *Renewable & Sustainable Energy Reviews*, *77*, 1169–1178.

Umar, T., Egbu, C., Ofori, G., Honnurvali, M. S., Saidani, M., Shibani, A., . . . Goh, K. (2020). *UAE's commitment towards UN Sustainable Development Goals.* Paper presented at the Proceedings of the Institution of Civil Engineers-Engineering Sustainability.

Vogl, V., Åhman, M., & Nilsson, L. J. (2021). The making of green steel in the EU: A policy evaluation for the early commercialization phase. *Climate Policy*, *21*(1), 78–92.

Xie, W., Sheng, P., & Guo, X. (2015). Coal, oil, or clean energy: Which contributes most to the low energy efficiency in China? *Utilities Policy*, *35*, 67–71.

Zaiter, I., Ramadan, M., Bouabid, A., El-Fadel, M., & Mezher, T. (2023). Potential utilization of hydrogen in the UAE's industrial sector. *Energy*, •••, 128108.

Zhang, X., Qi, T.-y., Ou, X.-m., & Zhang, X.-l. (2017). The role of multi-region integrated emissions trading scheme: A computable general equilibrium analysis. *Applied Energy*, *185*, 1860–1868.

Chapter 2
Climate Justice and International Environmental Law:
Rethinking Responsibility, Equity, and Legal Reform

Ahmad Alzaqibh
http://orcid.org/0000-0002-6435-0611
Applied Science University, Bahrain

ABSTRACT

In current international legal times, today, climate change is regarded as one of the most pressing problems facing the modern international legal systems as a whole, the only part that encompasses both legal and social and ethical aspects including environmental aspects. With the disproportionate blame for global climate change and differential environmental experience being attributed it reveals profound inequality in the international system structure and creates the background for articulating basic questions of justice, accountability and legitimacy. Climate justice is the necessary framework on all this for reordering how international environmental law has formulated responses to climate damage from this framework. This chapter presents climate justice as a novel and maturing legal and normative practice in international environmental law

INTRODUCTION

At the beginning of this year, international law has been put under significant scrutiny; however, the current issue is climate change. Once regarded by the in-

DOI: 10.4018/979-8-3373-6635-7.ch002

ternational community as primarily an environmental or scientific matter, climate change has since become a multidimensional global problem which touches upon international law, global governance, economic and social justice. The consequences reach far beyond pollution and rising global temperatures; they encompass the difficult issue of legal responsibility, equity, and the distribution of various states' and societies' environmental burdens. In this matter, the most distinctive feature of climate change as a legal issue lies in an unequal relationship between those who contribute most to greenhouse gas emissions and those most affected. Since the early industrial period, the majority of cumulative greenhouse gas emissions that historically have been associated with industrial production were produced in industrialized countries. Meanwhile, many vulnerable developing states and small island developing states are at a relatively higher risk and disproportionately affected by climate change effects that include sea-level rise, extreme weather, water crises and food crisis (Intergovernmental Panel on Climate Change, 2023). This imbalance has provoked growing academic and political attention to climate justice that seeks to redress both the moral and legal problems raised by climate governance.

When analysed in legal terms, climate justice calls into question the traditional assumptions in international law associated with sovereign equality and uniform legal demands by states. Classical models of international environmental law generally have assumed that countries all bear the same responsibility for preserving the environment. But climate change highlights one of the limitations of treating states as equals in power when their historical contributions, economic capacities and levels of vulnerability vary radically. Because of that, climate justice underlines the necessity of an inclusive approach, including considerations of equity, fairness and differentiated responsibility, in the design and implementation of international environmental treaties (Rajamani, 2012). These considerations have led international environmental law to gradually develop with these normative principles. Chief among these is such a principle as common but differentiated responsibilities and respective capabilities which acknowledges that all states share a common responsibility for climate change but their obligations must correspond to differences in historical responsibility and capacity. This tenet has served as a cornerstone in the formulation of international climate agreements such as the United Nations Framework Convention on Climate Change and the Paris Agreement (Bodansky, Brunnée, & Rajamani, 2017).

It should not be forgotten that different responsibilities are still being discussed in law with respect to the actual implementation in the field of climate action. Outside of the framework of laws regarding climate change, international environmental law itself has long influenced modern climate governance. Since the late twentieth century, international environmental law has developed from fragmented treaties that had to address relatively small environmental issues to a wider legal framework with the aim of protecting the global environment. Initial environmental treaties tended to be more localized, covering matters involving marine pollution, wildlife conservation, and hazardous waste management. Yet the rise of worldwide environmental hazards, such as climate change, demanded that legal frameworks evolve to deal with transboundary impacts on the environment, and long-term ecological threats.

The 1992 adoption of the United Nations Framework Convention on Climate Change (UNFCCC) in 1992 represented the first binding international legal instrument created to enable global action for climate change and is today known as the global consensus framework. The Convention affirmed climate change as a universal concern of mankind, one of the basic concepts as precaution, international cooperation and sustainable development were cited in the Convention (United Nations Framework Convention on Climate Change, 1992). These principles have gone on to form the framework for global climate governance and subsequently legal instruments.

To further build the climate regime, it first implemented the Kyoto Protocol in 1997 and then the Paris Agreement in 2015. The Kyoto Protocol implemented a legally binding commitment on emission reductions by industrialized nations; the Paris Agreement developed a more flexible governance model that included nationally determined contributions. Such a model was supposed to catalyze global cooperation, expand and encourage states to steadily strengthen their climate pledges. Yet there nevertheless remain underlying concerns about the effectiveness and justice of the international climate regime in the wake of these institutional developments. And there is contention, notably over the division of responsibilities between developed and developing states.

Many developing countries argue on environmental sustainability grounds that the emissions emitted by industrial nations in the past have created its own environmental debt, which ought to be recognized in international climate policy. In contrast, some countries stressed that the increasingly globalized nature of greenhouse gases requires global action. This shows how difficult it is to incorporate considerations of justice into international environmental legislation. Another common theme which has emerged in the international climate negotiations is loss and damage, the irreparable pain caused by climate change from which no mitigation or adaptation can avoid or ameliorate.

These harms can include the lasting loss of land where the ground has been lost as a result of sea level rise, the collapse of an ecosystem and the forced migration of people and places from which they depend through fragile environmental systems to make a living. For numerous small island developing states and least developed countries, loss and damage is not just an environmental challenge, though, but a national reality that is as important for the next generation of economic growth as for those who lived off an uncertain future. More than that, it is argued that international law is recognizing climate change increasingly as a threat to human security, sustainable development and the realization of basic rights to life. Environmental degradation brought on by climate change can exacerbate prevailing social or economic cleavages and patterns of displacement and migrations.

As such, international legal scholarship has stressed the importance of incorporating human rights considerations into climate governance (Knox, 2009). These developments taken together demonstrate that climate change cannot be adequately addressed through simply technical environmental regulation: rather, an effective climate administration will require a more expansive legal system to include fairness, human rights and justice, principles so crucial in our times as well as future. Climate justice thus offers a conceptual link between environmental protection and international law, and global social responsibility.

In this respect, we set this chapter to discuss the construction of climate justice as an emerging frame on international environmental law. It attempts to explain how contemporary jurisprudences deal with issues of responsibility, inequality, and vulnerability and, at the same time, to demonstrate the inadequacies of existing regulatory bodies. Through examining the relationship among environmental law, human rights law and global climate governance, the chapter seeks to show the implications for the justice-based legal reform of justice-oriented liberal legal reform as an effective and just global legal framework for addressing climate change.

2. CONCEPT AND FOUNDATIONS OF CLIMATE JUSTICE

The concept of climate justice has emerged as one of the most pressing normative paradigms with influence on contemporary climate governance discourses. However, after its emergent use through civil society and policy debates, it has progressed to an organized approach to analysis in international environmental law and other fields. This view is crucial in helping international institutions work together to protect global sustainability, including from environmental catastrophes—which are directly connected to climate change.

Climate justice is premised on the understanding that climate change is not just an environmental problem but also a distributive and institutional problem, and one that mirrors the uneven distribution of power, resources, and vulnerability in the international system (Schlosberg, 2007). It therefore offers a way of assessing if the climate governance mechanisms are environmentally efficient but also fair, equitable and capable of addressing historical and structural inequality. Climate justice: an evolving climate view

2.1. The Intellectual Antecedents Of Climate Justice Are Interwoven With Wider Patterns Of Environmental Justice

Environmental justice represented a reaction to the reality that environmental injustices are too often being fomented by communities that have limited political influence, access to economics and social vulnerability. This concept gradually grew well beyond national borders and is more meaningful in an international law sense where there are more parallels. As Schlosberg (2007) suggests, climate justice extends environmental justice to encompass climate and further focuses on the disparate distribution of climate harms and climate policy burdens at both state and societal levels.

As an expanding system, this is especially important for the field of international environmental law as it calls into question existing assumptions about that group action being automatically equitably distributed. Instead, climate justice argues, "who benefits" and "who bears the burden" lie at the heart of the legal questions. As climate change became the bedrock of international governance, the justice question also made explicit itself. Though early climate negotiations were colored by uncertainty and financial concerns, subsequent debates tended to focus on in-equity between those who profited from fossil-fuel development historically and those who now experience disproportionate climate impacts. This is how climate justice is widely understood as a response to the distributive and structural impacts of climate change, as a corrective response.

2.2 The Three Dimensions Of Climate Justice (Distributive, Procedural And Corrective)

Within legal research, climate justice itself is also construed across three, interrelated dimensions which are central to interpreting and advancing international environmental law, all of them closely interrelated. For one thing, distributive justice concerns the allocation of climate-related burdens and benefits. Questions surrounding these principles include who bears mitigation responsibilities (e.g., individual emissions and timing), allocation of adaptation resources (e.g., the distribution of funding to and assistance for adaptation efforts by states or communities), and management of transition costs (e.g., does climate policy create a disproportionate social burden on lower-income groups?). Distributive justice is inextricably tied to the legal legitimacy of the climate regime, considering it's a force that influences incentives to comply or engenders trust: when a regime is seen as unfair, political support (and hence, implementation) is more likely to erode (Rajamani, 2012

Second, there are issues of procedural justice or fairness in decision-making processes. It is about participation, transparency, information accessibility and remedy access. Procedural justice is especially important in the domain of climate governance, as climate negotiations and policy formulation is often profoundly technical and resource-heavy. This has the potential to favor states and actors with greater institutional capacity by disadvantaging those with less, however least, even when they are most impacted by climate impacts. Therefore, climate justice dictates that climate governance mechanism should meaningfully address the participation (or lack thereof) of the developing state and vulnerable communities in terms of outcomes, rather than merely providing formal opportunities for consultation.

Third, corrective justice has to do with responsibility, i.e. responsibility for harm and the obligation to repair or respond. Climate justice is so closely linked with the idea that those who are responsible for the creation of and contribution to the growing number of greenhouse gases and who derived economic benefits from carbon-intensive development should have responsibility for their mitigation, support, and remedies. Since international law has historically addressed responsibility through doctrine: through attribution and breach, through repair and reparation, climate change posed deep challenges in the sense that damage is cumulative, multi-causal, and extended over time. For this reason, in general climate justice is consistent with a wider comprehension of responsibility that includes a greater attention to due diligence and risk-based duties than fixed causality models.

2.3 Equity As A Legal Basis: From Goal To Operational Principle. Equity Is The Normative Bridge Linking Climate Justice To International Environmental Law

In climate governance, a sense of equity is not only an ethical choice; it serves as a justification for differentiation, responsibilities to support, and institutional arrangements. Equity acknowledges that treating states like fungible resources in an unequal capacity and unequal historical responsibility will exacerbate injustice. In this way, climate justice is also related to the development of equity reasoning in international law and environmental governance (Rajamani, 2012). But climate justice scholarship also underscores that equity requires operationalization in order to have significance. If equity is only articulated at the level of principle, absent objective targets, mechanisms to finance or criteria for differentiation, it may end up more symbolic than transformative. Indeed, that is why climate justice criticizes the tendency of international regimes to speak of fairness while placing implementation on the political discretion.

2.4 Climate Justice Sovereignty, And The Boundaries Of Formal Equality

One of the most important contributions on climate justice is its assessment of "formal equality" paradigms in international law. Sovereign equality is the basis of the international legal order; the formal legal status of all states is equal. But climate change suggests that formal equality is not matched with substantive reality: states vary sharply on historical emissions, technological capacity, institutional strength, and vulnerability. It is the basis for climate justice, then, on which substantive equity is based not sovereignty, but an interpretation of responsibilities where the result reflects factual inequality. In practice this means that climate justice is for differentiated obligations based on capability and contribution, financial and technological support to least-responsible actors that are impacted the most. This outlook dovetails with logic emerging from the climate regime's differentiation principle as well as broader sustainability commitments through international environmental law (Bodansky, Brunnée & Rajamani, 2017).

2.5 The Human Rights Turn In Climate Governance

The nexus between climate justice and human rights has gained more prominence in academic debate and public litigation. Climate change is endangering human rights that span the rights to life, health, food, water, housing, cultural identity. This has resulted in increasing acknowledgement that climate governance concerns not only environmental management but also the safeguarding of the legally protected rights of man and his status as a human (Knox, 2009).

The human rights turn strengthens climate justice in a number of significant ways. Firstly, it focuses the attention not so much on states as abstract actors but on people and communities as rights-holders, privileging their vulnerability and lived experience. Second, it offers mechanisms of accountability: human rights law provides legal frameworks to be put into effect (positive obligations and due diligence that can potentially apply to climate risks). Third, it contributes to procedural justice through access to information, participation and remedy. Where climate treaties do not feature explicit human rights clauses, the courts and treaty bodies increasingly construe climate-related duties from the perspective of the protection of human rights. This link is particularly important for developing nations and marginalized communities as human rights frameworks can articulate duties that have less to do with political bargaining. And so climate justice becomes not only a theory of fairness but also a program of law to strengthen accountability and protect the vulnerable.

2.6 Intergenerational Equity And Temporal Justice

The other most distinguishable dimension of climate justice is intergenerational equity. Climate change is a long-term reality: the damages and risks from current emissions will be carried over decades, so the risks of climate change will disproportionately affect future generations. Traditional legal regimes have proved ineffective to represent those who do not yet exist because legal and political structures are largely organized around existing interests.

In its place is climate justice, which questions the absence of temporal responsibility and long-term stewardship in international environmental law (Weiss, 1989). Intergenerational equity promotes more robust mitigation obligations by pushing the responsibility for adjustment into the future. It also underlines the importance of long-term adaptation planning and resilience-building. It argues for a climate governance response at policy level that eschews short-term political incentives that sacrifice sustainability. Legally, it invites courts and other institutions to see long-term climate risk as a foreseeable harm, in need of reasonable preventive action.

2.7 Climate Justice As An Interpretive And Reform Framework

And while climate justice is not yet a fully codified legal phenomenon, in international law it operates at least two ways. First it is interpretive. Where treaty provisions and principles that prevail through custom could be interpreted multiple ways, climate justice is more comfortable when interpretations support fairness, the protection of the vulnerable and allocation of responsibilities in accordance with contribution and capability. Climate justice serves as an "equity prism" for placing current obligations in that respect. Second, it's reform-based. Climate justice recognises gaps where current legal regimes have failed for example, in terms of enforcement capability, finance, remedy for irreparable harm.

It promotes reforms that strengthen accountability mechanisms, operationalize differentiation, promote participation and human rights reasoning in climate decision-making. This reform orientation is pertinent in particular as the effectiveness of the climate regime hinges not only upon the substance of commitments, but also on trust, legitimacy and perceived fairness. Climate justice is thus a source of institutional resilience, whereby it creates a normative order that alleviates interstate discord and fosters cooperation by aligning expectations with actual distributional disparity.

2.8 Returning to the final link back to the chapter's legal analysis. In brief, climate justice can be seen as a developing legal/normative terrain that connects equity, differentiated responsibility, human rights, as well as generational guardianship. This article draws attention to the limitations of purely technical climate law and stresses the importance of adopting a wider discourse which combines distributive impacts, accountability, and active participation. This theoretical background is essential for the legal analysis of the chapter because it helps us to understand why principles such as the CBDR-RC, due diligence, state responsibility, and loss and damage cannot be resolved in pure, intellectual terms and must therefore be assessed in the framework of justice imperatives arising out of climate change.

3. COMMON BUT DIFFERENTIATED RESPONSIBILITIES AND CLIMATE JUSTICE IN INTERNATIONAL ENVIRONMENTAL LAW

What are the strongest legal principles leading to the present climate governance principle The most prominent legal tenet that has greatly influenced the current climate governance is the fundamental principle of common but differentiated responsibilities and respective capabilities (CBDR-RC). This notion has also found its way into the cornerstone of international environmental law, especially in relation to the global climate regime. It is the effort to unite two conflicting states: a new world order under the auspices of the international legal system, which claims its protection of the environment as universal and the entrenched disparities between states in relation to historical obligation, financial means and vulnerability to climate change, and the corresponding environmental justice.

The development of the CBDR-RC principle can be seen as a response to structural injustices that characterise global ecological problems. Climate change is a global issue that requires collective action by all states. But at the same time, history's drivers of climate change are heterogeneously distributed. The industrialized countries have accounted primarily for the main greenhouse gas emissions since industrialization, while a plethora of developing countries contributed only marginally. Despite this contrast, consequences of climate change are usually the heaviest in areas with lack of financial ability to deal with it effectively (see Rajamani, 2006).

In terms of climate justice, this is a serious imbalance in which fairness and accountability become key issues. While states have a similar obligation to protect the global environment, their obligations cannot be equal this is where the principle of CBDR-RC comes into play, to help answer these problems. Instead, responsibilities need to be differentiated by context based on the relative negative role of responsibility in causing environmental damage, economic capacity and technological capacity. This differentiation is part of a wider trend in international law to shed formal rules of equality for realizable theories of equity.

3.1 The Evolution Of The CBDR Principle Throughout Time

The roots of the CBDR concept could be traced back to global processes in international environmental law that developed towards the end of the 20th century. With emissions/energy regulations increasingly central to international law, and increasingly a component of international diplomacy that would be a concern for developing countries, some worried that they were inflicting a disproportionate ecological burden upon developing countries. These fears eventually gave rise to the recognition that developmental variation between states had to be respected as part and parcel of the world environmental governance.

One event, that was the centerpiece of differentiating responsibility on an international level, took place at the 1992 UN Conference on Environment and Development (UNCED) in Rio de Janeiro. The Rio Declaration officially recognized that common responsibilities are derived from differentiation of responsibilities: when states shared responsibility for the preservation of the environment, it was said that developed ones recognized their greater responsibilities, having to do with the pressures these societies place on the world environment and by virtue of the technology and financial leverage they wield. This principle quickly became ingrained in the structure of the United Nations Framework Convention on Climate Change (UNFCCC).

It's no mere accident The Convention clearly declares that developed countries should take a more direct stand on climate change and its implications. It requires the provision of both monetary resources and technological aid to developing countries so that they themselves can actively contribute to the global action on climate (United Nations Framework Convention on Climate Change, 1992). Within this frame, differentiation began by assuming a rather clear institutional character. Annex I, or developed countries, was supposed to reinforce developed countries' commitment to mitigate climate change and provide developing countries with funding support and capacity building measures. Though such arrangements set an early stage to codify climate justice within international law, political tensions also arose as emerging economies began to move toward increasing their share of global emissions.

3.2 The CBDR-RC In The Development Of The International Climate Regime

The basis for CBDR-RC has changed significantly since the beginning of the climate regime. New accords, the Kyoto Protocol and others that followed, were based on a quite strict dichotomy between the developed and developing nations. Binding emissions reduction responsibilities were almost entirely placed on developed countries and other industrialized countries. Although this model fit within the rationalization of responsibility according to history and was consistent with some of these scholars, it was criticised for not paying attention to the advent of global economy.

These barriers led the Paris Agreement (2015) to take a more flexible approach to differentiation. Unlike countries that require emitters to make national targets for greenhouse gas emissions, the Paris Agreement is based on NDCs, under which each state imposes its own obligations. First and foremost of all, the Agreement reconfirms the principle of the CBDR-RC, acknowledging that differentiation still plays a role in how climate responsibilities are carried out under certain national conditions (Bodansky, Brunnée and Rajamani, 2017). This evolution also marks progress and constraint from a climate justice perspective. On the one hand, however, the Paris framework also allows countries to play in climate action actively through incorporating all states with mitigation obligations. However, this flexible nature of the NDC system begs the same questions, as to whether it's possible for differentiated responsibility to address past inequities. Without rigorous accountability mechanisms, climate aspirations might continue to be politically motivated rather than grounded in objective markers of accountability and competence.

3.3 Climate Justice And Discussion On Historical Responsibility

A key aspect of climate justice is a concept called historical responsibility. The idea is grounded in the knowledge that today's climate is largely the result of decades of greenhouse gas emissions that have accumulated over time. States that experienced early industrialization had a substantial part to play in shaping the environmental conditions that now characterize the world. Historical responsibility has been the subject of much discussion in the legal scholarship. Some scholars argue that those states which were among the largest contributors to historical emissions bear a moral and legal responsibility to take on more responsibility to mitigate and support states that bear some responsibility.

Others warn that scapegoating historical emissions leads to fraught problems of both causation and equity, from that of the intergenerational nature of responsibility and that it involves issues of causation, with respect to non-state causes of global emissions. For all these arguments however, the concept of historical responsibility is at the heart of climate justice talks. It provides support for suggestions for better financial support, transmission of technology, and adaptation aid for the developing world. And it underscores the message that global climate governance should not just consider future emissions, but also the cumulative environmental costs of economic development in the past.

3.4 Differentiation, Vulnerability, And Climate Equity

A second key dimension of climate justice relates to vulnerability. As historical responsibility focuses on the causes of climate change, vulnerability focuses on the consequences. Some states especially small island developing states, least developed countries and countries in parts of the world where water is scarce or climate-sensitive face disproportionate risks from climate threats like sea-level rise, drought and extreme weather events. Climate justice provides that legal and policy frameworks must consider this uneven vulnerability.

Differentiation needs to be based, therefore, not just on emissions responsibility and economic capacity, but on exposure to climate risks. In other words, climate finance, adaptation assistance, resilient building programs should prioritize those countries and communities who most face the impact of climate change. This aspect of climate justice has become increasingly recognized by international bodies. Adaptation finance, capacity building and technology transfer mechanisms have been planned to better support vulnerable states to adapt to climate change impacts. However, plenty of researchers insist that the levels of funding to date fall far short of the scale of issues confronting developing regions.

3.5 Limits Of CBDR-RC And The Requirement Of Enhanced Accountability

While the cornerstone principle of CBDR-RC is perhaps the most ambitious effort to enshrine climate justice in international law practice, its implementation has been debated. Perhaps the most obvious criticism is the lack of enforcement structures to make clear how this can be enforced. Global climate treaties depend more extensively on voluntary compliance and peer review and political pressure than on binding legal sanctions. According to the climate justice analysis this is a weakness in the institution and as such, it is to blame for the problem with respect to accountability. Should distinct responsibilities go unmonitored and unenforced, states could adopt climate pledges that are politically and not environmentally beneficial. Another issue pertains in relation to the ebb and flow of worldwide economic growth.

And the fast-paced economic rise of some emerging economies has obfuscated the distinction between developed and developing economies. Given this shift in emissions models and patterns, identifying appropriate levels of responsibility becomes murkier. The challenge for climate justice is therefore that it depends on adjustable frameworks to adapt to changing global realities while remaining grounded on principles of equity.

3.6 Climate Justice As An Axiom For Future Climate Governance

Nonetheless, the fundamental of CBDR-RC remains at its core in the normative framework of international climate governance. Climate justice helps underscore the significance of this principle by arguing that effective climate cooperation has to do with not just what is scientifically necessary, but what constitutes fair treatment of the participating states. When states see the international climate accords as fair, they will be more likely to take part and fulfill their obligations. In contrast, perceptions of injustice can undermine cooperation and undermine the efficacy of global climate governance.

Climate justice serves, as such, as both a moral framework and pragmatic basis for international cooperation on the environment. Climate justice is likely to lead to greater differentiation in future law. Some of these models may incorporate hybrid models where historical responsibility, economic capacity, and risk factors are used to form the basis of individual or collective climate duties. These methods might further enhance the credibility of international climate agreements, and consequently serve to make the global climate battle even more successful. Eventually, the principle of common but differentiated responsibilities reveals the extent to which international environmental law has sought to moderate the universal nature of climate change with the uneven realities of global development. Climate justice further develops this principle by arguing that legal frames should continue changing in line with the ethical, social and economic facets of climate governance.

4. CLIMATE JUSTICE, HUMAN RIGHTS, AND INTERNATIONAL LEGAL OBLIGATIONS

Climate justice has become increasingly defined in the context of international human rights law in recent years. This move is a sign of more understanding that climate change is not just posing a danger to ecosystems or economies, but also to the delivery of the most basic human rights. Warmer temperatures, extreme weather conditions, sea-level rise, food insecurity and limited availability of water, are all direct threats to people and communities' quality of life and the ability to live in dignity and security. As a result, the interplay between climate change and human rights has become one of the most dynamic fields of modern international law (Knox, 2009). Historically, international environmental law and international human rights law emerged as two parallel fields.

Environmental law was focused initially on the regulation that states need for environmental and natural resources and the environment protection, and human rights law was largely concerned with protection of individual people from abuses of state power. But the increasingly dire plight of global environmental issues especially climate change is blurring the lines between these legal domains.

In conclusion, environmental degradation poses a growing threat to both the fulfilment of human rights, and frameworks for human rights offer a legal mechanism for the implementation of those duties against environment-related harm. From the place of climate justice on the other hand, this convergence is important because it brings the focus of climate governance into the domain of vulnerable individuals/communities. The climate justice framework recognizes that climate change impacts are not distributed in sufficient ways. Marginalized groups, indigenous communities, small island states, and low-income populations often bear the brunt even though they contribute the least to global emissions. The legal and normative framework of human rights thus gives an important normative framing against which to address these inequalities.

4.1 Climate Change As A Human Rights Issue

The acknowledgement of climate change as a human rights issue has grown considerably over the past two decades. Academics and policy-makers growingly admit that the environment impacts a variety of internationally protected human rights. These include the right to life, right to health, right to food, right to water, right to housing and right to a standard of living (Boyle, 2012). Even climate-related disasters like hurricanes, floods, and heatwaves could directly impact the right to life as well. By the same token, the shifting environmental conditions lead to the transmission of diseases, food insecurity, an inability to make ends meet, and water scarcity, and the rights to health and to an adequate standard of living. In coastal areas and island states, sea-level rise poses an existential threat to entire communities on a basis of displacement, bringing into sharp focus the difficult task of climate migration and the responsibility of defending uprooted communities. The need to recognize these interlinkages has led to some international organizations starting to treat climate change as a human rights issue. The United Nations' Human Rights Council has adopted a number of resolutions recognizing that climate change threatens the full enjoyment of human rights. Furthermore, the UN Special Rapporteur on Human Rights and the Environment has played an important role in articulating the legal relationship between environmental protection and human rights obligations. From a climate justice perspective, the human rights approach strengthens the legal relevance of vulnerability. In order to shift our focus from climate change as an environmental conflict for states to those who bear the consequences of this action, frameworks for human rights emphasize the voices of individuals and communities.

4.2 State Responsibilities And Due Diligence In Climate Governance

The incorporation of human rights into climate governance also bears significance to state responsibilities under international law. Human rights law obligates States to avoid violating rights in the clear and mean and to do everything possible to protect people from harm that can be foreseen. This notion of positive obligations is very much integrated in the due diligence doctrine, which holds states responsible for not doing harm to the environment beyond the sphere of their jurisdiction or authority. With respect to climate change, however, due diligence obligations mean an obligation to implement sound mitigation and adaptation n measures in order to protect vulnerable groups from known climate-related threats. Negligence on the part of states to act (re)act about climate change will thus be seen as a breach of international obligations. International jurisprudence has increasingly acknowledged the importance of due diligence. The International Court of Justice (ICJ) stressed the significance of due diligence in environmental protection, not least in transboundary cases of environmental damage. Although these rulings were not specifically dealing with climate change, however, they have helped to establish legal standards to rely upon in disputes that arise over climate change.

There is an additional layer of climate justice in this argument that has reinforced on which case due diligence obligations should be read to include both environmental protection imperatives and human rights concerns alongside environmental responsibilities. States must therefore take a proactive role in curbing emissions, improving resilience and protecting vulnerable people from environmental consequences of climate change (and other climate risks).

4.3 Climate Litigation And Court Accountability

One of the more telling indicators of climate justice-human rights-law connection has been the fast-growing development of climate litigation. For the past 10 years courts have been called on to question whether governments and corporations have complied with their legal obligations under the law on climate change in many locations. Most climate litigation is a hybrid of environmental law, constitutional law, and other human rights principles that must be reached between the parties. Plaintiffs often contend that lax climate policies violate fundamental rights because they imperil people by putting them at risk of suffering foreseeable harm from its impacts. Courts are now more willing to wrestle with these claims, leading to more and more jurisprudence on climate responsibility.

The Netherlands experienced one such case in Urgenda, when the Dutch Supreme Court decided that the government had a legal responsibility to reduce greenhouse gas emissions in a manner that is safe for people's human rights. The court drew on the European Convention on Human Rights, especially the right to life and private life when seeking to rationalise the need for more drastic climate action. In countries like Germany, Colombia and France, among others, some judges have also ruled that states should enforce tougher climate policies that are consistent with its own. Such decisions are the beginning of the broader trend of the judicial recognition of the climate justice cause as an argument in domestic courts.

Litigation on climate also extends well beyond the domestic courts. The cases of climate change have attracted increasing attention and power from regional human rights organizations. Two courts with climate change effects on human rights are the European Court of Human Rights and the Inter-American Court of Human Rights. This is yet another area in which the jurisprudence development continues; but in these cases we can see that the development of the impact of the human rights regimes can be seen as more significant in putting into place a climate governance framework.

4.4 Climate Justice, Vulnerable Groups, And Loss And Damage

A different important aspect of climate justice concerns the security of especially vulnerable populations. Not all communities suffer equally from climate change. Geographic position, economic resources, political resources, social marginalization, all influence the proportion of communities exposed to climate risks. Even small island developing states are the starkest examples of climate vulnerability. Most of these states face existential threats from rising sea levels, even though they represent only a negligible share of global greenhouse gas emissions. Climate justice insists that the international community has an obligation to assist such states in responding to climate impacts and to address irreversible losses. Loss and damage is a topic which attracted many of this attention with regards to the framework of international climate negotiations.

Loss and damage are the climate damages that cannot be avoided through mitigation or adaptation. We may see examples of that in other areas of climate justice work such as the loss of land because of sea level rise, the loss of ecosystems and communities, displacement and loss of life with loss and damage. While legal frameworks that would limit their scope to loss and damage are scantily developed, the concept has come to have considerable political and normative traction. Climate justice activists insist that loss and damage are critical in ensuring fairness within a global climate regime for equitable functioning.

4.5 The Future Role Of Human Rights In Climate Governance

An emerging synthesis of climate justice and human rights law indicates that human rights frameworks could be an increasingly significant element of determining future climate governance. Human rights law offers ways of holding parties to account, of including citizens, and of providing remedy. These are rarely found in the traditional ways we have been talking about environmental agreements. Climate justice, framing climate change as a human rights issue, reinforces, for lawyers and otherwise, the legal and moral underpinnings of climate change action.

It also strengthens the case that to defend the environment is not only a policy objective in the strict sense, but is also a matter of law tied to the preservation of human dignity. However, challenges remain. Human rights frameworks were not conceived to handle problems of a global environmental nature; for instance climate change. Issues of jurisdiction, causality, and attribution still pose challenges to the application of human rights to damage in relation to climate change. However, the swift unfolding of climate litigation and human rights jurisprudence indicates that courts and legal bodies are willing to think through these issues more than ever before. Climate justice can be positioned as a bridge between environmental protection and human rights law. It integrates these fields and contributes to a more comprehensive perspective of state responsibility with respect to global environmental change.

5. CLIMATE JUSTICE, CLIMATE FINANCE, AND GLOBAL GOVERNANCE REFORM

Climate justice is a key focus of dialogue among many in the international discussions on global climate governance and the restructuring of international environmental cooperation. One theme of particular emphasis in this debate is climate finance, one of the main tools of the international community to resolve inequality brought on by climate change. Climate finance is more generally framed as the financial resources mobilised to facilitate mitigation and adaptation – particularly in the developing world, where economic resources are difficult to mobilise for adaptation and mitigation to the climate crisis where there may be limited economic resources to respond adequately and effectively to climate problems (Newell & Mulvaney, 2013). Climate finance must be seen not only as a technical political tool but also as one of the key aspects of issues relating to fairness, responsibility, and international solidarity. But developing country governments have maintained for decades that the expense of climate mitigation and adaptation shouldn't be disproportionately saddled on several states with little to no impact on the historical build-up of greenhouse gas emissions. Instead, climate justice means rather that richer, historically high-emitting states should be asked to play a leading role in investing resources to fight the impacts of climate change.

5.1 Origins And The History And Development Of Climate Finance

The idea of climate finance developed slowly within the framework of the international climate regime. Early climate agreements acknowledged that effective cooperation between countries worldwide would require economic assistance to assist developing nations, especially in the fields of technology transfer, capacity, and planning for adaptation. The United Nations Framework Convention on Climate Change (UNFCCC) set an overall framework for financial assistance, stating that developed countries should provide finance to developing states to meet their climate obligations (United Nations Framework Convention on Climate Change, 1992). These commitments grew into more formal mechanisms of finance over time. The most significant event is the Copenhagen Conference of the Parties (COP15) of 2009, when developed countries committed 100 billion US dollars a year to provide assistance for climate action in developing countries by 2020.

Although this was a massive political undertaking, the realization of this commitment has been inconsistent and the extent of actual finance inflows is said to be extremely low relative to the severity of global climate challenges by many analysts (Roberts & Weikmans, 2017). Later negotiations sought to strengthen the institutional framework for climate finance. This was a major step in such a direction given that the Green Climate Fund (GCF) was also set up. The GCF has been to provide the funding to finance projects that help finance the mitigation and adaptation of developing countries. Despite the fact that a fund has been able to promote a slew of types of projects worldwide, the sufficiency of climate finance instruments, their availability as well as their ability to survive over time is uncertain.

5.2 Climate Justice And The North-South Divide

Climate finance cannot be separated from debates about the global economic order and a history between developed and developing powers. Climate justice exposes deep-seated inequalities in the international system, in particular the economic and technological advantages enjoyed by industrialized countries. Developing countries will claim that our crisis of climate is directly related to industrial development systems established back in the days when global economic expansion was the norm. These economies had large-scale fossil fuel consumption during the nineteenth and twentieth centuries, which contributed to the boom in economic growth while imposing large environmental externalities.

Many developing countries struggle with structural development issues while dealing with climate impact. Climate justice therefore highlights the importance of climate governance addressing not just the environmental but also developmental inequities. Rather, financial assistance, technology transfer, and capacity building schemes are viewed more broadly from a structural perspective, and are viewed not necessarily just as methods of implementation but rather as actions to redress structural imbalances within the international system (Okereke, 2008). This approach is particularly evident in the negotiations about the Global South, where many states emphasize that climate action should only be a part of broad development projects. If emissions reduction is imposed without adaptation, and not in terms of meeting development imperatives, environmental conservation will likely lead to conflicting demands between economic growth and environmental protection.

5.3 Adaptation Finance And The Protection Of Vulnerable States

Whereas in early years climate negotiations primarily dealt with mitigation cutting greenhouse gas emissions the importance of adaptation finance has begun to receive more emphasis in recent conversations. Adaptation means to work on reducing the susceptibility of communities and environments to climatic influences. These include such measures as improving coastal defences, improving water management systems, developing climate-resilient agriculture and upgrading disaster preparedness.

This is particularly important for fragile states which are facing severe climate risks but contribute relatively little to global emissions - adaptation finance is therefore key. For instance, small island developing states are faced with existential threats from sea-level rise, and many countries in Africa and Asia are now experiencing an increase of drought frequency and extreme weather events. Several international organizations have tried to tackle these problems by providing financial support to the development of adaptation initiatives specifically for developing countries (for example, the Adaptation Fund and the Green Climate Fund). However, there remain deep funding gaps. Current levels of adaptation financing fall far short of what is needed to adequately meet the estimated magnitude of climate impacts (UNEP, 2022) according to a multitude of international assessments. The critical issue of closing this adaptation gap is crucial from a climate justice perspective. In these vulnerable states may have little means to protect their people and ecosystems from climate change. Climate justice thus positions adaptation finance as a moral imperative as well as a practical necessity.

5.4 Loss And Damage And The Development Of New Financial Instruments

One of the fiercest themes of controversy in climate negotiations pertains to loss and damage. Unlike mitigation and adaptation (both strategies to prevent or tackle the effects of climate change) the loss and damage concept aims to address such harms as they are inevitable and/or irreversible. These could be the loss of permanent territory that result from sea-level rise, destruction of cultural heritage sites or displacement of communities forced to leave their homes. For plenty of developing countries, especially s

mall island states, loss and damage is one of the most urgent aspects of climate justice. These states say that international law has to acknowledge the responsibility of the rest of the world to deal with harms that are not prevented by conventional climate policies. The creation of the Warsaw International Mechanism for Loss and Damage in 2013 represented a significant advance in institutionalizing this idea within the framework of the climate regime. In short, it was international negotiations that took place, and led to the establishment of a Loss and Damage Fund for the funding of countries enduring damage from extreme climate impacts. While this fund represents a great political success, many questions of its governance, funding and operating functions have yet to be explored. But critics argue that the loss and damage framework with no large-scale financial commitment from developed states is unlikely to catch up with the aspirations of fragile countries.

5.5 Climate Justice And Reform Of Global Environmental Governance

In addition to various financial instruments, climate justice brings broader issues about the architecture of global environmental governance to the fore. International climate regime of today is to be very much based on voluntary commitments, peer review and negotiations. And while these measures have enabled mass participation, they are frequently poorly enforced and do not represent a reliable mechanism of enforcing climate pledges. Some academics suggest that these structures might rely on stronger institutional processes, to include: better monitoring and accountability systems, and better coordination among international institutions. Climate justice reinforces such reforms, as the principle that equitable governance arrangements are necessary to sustain trust in states in these interactions is stressed by climate justice. The other dimension is that climate justice has led to broader participation of non-state actors in climate governance. Civil society, indigenous groups, academia, and private entities increasingly have a significant role in climate policy and advocating for environmental justice/accountability. Their participation promotes transparency and can support the development of climate policy that is open to stakeholders so that people affected by the climate crisis are at the heart of the matter.

5.6 Climate Financing As The Basis For Fair Climate Governance

In the end, climate finance sits atop the broader umbrella of climate justice. Financial mechanisms represent one of the fundamental ways in which the international community seeks to translate a set of principles of equity and responsibility into actual policy results. But the success of climate finance depends not just on how much money has been mobilized but on how these assets are assigned, managed and observed as well. Transparent governance mechanisms, fair distribution of resources and the involvement of the recipient communities are all important considerations for real climate finance to advance true climate justice. Climate justice thus offers as both a normative and policy anchor for future global environmental governance reforms. Climate justice promotes fairness, responsibility, solidarity, and transparency, calling for financial, and institutional, mechanisms to address the multifaceted inequalities in relation to climate change, and other complex social questions within the international community.

6. CLIMATE JUSTICE, CLIMATE LITIGATION, AND EMERGING JUDICIAL DEVELOPMENTS

Climate litigation is among the biggest legal events that have been shaping the evolution of climate governance and environmental care over the past few years. Courts across the globe are increasingly being called on to adjudicate whether governments and corporations have met their obligations to address climate change. These courts demonstrate a larger shift by which climate justice arguments are transitioning from literature to the courts. As such, climate litigation forms one significant method through which the principles of climate justice (namely accountability, equity, protection of vulnerable populations) can be translated into binding legal responsibilities (Peel & Osofsky, 2015). Several factors contributed to the expansion of climate litigation. First, the accelerating effects of climate change have increased public pressure to respond legally. Second, the perceived inadequacy both of political and diplomatic processes has led activists, civil society organizations and impacted communities to seek redress in the courts. Third, international human rights law and environmental law developments have created new legal frameworks through which courts can creatively interpret climate obligations. All of these aspects coalesced into a rapidly growing climate law field that spans multiple jurisdictions.

6.1 Climate Litigation as an Instrument for Climate Justice

Climate litigation has a range of critical roles to play in climate justice. Above all, it serves as a way to make governments and private investors accountable for bad climate policy. World climate treaties are, in many cases, based less on explicit commitments and more on voluntary or negotiated agreements, and thus less binding. Judicial cases, on the other hand, permit people, and even organizations, to challenge the behaviour of states which threaten environmental protection or violate individual rights. Climate litigation can also help establish legal norms to specify the extent to which states should respond in various ways. Courts can also advance and extend existing, general legal concepts like due diligence, precautionary action, or human rights in response to climate related ills through judicial readings. Climate litigation therefore has a real influence on how international environmental law is further shaped. Likewise, litigation over climate can amplify the perspectives of marginalized communities whose political will is frequently marginal to the mainstream. Indigenous groups, youth activists and people living in poor lands around the world have increasingly turned to the courts to contest insufficient policy around warming. These cases are social and distributive, and bring some of the fundamental arguments in climate justice.

6.2 The Urgenda Case and New Climate Litigation Avenues

it's all based on rights. Urgenda Foundation v. State of the Netherlands is one of the most significant climate litigation cases in recent years. This landmark judgment by the Dutch Supreme Court established a legal obligation of the Dutch government to adopt stricter emissions-reduction steps to safeguard the rights of citizens under the European Convention on Human Rights. It ruled that too little on climate could violate rights to life and private life, and set an important precedent for rights-based litigation on climate. The Urgenda ruling reinforced the view that the courts will apply a set of human rights obligations that obligates governments to respond to climate change responsibly. The case's legal substance is strengthened by linking environmental protection to human rights obligations (Setzer & Higham, 2021). And the case of Urgenda is also relevant beyond the Netherlands. Its reasoning has been adopted by courts in many other places, and ever more lawsuits, grounded in law, show the connection between climate policy and the protection of human rights.

6.3 Climate Cases and Intergenerational Justice in the Court of Law

Another important aspect of climate litigation has been the search for constitutional challenges to insufficient climate practices. The courts have long recognized the way that constitutionally guaranteed rights (such as the guarantee of fundamental rights and the responsibility of the state to safeguard the environment) could underwrite obligations of governments to take stronger climate actions. One such example is the German Federal Constitutional Court's ruling on the Climate Protection Act of 2021. But in this case, the legal framework the government adopted in its passing of climate law failed, the court found, because it imposed too much cost on future generations. It emphasized the notion of intergenerational equity in its decision, and it said that laws for today should not take away future citizens' rights by deferring key cutbacks on carbon emissions. This decision is a significant step forward in a field of climate justice jurisprudence that recognizes those people's rights when constitutional law comes into force through this decision. It was significant because, in emphasizing long-term environmental responsibility, the court reaffirmed that whatever climate governance development state emerges by "the time and conditions of today", must account for both today and tomorrow's interests.

6.4 Courts for Regional Human Rights and Climate Litigation

And regional human rights organizations have also begun to accommodate some climate-related claims. While the jurisprudence on this issue is limited, regional courts' decisions could shape climate law to a significant extent, as of now. The European Court of Human Rights (ECtHR) has heard multiple cases regarding the human rights-related issues concerning climate change. These are cases that pose the question in today's climate-focused world of whether governments have properly met their obligation to protect citizens against foreseeable environmental harm. As well, if it indeed accepts climate-related human rights abuses, these judgments could offer hopeful legal precedent in climate governance across Europe. Similarly, when it comes to environmental protection, the Inter-American Court of Human Rights has ruled on climate change-related cases.

The Court has reiterated its explicit acknowledgement of these and other human rights issues as important areas of environmental protection, including, with respect to Indigenous peoples and other vulnerable groups, on the Court in its advisory opinions and on its judgments. These readings suggest that regional human rights systems may increasingly shape climate justice law and jurisprudence (Savaresi & Auz, 2019).

6.5 Corporate Responsibility and Legal Conflict Over Climate

While many climate cases have focused on the blame of the federal government, lawsuits against corporations have also been on the rise. For years fossil fuel companies, large emitters and a great number of other major emitters have been responsible for causing great harm to the environment and failing to disclose the risks brought about by climate change on an environmental level. Various suits have sought to hold corporations to legal account for breaches of environmental health stemming from fossil fuel production. These issues often go hand in hand in a patchwork of legal theories tort law, consumer protection law and securities regulation. Results vary from jurisdiction to jurisdiction, but a legal lesson from corporate climate litigation demonstrates a broad realization that today's world understands that taking responsibility for climate change is a responsibility of a wider range of actors than just states with notable power. Corporate accountability is, from the lens of climate justice, critical in a world where much of the driving force behind climate change is performed by private actors that operate independently on global markets. So, lawsuits targeting corporate emissions can fit nicely amid broader efforts to address structural causes of damages to the environment.

6.6 Climate Litigation and Environmental Accountability

The rapid growth of climate litigation indicates that courts are becoming an even more central arena for fighting environmental responsibility. Judicial decisions can't replace political negotiation or international collaboration, but help them build clearer legal commitments and better mechanisms for accountability. In fact, the Court's cases of climate justice would perhaps have had an important influence on these legal changes. Climate justice can frame how to understand justice to the courts with respect to the climate crisis as a normative matter that entails fairness, accountability and preservation of that world's most vulnerable communities and individuals. Indeed, given climate impacts are increasing, this should have implications in multiple jurisdictions and contribute to climate litigation, too.

Courts will also be more increasingly urged to consider whether governments and companies have taken sufficient steps to address climate risks. And in doing so, they will move towards the step-wise shaping of the criteria of international law on climate responsibility. Climate litigation, meanwhile, represents a crucial link between legal theory and practice in this respect. The abstract principles of climate justice are thus operationalized and turned into actual legal claims, having the potential to influence policy decisions as well as future international environmental law direction.

7. FUTURE LEGAL DEVELOPMENT AND REFORM OF INTERNATIONAL CLIMATE GOVERNANCE

From the top to the bottom, in an era characterized by an ever-worsening climate change crisis, the shape of future international environmental law will depend on how accurately lawmakers apply the tenets of climate justice as instruments for governance. But although climate justice has already played a role in the discussions of responsibility, finance and human rights and litigation, its systemic impact on reform of global environmental governance has not been defined firmly yet. The future of the international climate law terrain will evolve along numerous lines of institutional reform, doctrinal solutions towards accountability from an accountability and state accountability standpoint (Bodansky et al., 2017).

The greatest challenge to the international climate regime lies in the chasm between political aspirations and the implementation. International agreements, particularly the Paris Agreement, established a global benchmark for climate cooperation, but these agreements also place great reliance on voluntary commitments by states. While this framework has allowed for wide participation and transparency, it has also exposed the global inadequacy of climate commitments to date. Climate justice therefore encourages the establishment of governance mechanisms that can lead to the establishment of more transparent, efficient and accountable international climate governance (regulatory, monitoring of compliance and holding the international climate regime to account).

7.1 The New Formative Stage of the International Courts and Advisory Opinions

The influence of international courts and advisory opinions on climate governance. More and more international legal institutions have started to address questions of climate (and environmental) impacts in recent years, as there is increasing acknowledgment that climate change leads to nuanced issues of international responsibility. An important recent development is the call for an advisory opinion from the International Court of Justice (ICJ) on states' obligations with respect to climate.

This collaboration, which involves a coalition of small island developing states and other vulnerable states, aims to provide clarity as to the obligations of states as defined in international law to protect the planetary system, which is affected by climate change. If ICJ gives an advisory opinion ultimately regarding these issues, the decision could have far-reaching implications for the establishment of international environmental law.

In a similar light that contentious judgments generate binding obligations, an advisory opinion does not necessarily create binding obligations, and therefore can only take the form of a piece of the bargain in international legal practice (Voigt, 2016). However, from this perspective of climate justice, the process at the ICJ serves as a reminder that very vulnerable states are looking for a legal clarification on what significant emitting countries are doing. Such clarification could bolster normative expectations about mitigation, adaptation assistance, and climate finance.

7.2 Strengthening Climate Governance (CCG) Institutions Around the World

Reform in another aspect in future relates to the institutional architecture of global climate governance. The existing global climate regime exists largely under the United Nations Framework Convention on Climate Change (UNFCCC), comprising COP-system. Although this system has made international disputes simpler, it is described as slow in taking decision and harder to find consensus among states with competing interests. Global climate peace and justice academics contend that enhancing international climate governance may necessitate innovative administrative tools that promote an increased level of cooperation and accountability.

Possible reforms entail better tracking of national goals and responsibilities related to climate, tougher reporting requirements for their implementation and clearer frameworks for gauging the extent to which national policies align with targets for global temperatures. Greater coordination among international organizations might also be needed. Climate change cuts across trade and development policy, human rights law, and international finance and is governed by various global regimes. Effective climate governance thus depends as much on collaboration between groups including the United Nations, the World Bank, regional development financial institutions and global environmental NGOs as it does on state power.

7.3 Climate Justice and Sustainability

Yet the future evolution of international climate law also depends on whether its frameworks fit within wider frameworks of sustainable development. It's linked inextricably to poverty reduction, food security, energy access, and economic development. As such, climate change policies will need to strike a balance between what is needed to address climate change, its impact on land use, and to those who are most vulnerable to global warming, both socially and economically. Sustainable development strategies should balance the needs of vulnerable populations with long-run sustainability for the environment, Climate Justice argues.

The United Nations Sustainable Development Goals (SDGs) provide an important platform by aligning climate action with other development needs. Specifically, SDG 13 (Climate Action) reinforces that this is a matter of urgent concern for action on climate change and its immediate harms, and when cooperation is to occur at an international level, it must occur simultaneously (United Nations, 2015). Incorporating climate justice in sustainable development models may guarantee that climate policies drive shared economic transformations rather than replicate disparities. This could translate into, for instance, improved access to clean technology, financial channels and resource flows for climate finance (if relevant for the developing world) and capacity development that can underpin sustainable growth:

7.4 The Increasingly Prominent Role of Non-State Actors in Climate Governance by NGOs and Non-State Actors in the Future

There's also a stake in this for non-governmental organisations and the growing number of non-state actors. Climate policies are increasingly crafted by cities, local governments, private companies, civil society organizations and academic institutions. Several non-state actors help shape climate governance. Private companies invest in renewable energy technologies and sustainable production systems. Cities have climate adaptation and greenhouse gas reduction strategies. Civil society groups call for better environmental safeguards and pressure governments to honour environmental promises.

According to a climate justice theory, the institutional engagement of those institutions should be recognised, and reinforced, in forms of governance in an inclusive manner. More transparency and public involvement can strengthen the legitimacy of climate policies and the extent to which environmental solutions are in the best interests of the communities affected by them.

7.5 Struggling for a Fairer Global Climate Order

At the end of the day, international environmental law's future will be whether it can combine (or more appropriately, reconcile) that scientific imperative and climate change reality with the political and economic realities of a global order. Climate justice gives a framework that must serve as a guide for conduct of actions and addressing this crisis. Climate justice serves as a key to the challenge of global cooperation for fairness, responsibility, and solidarity which are the basic guiding values of the world and is why climate justice offers up essential principles to guide us through this road to success. These principles of climate justice-oriented legal reforms could yield better mechanisms for climate finance, clearer standards through which states can be held responsible, and increased protection for vulnerable communities from harm.

These are not political devices that will bring down the curtain on international climate negotiations as they tend to; what they would accomplish, in the end, is ensure that world-wide climate governance develops according to what works and to equitable principles. Given that climate change is already transforming the way global environmental, economic and social systems work, international law will remain key in the space through which responsibility and cooperation will be explored. Climate justice thus is neither just an idealizing principle as to the future nor its existence, but more of a practical blueprint for what global governance of climate ought to comprise.

CONCLUSION

The concept of climate justice has emerged as one of the key frames by which to understand legal, ethical and institutional challenges stemming from global climate change more broadly. As the chapter shows, if climate change was to be eradicated with technical environmental regulations and a narrowly defined policy instrumentation alone, a comprehensive legal approach is needed. It acknowledges the unequal distribution of responsibility, capabilities and vulnerabilities among states or communities. Climate justice, then, offers an important normative framework through which international environmental law can critically assess current governance institutions and point toward new reforms.

It is the case throughout the chapter that climate justice challenges the very assumptions upon which international environmental law rests. Formerly, environmental governance, in particular, depended on the idea of formal equality between states, under which it was assumed that states were responsible for meeting common responsibilities to address environmental challenges. But the global climate crisis has the downside of laying bare the limitations of such an approach. States have given uneven responses to greenhouse gas emissions, have radically different economic and technological capabilities, and are subject to climate effects in profoundly unequal ways. As such, effective climate governance requires incorporating equity and differentiated responsibility (Rajamani, 2012).

One of the key instruments through which climate justice had an impact on international environmental law was that of the principle of common but differentiated responsibilities and respective capabilities (CBDR-RC). It acknowledges that all states should, as a group, share a common obligation to safeguard the global climate system and yet that that responsibility has to be based upon historical emissions, economic means and degree of risk. While this principle has been embedded in international climate agreements for many decades, this principle is not consistently enforced everywhere. We must reinforce the operationalization of differentiated responsibilities as part of future work in the international climate governance. The chapter also has drawn attention to increasing intersection of climate justice and international human rights law. Climate change presents direct challenges to essential human rights such as the right to life, health, food, water, and housing.

As a result, climate governance discourses and litigation have gradually utilized human rights frameworks. The judicial rulings in several courts suggest that the courts are also starting to construe the legal obligations currently established in the interest of compelling governments to take more aggressive climate measures. Such trends suggest that human rights law itself will become the foundation to climate accountability. A related central aspect of climate justice is its closely related key component climate finance and global economic inequality. International climate policies take the position that developing countries need financial and technological assistance if they are to confront climate change.

The Green Climate Fund and other adaptation finance schemes are effective instruments that address these concerns. But substantial funding gaps remain and whether climate finance mechanisms are reliable mechanisms of climate finance remains a major point of discussion. Climate justice emphasizes such financial support cannot be framed from the vantage point of voluntary contribution not as a tool of charity, but as a structural part of a broader framework of responsibility grounded on historical emissions and global equity. Additionally, this chapter has examined increasing significance attached to climate litigation on these issues as well as the emergence of climate justice. In many jurisdictions, courts have been asked whether governments and corporations have taken sufficient action to combat climate change.

In landmark events such as Urgenda, and in a number of constitutional climate decisions, legal institutions can further this goal of enhancing environmental accountability. Judicial processes could never do the disjunction business associated with political negotiation, but they also have the potential to build legal norms in relation to climate responsibility (Peel & Osofsky, 2015). By way of doing so the chapter not just addressed the law, but also the wider implications of climate justice's relevance for the reform of global environmental governance. International climate cooperation today is a networked affair, comprising treaties among countries, institutions to help regulate and enforce that cooperation, as well as treaties they broker and negotiate; treaties between states.

This system has come a long way towards global climate action, but has faced tremendous challenges against enforceability, accountability, and equitability. Thus, climate justice advances reforms for better transparency, strengthened governance, robust monitoring, and empowering the most vulnerable to contribute adequately to decision-making. International climate law might be an evolving field, possibly with broader involvement, too, by international courts and advisory societies. Recent calls to the International Court of Justice and a number of other institutions to provide advisory opinions indicate more efforts are being made to determine the legal responsibilities that states have when considering climate change. These developments may make the move toward the development of more precise legal obligations concerning environmental issues at the international level (Voigt, 2016).

Climate justice, by contrast, helps us see that the legitimacy and efficacy of the international climate governance of states does not only hinge on scientific knowledge or economic concepts alone but also on a measure of justice and state kinship. No global collaboration on climate change is likely to work unless international institutions remember the deep, societal inequalities on which they've been based.

Combining alongside those objectives, climate justice in the scope of international environmental law is therefore not an optional moral wish; rather, it is an ontological truth for the world to become sustainable in terms of global climate governance. Legal systems that rest on principles of equitable, accountable and participatory governance are more likely to secure long-term cooperation and to make a difference in policy and results. Accordingly, in a change vision, climate justice is a paradigm change for reconsidering the applicability of international law in responding to the climate crisis.

Concentrating on fairness, equity, accountability and the rights of the least favoured, climate justice serves one the purposes of buttressing, independently, the normative foundation for environmental politics, but also of generating institutional realignment to deal with the most difficult problems of the twenty-first century. As international environmental law grows, however, the concepts of climate justice will remain a focus of both scholarly and policy work in pursuing the development and sustainment of cooperation to address the climate system in the world through the multilateral process that supports it, whether at home, abroad, or by means of the worldwide climate crisis.

REFERENCES

Bodansky, D., Brunnée, J., & Rajamani, L. (2017). *International climate change law*. Oxford University Press.

Boyle, A. (2012). Human rights and the environment: Where next? *European Journal of International Law*, *23*(3), 613–642.

Intergovernmental Panel on Climate Change. (2023). [*Synthesis report*. IPCC.]. *Climatic Change*, •••, 2023.

International Court of Justice. (2010). *Pulp mills on the River Uruguay (Argentina v. Uruguay)*, Judgment, I.C.J. Reports 2010.

Knox, J. H. (2009). Climate change and human rights law. *Virginia Journal of International Law*, *50*(1), 163–218.

Newell, P., & Mulvaney, D. (2013). The political economy of climate justice. *Review of International Political Economy*, *20*(1), 132–158.

Okereke, C. (2008). Equity norms in global environmental governance. *Global Environmental Politics*, *8*(3), 25–50.

Peel, J., & Osofsky, H. M. (2015). *Climate change litigation: Regulatory pathways to cleaner energy*. Cambridge University Press.

Rajamani, L. (2006). *Differential treatment in international environmental law*. Oxford University Press.

Rajamani, L. (2012). The principle of common but differentiated responsibilities and respective capabilities. *The International and Comparative Law Quarterly*, *61*(2), 493–523.

Roberts, J. T., & Weikmans, R. (2017). Postface: Fragmentation, failing trust and enduring tensions over what counts as climate finance. *International Environmental Agreement: Politics, Law and Economics*, *17*(1), 129–137.

Savaresi, A., & Auz, J. (2019). Climate change litigation and human rights. *Journal of Human Rights and the Environment*, *10*(1), 76–101.

Schlosberg, D. (2007). *Defining environmental justice: Theories, movements, and nature*. Oxford University Press.

Setzer, J., & Higham, C. (2021). *Global trends in climate change litigation*. Grantham Research Institute, London School of Economics.

United Nations Environment Programme (UNEP). (2022). *Adaptation finance gap report*. UNEP.

United Nations Framework Convention on Climate Change. (1992). *United Nations Framework Convention on Climate Change*.

United Nations. (2015). *Transforming our world: The 2030 agenda for sustainable development*.

Voigt, C. (2016). Equity in the Paris Agreement. *Climate Law*, *6*(1–2), 50–69.

Weiss, E. B. (1989). *In fairness to future generations: International law, common patrimony, and intergenerational equity*. United Nations University.

World Commission on Environment and Development. (1987). *Our common future*. Oxford University Press.

Chapter 3
Climate Justice and Textile Industry in the Global South:
A Constitutional and International Law Perspective

Pallavi Pallavi
http://orcid.org/0009-0009-3217-3583
Central University of South Bihar, Gaya, India

Ashok Kumar
http://orcid.org/0000-0002-9204-6051
Central University of South Bihar, Gaya, India

ABSTRACT

In today's globalised environment, where there is constant development in law and technology, the lines surrounding the border between each nation have been diminished as far as environmental problems are concerned. Among the many factor of pollution, the textile & fashion industry is a major contributor, from growing materials in one country, their manufacturing in another, and their point of sale in other country, making it an absolute necessity to focus on it in relation to legal implications, climate change, and sustainability. This Chapter explores how courts have played a part in invoking constitutional rights to enforce environmental justice. It will also evaluate how environmental rights and international climate justice frameworks can be utilised to ensure the sustainable and equitable regulation of the textile industry in the Global South. The researcher will also highlight the issue of the unbalanced distribution of the contribution and burden of environmental problems.

DOI: 10.4018/979-8-3373-6635-7.ch003

1. INTRODUCTION

The richest countries, the "Global North," are responsible for nearly 50% of the emissions generated since the Industrial Revolution, as estimated by the World Inequality Database (Intergovernmental Panel on Climate Change [IPCC], 2012). The comfortable living standards of European, North American, and many other "Global North" country citizens create a footprint 100 times greater than the impoverished living standards of the impoverished nations (Chancel, 2022). The top 10% of the world's biggest polluters, which is equal to 771 million people, were responsible for nearly 48% of the emissions generated globally, whereas the bottom 50%, which equals 3.8 billion people, were responsible for nearly 12% of the emissions generated globally as of 2019 (Chancel, 2022).

This shows that the least developed countries, which are mostly from the "Global South," contribute the least to the increase in the earth's temperature. This also proves that the least developed countries have received a lesser share of the direct benefits of the use of fossil fuels, which are energy consumption.Climate change is often analysed, explained, and understood as an environmental issue, but it also symbolises climate justice, which emphasises the imbalance in climate debts and rights among governments, societies, and people. Climate justice emphasises that countries and people that contribute little to global greenhouse gas emissions are most at risk from the devastating effects of the climate catastrophe. Justice in the Climate Disasters is anchored in equity, responsibility, and human rights (Dolšak & Prakash, 2022). Climate justice essentially draws attention towards the inequitable nature inherent in the state of the global climate system, where the developed countries (Global North) are essentially responsible for the concentration of atmospheric carbon and have been benefiting from their development; in contrast, the Climate Southerners have been facing higher prospects (Füssel, 2010). The risks include rising sea levels, more frequent and severe weather events, failed agriculture, water shortages, loss of biodiversity, food insecurity, as well as displacement. Vulnerabilities are heightened by the existing conditions of poverty, inadequate infrastructure, and adaptive capacity.

While their effect on global warming is insignificant, least developed countries (LDCs) are often underrepresented in international arenas of global warming decision-making, thereby further diminishing their impact on setting the global warming agenda. Existing global climate governance frameworks often fail to adequately consider historical obligations, economic inequalities, and the distinct needs of at-risk groups (Wang & Fang, 2024).

The objectives of this chapter include the following: First, it explores the interconnection between constitutional rights and environmental rights within the Global South jurisdictions. Second, it analyses the role of the courts in advancing environmental justice, especially in the textile industry. Third, it assesses the tension between industrial growth and sustainable development within the textile industry of the Global South. Fourth, it analyses the relevance and limitations of climate justice frameworks for the textile industry.

An unbiased and equitable approach, free of boundaries, is required to transform the textile industry into a sustainable sector that upholds environmental protection and equitable burden-sharing.

2. METHODOLOGY

The research methodology for the chapter is doctrinal. It involves a comprehensive literature review to synthesise existing knowledge on Sustainability, Climate Change, and Environmental rights. There will be a structural analysis of existing legal rules, principles, and judicial pronouncements relevant to environmental rights, sustainability, and the textile industry. Case studies from various parts of the global south will be analysed to understand the real problem. Scholarly articles, books, policy papers, and reports of international organisations are used to provide critical insight and contextual understanding. Ethical considerations will be prioritised throughout the research process, and an interdisciplinary approach will integrate insights from environmental science, sociology, and ethics. This methodology aims to produce rigorous empirical evidence and theoretical insights, and to contribute to a comprehensive understanding of the relationship among Sustainability, environmental protection, and development.

3. LITERATURE REVIEW

The academic literature on climate justice has shifted from the early environmental protection stories to a thriving academic field that encompasses international law, human rights, political economy, and ethics. Among the first major contributions to the climate justice literature are the works of Henry Shue, for instance, "Climate Justice: Vulnerability and Protection," wherein the author argues that climate change is a justice issue because those who are causing the emissions are the ones who are most vulnerable to its impacts. The concern for historical responsibilities and moral obligations, as seen in the works of Shue, strikes a chord with the later works on climate justice. Similarly, the works of Simon Caney, for instance, "Justice Beyond Borders," frame the issue of climate change within the context of global distributive justice, wherein states are expected to share the burden equitably according to their capacity.

In international environmental law, Philippe Sands has made a significant contribution in shaping the concept of how environmental principles, such as sustainable development, precaution, and polluter pays, have evolved. His contributions have shown how these principles, although recognized in international law, are not enforceable, especially in instances where economic interests are at play. Edith Brown Weiss's theory on intergenerational equity further strengthens the climate justice movement in its assertion that the current generation is a custodian of environmental resources for future generations, a notion that strengthens constitutional environmental law in the Global South. The concept of CBDR has been subject to scrutiny by several scholars, such as Lavanya Rajamani, who focuses on the UNFCCC and Paris Agreement, critically assessing how equity has been undermined in these global climate change frameworks. Rajamani argues that, although the Paris Agreement makes rhetorical commitments to equity, its focus on voluntarism places an unfair burden on developing nations. In industries with high levels of climate change, such as textiles, economic integrity in the Global South may be undermined if costs are a barrier.

The connection between climate change too human rights has been another key area of engagement for current scholarship. John Knox, former United Nations Special Rapporteur on Human Rights and the Environment, is one of the key individuals who have played an important role in foregrounding the normative connection between the environment and the length of human rights. Sumudu Atapattu's research on the connection between environmental justice and human rights explores how the experiences of developing countries are marginalised in the context of industrial pollution and climate change through international law. In the constitutional and domestic environmental law tradition, scholars such as Shyam Divan and Armin Rosencranz have written extensively about the role of the judiciary in advancing environmental rights through public interest litigation. Their works document how the right to a healthy environment under Article 21 of the Indian Constitution has been judicially construed to keep polluting industries, such as the textile sector, in check. Upendra Baxi's critical approach to environmental movements warns against the dangers of judicial activism and the symbolic provision of environmental justice when fundamental economic disparities remain untouched.

The literature on specific sectors, including the textile and fashion sector, has been supplemented by interdisciplinary scholars/activists. Kate Fletcher's scholarship on sustainable fashion examines fast fashion's business model and highlights its lack of sustainability. Another example is Safia Minney, founder of People Tree, whose scholarship and activism on sustainable fashion advocate for ethical fashion and sustainable trade. In scholarship on the Global South, authors such as Mark Anner examine labour exploitation, supply chain governance, and other aspects of garment manufacturing, especially in Bangladesh, highlighting the connection between environmental and worker exploitation. Activism has also been very important in defining climate justice talks, alongside academic writings. Vandana Shiva, an activist and academic in the field of environment, has been instrumental in discussing anti-industrial agriculture, corporate globalisation, and resource extraction, including cotton farming and textile manufacturing. Naomi Klein, through books like "This Changes Everything," among others, has been crucial in defining a response to climate change that cannot be discussed without addressing economic inequalities.

Recent research has also highlighted the significance of legal and institutional strategies for achieving sustainable development goals beyond environmental regulation. For example, Alqodsi and Abdallah (2024) have highlighted the significance of legal frameworks in achieving sustainable development goals alongside educational and institutional reforms. The analysis in this article highlights the role of sustainability goals in achieving environmental sustainability, thereby reinforcing the theory for this chapter by emphasising the importance of climate justice within broader legal frameworks for sustainable industrialisation.

Nevertheless, the current literature still presents certain knowledge gaps. Most literature on the topic addresses climate justice or environmental law on the one hand, and industry regulation on the other. Interventions on the topic remain exceptionally limited in the legal literature, exploring the textile industry within the rubric of the combined constitutional and international climate justice paradigm as conceived in the Global South. This paper addresses the identified knowledge gaps to present a more integrated picture of sustainability within the textile industry.

4. THEORETICAL FOUNDATIONS

"The history of industrialisation, and in particular the history of the North, bears a profound connection to the origins of climatic injustice" (Page, 2008). In the 18th and 19th centuries, nations across Europe, North America, and parts of East Asia initiated the process of industrialisation. These developing countries relied mainly on fossil fuels such as coal, oil, and gas to power industrialisation. Due to their long reliance on fossil fuels, they achieved unprecedented economic growth and technological innovation (Page, 2008). Notwithstanding, this progress was accompanied by a massive toll on the environment, including the accumulation of greenhouse gases in the Earth's atmosphere, which has been identified as the main driver of climate change (Intergovernmental Panel on Climate Change [IPCC], 2012).

On the other hand, there are many nations in the Global South, including a large part of Asia, Africa, and Latin America, in addition to some island nations, that either remained under colonization in the age of the Industrial Revolution or did not have enough means to pursue a high-carbon development trajectory and have made a very insignificant contribution to climate change in the whole history, despite being the ones most impacted by this effect, while among the biggest worldwide imbalances there is this: those who contribute least to the creation of climate change are those who suffer the most from its effects (IPCC, 2023). To address and equalise such an imbalance, two key principles have emerged for the realm of international environmental law, namely:

Polluter Pays Principle- This concept holds that the burden of paying for the costs of managing and remedying environmental damage ought to be borne by those countries, companies, or individuals responsible for the damage (Khan, 2015). It implies that nations and companies that were bigger polluters in the past ought to be held responsible for the environmental damage caused by those emissions. The process of taking responsibility for those damages should be achieved by reducing the environmental damage they now cause, as well as compensating those nations for environmental damage occasioned by those emissions (Heine, Faure, & Dominioni, 2020).

CBDR-RC - The aforementioned was first enshrined in the UNFCCC 1992 and has been reiterated in various agreements thereafter, recognizing that while all countries share a Common Responsibility to deal with climate change, they do this in light of their different capabilities and responsibilities in this matter in the past and in the present (Wang & Gao, 2018). It is expected of developed nations to lead in reducing emissions and to provide economic and technological assistance to developing nations to help them adapt to the impacts of climate change.

These guiding principles, considered together, are the philosophical foundation for the concept of Climate Justice (Roberts & Parks, 2007).

5. CONSTITUTIONALISING OF ENVIRONMENTAL LAW IN THE GLOBAL SOUTH

A. India

Among the first priorities after attaining independence included the promotion of economic development. Nothing in the Constitution protected the environment from accelerated deforestation and river pollution; at the same time, air quality was steadily deteriorating (Divan & Rosencranz, 2001). The IFA was declared inadequate because the courts were not empowered to enforce it fully. The "Lex non cogit ad impossibilia" was cited. Without the backing of the constitution, protecting the environment was not feasible. The inadequacies of the legislative system were not remedied; instead, the deterioration worsened (Divan & Rosencranz, 2001).

India in 1972 was a participant at the UNCHE marking a historical milestone in the global setting. It recognised the right to a healthy environment (Baxi, 2010). Principle 17 establishes a connection between environmental protection and human rights. The idea was warmly received in India. It is also committed to promoting sustainable development. The Declaration significantly influenced domestic policy. The document ushered in constitutional reforms and even spurred judicial activism, thereby giving rise to the prominence of salus populi suprema lex (Kingsford et al., 2021).

The Declaration laid the foundation for constitutional amendments. The Declaration emphasised the urgency of ecological preservation to which India pledged unambiguously (Kingsford et al., 2021). In the year 1976, the Parliament of India translated the commitment made by the then Prime Minister, Mrs Indira Gandhi, at the United Nations Conference on Human Environment held in Stockholm into action. The Stockholm Conference highlighted the various perspectives that formed a core part of India's environmental philosophy, as reflected in subsequent government policy declarations (Breggin & Pendergrass, 2007). These provisions are stated as under:

- Article 48A states: "Protection and enhancement of the environment, as well as the preservation of forests and wildlife." The State shall make every effort to preserve and enhance the environment and to conserve the country's forests and wildlife (Sharma, 2025).
- Enacted the Constitution (Forty-second Amendment) Act, 1976. The Forty-second Amendment Act sought to make certain provisions of the Indian Constitution more explicit. The Amendment introduced two major provisions under the Indian Constitution: Article 48A and "Fundamental Duties" (Sharma, 2025).

B. Bangladesh

Although the Constitution of Bangladesh did not contain an explicit environmental right, the judiciary held that the constitutional right to life under Article 32 included the right to a healthy, pollution-free environment. Through PIL, the courts dispensed with traditional rules of locus standi and allowed environmental activists and civil society organisations to approach the judiciary on behalf of affected communities and natural resources (Mohammad, 2012). One of the most significant developments is the granting of juridical personality and legal rights to rivers, and the establishment of the National River Protection Commission with juridical guardianship (Khalequzzaman, 2019). This legal standpoint not only depicts an eco-centric approach, effectively superseding an anthropocentric basis for environmental perspectives, but also emphasises grave directives concerning unauthorised occupation, pollution, and harmful development projects in an eco-centric context.

C. Pakistan

Pakistan's Judiciary's Contributions Towards Environmental Conservation Through Judges' interpretation of Article 9 of its Constitution, Pakistan's Judiciary has substantially developed environmental protection (Khan, 2025).

D. Nepal

The Constitution of Nepal expressly recognises the right to a clean and healthy environment as a fundamental right, making the same directly enforceable before the judiciary (Yonzon & K. C., 2021). The Supreme Court of Nepal, through PIL, has extensively engaged with issues of deforestation, pollution, biodiversity loss, and unsustainable development (Yonzon & K. C., 2021). The courts regularly apply the principles of sustainable development, intergenerational equity, and the precautionary principle when examining state actions and development proposals. In this backdrop, judicial interventions in Nepal reflect a rights-based, ecologically sensitive framework for balancing environmental preservation with socio-economic development. In Nepal Court upheld the validity of the governmental measure because personal freedom to conduct business must give way to environmental rights under the constitution's "right to life": "every human being has an inherent right to live in a healthy environment" (Bhattarai, 2019).

It must be emphasised that the effectiveness of constitutional environmental rights ultimately depends on mechanisms of administrative accountability and institutional guarantees (Kozlowski et al., 2018), while examining the legal guarantees from the perspective of public sector accountability within the United Arab Emirates, also indicates the importance of objective regulatory oversight and performance evaluation for the enforcement of legal rights. Even though the context of the jurisdiction would be different from the one under consideration, the relevance of the argument would be similar to the one propounded from the perspective of the law.

6. JUDICIAL RESPONSE TO ENVIRONMENTAL INJUSTICE IN THE GLOBAL SOUTH

6.1. Judicial Activism

A. India

It is observed that the term "Environment" is extensive in its scope, and it is exceedingly challenging to delineate it in its definitive form. Even eminent environmentalists who have contributed to environmental conservation have not defined the term' environment' with the necessary precision (Price, 1985). Environmental conservation is an intrinsic foundation of India's legal system and is grounded in the principles of environmental constitutionalism. The Supreme Court has emphasised the importance of the "polluter pays principle" in the case of Vellore Citizens Welfare Forum vs. Union of India, describing it as a "broad principle" applicable to environmental laws. According to the decision in this case, industries that cause environmental pollution are required to pay for cleanup efforts, demonstrating the judiciary's commitment to holding polluters accountable for their actions. As such, in the matter of Rural Litigation and Entitlement Kendra v. State of Uttar Pradesh, the Supreme Court stepped in to stay the quarrying of limestone in the Mussoorie hills, recognising that it would cause irrevocable harm to the environment. In these situations, it is very clear that the Indian courts are ready to give precedence to the cause of protecting the environment, even at the cost of economic growth. The Indian courts have expansively interpreted the scope of environmental rights by incorporating the rights of the marginalized sections of society. In the case of Narmada Bachao Andolan v. Union of India, the Supreme Court recognized the rights of the displaced communities affected by major projects, including dams and reservoirs.

Furthermore, the judicial system has been very active in dealing with new issues that have raised environmental concerns, such as global warming. For example, in the context of air pollution in India's National Capital Region, the Supreme Court in India decided in the case of "MC Mehta vs. Union of India" that the government should take urgent measures to address air pollution. The aforementioned cases show that the judiciary has played a very important role in the development of environmental governance in India by incorporating the principles of sustainable development, intergenerational equity, and environmental justice in the Constitution.

B. Bangladesh

Bangladesh is one of the very important jurisdictions of the Global South, where the judiciary has acted vigorously and transformationally to protect the environment.

Confronted with serious ecological vulnerabilities arising from climate change, river pollution, deforestation, and unregulated industrialisation, the Supreme Court of Bangladesh has adopted an expansive constitutional and rights-based approach to environmental governance (Lie & Alam, 2025).

The scope of PIL expanded with Dr Mohiuddin Farooque v. Bangladesh. BELA petitioned the High Court on behalf of Tangail inhabitants in 1994 to evaluate FAP activities. In some parts of Tangail district, the petition alleged environmental degradation, ecological imbalance, and law violations. It was initially denied for lack of locus standi (Razzaque, 2004). Significantly, when an appeal was made against the refusal, the court explained 'Any person aggrieved' and 'Sufficient interest'. Thus, the appellant, an environmental association of attorneys, was found to be 'a person aggrieved' since its bona fide cause impacts an indefinite number of people on a major public issue. The Court used Principle 10 of the Rio Declaration to evaluate "sufficient interest" and Lucas Standi (Hossain, 1997). The FAP 20 Case not only widened the scope of the locus standi but also created a judicial precedent to recognise environmental matters as a justifiable ground for PIL. Moreover, the Court widened the ambit of the "Right to Life" as was stipulated in Article 31 of the Constitution of Bangladesh. Most significantly, it was pronounced that "Right to Life" involves within its ambit the preservation of the environment free from air and water pollution, ecological balance, and proper sanitation, without which life can hardly be enjoyed."

C. Pakistan

Pakistan's Judiciary's Contributions Towards Environmental Conservation through PIL: important decisions, like Leghari v. Federation of Pakistan, hold governments liable for issues of Climate Change.

Ashgar Leghari vs. Federation of Pakistan is a landmark case in the middle of this wave of climate change litigation; importantly, it is the first climate change case from the global South to receive international media attention. In this case, Ashgar Leghari is a law student and an agriculturalist who has filed a PIL, claiming that it violates the constitutional rights to life and dignity due to "negligence and inaction" by the government of Pakistan in relation to the National Climate Change Policy. The Green Bench, having thereby defined the idea of 'Climate Change from a national perspective,' referenced the Constitution of Pakistan and international principles, including rights to life and dignity, intergenerational equity, and the precautionary principle, to call for a 'transition to Climate Change Justice.' It also established a Climate Change Commission to assist the Court in supervising the development.

D. Nepal

The judiciary in Nepal has adopted a progressive approach to environmental protection, aligning with the Global South and integrating constitutional guarantees, human rights, and principles of international environmental law. In Suray Prasad Sharma Dhungel v. Godavari Marble Industries and others (1995), the Nepalese Supreme Court observed that since a clean and healthy environment is a fundamental part of human life, the right to a clean and healthy environment is inalienably an integral part of the Right to life. Therefore, the absence of an appropriate environment reduced human life.

In the case of Advocate Kedar Bhakta Shrestha versus HMG Department of Transportation Management, it was submitted by the petitioners that by prohibiting the use of "tempos," three-wheeled diesel-powered vehicles and one of the main causes of air pollution in Kathmandu, the government denied them their right to carry on trade or commerce. The Court upheld the validity of the governmental measure because personal freedom to conduct business must give way to environmental rights under the constitution's "right to life": "every human being has an inherent right to live in a healthy environment."

6.2. Structural inequality and judicial response in the global south

The textiles industry is one of the strongest examples of climate injustice in the current global economic system. While climate change is a global issue, the costs of producing goods using environmental resources are unevenly shared, even though they are borne by the entire world. The Global South, especially India, Bangladesh, and Vietnam, have become the hub for manufacturing in the global textiles and fashion industry, but they bear the costs of this process (Niinimäki et al., 2020). This is not a random element but is organized along the global supply chain. There, the costs of the damages are externalized in the Global South, while all the benefits are reaped in the Global North..

"Economic dependency, in turn, further accentuates this imbalanced situation." The textile industry is an important sector for employment, exchange earnings, and industrial development in developing countries (Muthu, 2014). This further limits the capabilities of governments in controlling the regulations in these countries. The reason for this is that stringent environmental regulations would only serve as a barrier to economic development. This economic struggle for survival, in turn, affects the environment. This further highlights the failure of climate governance in extending support to the changes in these countries in the Global South, as there can be no punitive measures for sustainability without actual support. Big Fashion has been putting pressure on prices, timeframes, and the quality of the production process. The Southern manufacturers have no leeway to absorb the costs of environmental regulations. Sustainability labels and environmental regulations are in high demand by the big players in the international market. The Southern manufacturers are left to bear the costs of complying with the principles that these regulations and labels advocate. The truth is that climate justice is not very practical and tangible in the absence of a duty on corporations to bear the costs of climate changes and environmental degradation.

The case of Tiruppur, a textile hub in India, brings these structural biases into focus. While Tiruppur has remained a key hub for the export of knitted fabric, thereby making important contributions to economic development, rapid industrialisation has resulted in the dangerous pollution of the Noyyal River, rendering water sources unfit for agricultural or household use. However, intervention by judicial bodies, on the basis of constitutional rights regarding environmental matters, alongside the implementation of the "polluter pays principle," made it necessary to set up "Zero Liquid Discharge Systems" within textile mills. Even while emphasising that this has made important strides towards holding industries accountable towards environment-related standards, serious biases have remained, since smaller-scale mills remained unable to pay economic costs, thereby again stressing an important role for governments, alongside international camaraderie, towards keeping the burdens of environment protection laws below those of poorer producer nations.

A case can also be made for Bangladesh, where the garment and textile industry sustains the country's economy. Here, the industry has a significant negative environmental impact, with untreated effluent contributing to pollution of the Buriganga and Turag Rivers (Sakamoto et al., 2019). Environmental degradation does not occur in isolation, as the industry, due to climate-related risks such as flooding and water shortages, specifically targets industrial areas in urban Bangladesh (Sakamoto et al., 2019). This seems to be a challenge to environment, especially given the availability of environmental protection statutes and effluent discharge standards. Here, the tragedy of Rana Plaza, which began as a labour rights issue, has reflected the failure of environmental protection in this industry, where economies have played a crucial role in contravening these rights (Mahmud et al., 2025).

The Vietnam case provides a reform-oriented, though no less complex, picture. In fact, Vietnam, as a growing textile manufacturing center, has taken steps to reform regulations in line with Maastricht environmental requirements, driven by trade agreements and environmental conventions (Doan & Vu, 2023). Though these are promising signs for the adoption of clean tech and a better environmental management regime, the ground-level issues remain a concern, especially for SMEs that cannot draw on financial, knowledge, or infrastructural resources that could be deployed for compliance. The dynamic character of the industrialization process in

Vietnam has intensified water pollution and waste management problems, which point to the limitations of environmental regulatory reform in Vietnam without further international support (Doan & Vu, 2023). The case of Vietnam thus proves that environmental regulatory reform needs to be complemented by economic policies that reduce the scope for sustainability-induced widening of gaps between people (Fadly, 2020).

Nevertheless, despite these varying realities in the Global South, some commonalities can be distinguished in relation to the challenges that are faced in the realisation of climate justice. The aforementioned factors, namely the lack of enforcement, dependencies, and corporate power, all add up to the challenge of climate justice in the textile industry. Legal enforcement is, of course, inevitable but is seen as an isolated action that does not address the imbalances that have been established in global supply chains. The enforcement of legal measures is still locally based, failing to take into account transnational patterns of consumption that affect local environments. The net impact of all these dynamics is a form of environmental justice in which the Global South itself becomes a site of production as well as pollution, and the benefits of decision-making, along with its costs, lie elsewhere. The issue of climate justice, in this sense, calls for a redistribution of responsibility that takes into account historical emissions, corporate responsibility, and the developmental needs of countries that produce textiles. International law needs to go beyond mere promises of action to ensure that the responsibilities of climate action are shared equally.

7. ENVIRONMENTAL AND SOCIETAL IMPACTS OF THE TEXTILE INDUSTRY

The fact that the fashion industry is having an increasingly negative influence on the environment in terms of pollution, water consumption, carbon emissions, human rights, and gender inequality makes a clear and present necessity for a transition to sustainable fashion. There is a growing number of organisations in Geneva and beyond that are joining forces to drive change in the fashion sector and encourage international cooperation on the matter (Fang, 2023).

A 1.3 trillion US dollar apparel industry accounts for a substantial part of our economies and employs more than 300 million people along the value chain (Niinimäki et al., 2020). Global fibre output has nearly quadrupled over the past two decades, increasing from 58 million tonnes in 2000 to 116 million tonnes in 2022 (Textile Exchange, 2023). If this trend continues as it is, projections indicate it will reach 147 million tonnes in 2030 (Textile Exchange, 2023). In 2014, people bought 60 per cent more garments than in 2000; yet they kept the clothes for only

50 per cent as long. Although the fashion industry witnessed tremendous growth, awareness of the wide range of environmental problems it caused has also started to rise (Messe Frankfurt, n.d.). Apart from being the second-largest consumer of water, the fashion industry accounts for 2-8 per cent of global carbon emissions (Textile Exchange, 2023).

Fast fashion refers to the production of inexpensive clothing at an unprecedented rate, which is expected to be disposed of after a few uses (Bick et al., 2018). Fast fashion is characterised by fast production cycles, poor textile quality, and poor working standards. Fast fashion is a business model that encourages overconsumption and breeds a "throwaway culture" (Bick et al., 2018). This section of the economy is characterised by rapid production and consumption cycles that often harm the environment and those who are most vulnerable. Retailers such as Zara, Forever 21, H&M, FashionNova, and Shein are among the most prominent examples of fast fashion. The fashion industry supply chain results in an ecological footprint from the procurement of raw materials through the manufacture and distribution of the clothes to their disposal. The fashion industry consumes significant amounts of fossil fuels, uses considerable amounts of water in production, and employs chemical-intensive manufacturing practices, all of which contribute to carbon emissions, water pollution, and resource depletion. In addition, the disposal of items considered "fast fashion" typically has negative environmental impacts. This is because the clothes considered "fast fashion" and disposed of end up in dumpsites or are burned, releasing toxic materials into the atmosphere and the soil (Bick et al., 2018). In addition, the fashion industry tends to exploit workers in countries with low wage rates and weak environmental and labour regulations. Many textile workers have to work in dangerous conditions, for too long, and for too little pay. Most of these workers come from neighbourhoods that don't receive adequate services, which makes them more likely to be affected by climate change and less able to adapt to its impacts (Joy et al., 2012). Not only is the fashion sector accountable for 8-10% of the world's greenhouse gas emissions, but it is also accountable for more than 20% of the world's water pollution, and it generates 92 million tonnes of textile waste every year (Bick et al., 2018).

To shed light on the real-world repercussions of climate injustice in the fashion industry, let's look at some culturally relevant examples. The Barbie movie is an example of pop culture (Bobaru, 2024). In spite of the fact that it conveyed a message of equality and equity, the frenzy that resulted from it created a phenomenal increase in the number of purchases of fast fashion, which had a severe impact on the garment workers who were employed in the initial stages of the supply chain (Bobaru, 2024). Additionally, this is in addition to the indirect damages that were caused by the large amount of pink fast-fashion that was thrown away in landfills after the trend had passed. This example, one of many, illustrates the impact of fast-paced fashion trends on climate change and climate justice.

8. INTERNATIONAL LAW AND CLIMATE JUSTICE FRAMEWORKS

One such normative framework that has arisen within international environmental law regarding climate change is climate justice. This is due to the observation that climate change is a global problem that does not respect territorial boundaries and has been distributed unevenly across both causes and consequences. The Global South, which includes large textile-producing countries such as India, Bangladesh, Vietnam, and Sri Lanka, is responsible for only insignificant past greenhouse gas emissions but is bearing the brunt of climate change consequences (Chen et al., 2024). International law attempts to address this disparity in respect of equity, responsibility, and sustainability, as underpinned by a number of frameworks in areas such as the environment, human and labour rights. The efficacy of such frameworks in moderating highly materialised industries, such as the textiles industry, remains in question (Chen et al., 2024).

The roots of climate justice as a concept lie in the nascent stages of international environmental law and the need to harmonise environmental protection and development. This was especially the case with the adoption of the Stockholm Declaration on the Human Environment in 1972 (Jensen & Whitfield, 2022). This marked the beginning of climate justice, in which the capacities of states to address environmental issues were formally acknowledged while development was taken into account. However, climate justice was not expressly formulated during this time. The Rio Declaration on Environment and Development further expanded the concept, introducing equity and fairness into the principles of environmental governance. The third principle stresses the "right to development", whereas the seventh introduced the concept of "Common but Differentiated Responsibilities" (Jensen & Whitfield, 2022). This tenet recognises that, although all nations have equal duties in addressing environmental degradation, greater responsibilities lie with those nations that have been historically major contributors to greenhouse gas emissions and possess greater capacity.

However, climate justice received clearer articulation within the UNFCCC, where CBDR has been recognised as one of its tenets (Rajamani, 2012). The above fact remains pertinent to textile-producing countries, especially in the Global South, where they depend on industrialisation for employment creation to alleviate poverty, while at the same time, they are under pressure to abide by environmental standards that developed nations violated when they were still developing.

The Paris Agreement of 2015 marks a dramatic change in international climate governance, adopting a flexible approach to reducing GHG emissions through Nationally Determined Contributions. Although the agreement emphasised the value of CBDR & Equity, the regulatory framework was comfortable from a binding GHG reduction regime to voluntary one (Pauw et al., 2020). From a Climate Justice perspective, it is important to recognise that, under the Paris Agreement, there is a need for financial support and technology transfer to enable developing countries to pursue low-carbon pathways. Under Article 4 of the Agreement, there is also consideration that developing countries will need longer timelines in order to peak their greenhouse gas emissions (Pauw et al., 2020). This is particularly crucial for textile-dependent economies, as a sudden change in the regulatory framework could lead to widespread unemployment and social unrest. However, the Agreement does not impose any binding obligations on multinational companies with global supply chains. The textile companies based in the Global North are still offloading environmental costs onto manufacturing centres in the Global South. Consequently, the Paris Agreement signifies normative progress, but it lacks the effectiveness to address environmental injustice.

The concept of "Common but Differentiated Responsibilities" is fundamental to understanding the inequities within the global textile supply chain (Bick et al., 2018). The textile industry involves significant energy use, excessive water consumption, and chemical emissions, all of which contribute to global warming and environmental degradation. CBDR requires developed countries to lead in mitigating emissions and to promote sustainable transition industries in developing countries (Bick et al., 2018). Specifically, within the textiles industry, it means the following responsibilities: financing eco-friendly technologies, ensuring fair trade, and ensuring supply chain responsibility. However, in practice, developed countries often use trade-based environmental regulations that shift compliance costs onto manufacturers in the Global South, with no accompanying financial or technological support. Production countries in the textile industry are forced to decide between economic survival and environmental protection, a dilemma that international law has so far failed to resolve.

There is an increased understanding of climate justice through the human rights frame, recognising that environmental degradation directly impacts basic rights to life, health, livelihoods, water, and a healthy environment. Normative support for linking environmental degradation with violations of human rights is given by international human rights instruments such as the International Covenant on Economic, Social and Cultural Rights and the Universal Declaration of Human Rights. In many textile-producing areas, environmental pollution commonly causes water contamination, respiratory diseases, occupational hazards, and community dislocation. Recognition that environmental degradation is also a human rights violation strengthens the legal framework for accountability at the state and corporate levels under international law. The United Nations Human Rights Council has increasingly recognised the nexus between climate change and human rights, leading to the recognition of the right to a clean, healthy, and sustainable environment as a universal human right. The implications of such a shift for the textiles industry cannot be underestimated, as state duties regarding private actors' environmental impacts expand.

Climate justice in the textile industry cannot, therefore, be discussed without labour justice. The textile industry is marked by precarious work, dangerous working conditions, and gendered exploitation at work. Labour law at the international level, through the International Labour Organisation, sets minimum standards for safety at work, remuneration, and working conditions. Climate change worsens these weaknesses by leaving workers vulnerable to hazardous chemicals, dangerous factory conditions, and climate-related threats such as heat stress and flooding. Climate justice must thus consider labour rights and ensure that environmental sustainability will not threaten them. The nexus between labour rights and environmental protection is also reflected in international law documents, such as the UNGPs (Ruggie & Sherman, 2017). The UNGPs include the obligation of all firms to respect human rights, including environmental rights. It can be noted that, even though UNGPs are soft law, they have shaped recent regulatory developments to increase corporate accountability (Ruggie & Sherman, 2017).

With growing concerns about businesses' environmental accountability, numerous global and regional efforts have promoted accountability across supply chains (Bright et al., 2020). Tools such as the OECD Due Diligence Guidance for Responsible Business Conduct and the Corporate Sustainability Due Diligence Directive (CSDDD) of the European Union promote mandatory due diligence for environmental and human rights (Bright et al., 2020). Such regulations require companies to identify and mitigate environmental impacts in their supply chains. From a climate justice viewpoint, these sets of approaches are best combined with positive actions like financial, technological, and capacity-building assistance; otherwise, these sets of approaches in supply chain regulation could result in more inequality.

Apart from traditional command-and-control approaches, economic approaches like environmental taxation have emerged as relevant approaches in pollution mitigation regulation schemes. Alshawabkeh (2025) examined the viability of environmental taxation in the UAE as a regulation approach and found that fiscal policy could be a viable regulation approach if well-designed within constitutional and administrative parameters, which is akin to the polluter-pays approach discussed earlier and could be a useful case study for states in textile production caught in a web of economic interests and climate justice concerns.

The Sustainable Development Agenda for 2030 has an overall Framework for addressing climate justice, incorporating various goals in an interrelated manner

(Kozlowski et al., 2018). The text indicates that the goals closely associated with the issue of climate justice and the textile sector are SDG 12 (Responsible Consumption and Production), SDG 13 (Climate Action), SDG 6 (Clean Water and Sanitation), and SDG 8 (Decent Work and Economic Growth)(Kozlowski et al., 2018). The SDGs include calls for sustainable industrialisation, resource efficiency, and sustainable economic growth. In textile-producing countries, achieving these requires that local policies be brought into line with international commitments and that policies supporting sustainability transitions do not worsen poverty or unemployment. Finally, climate justice involves cooperation over imposed regulation.

- Limitations and Challenges of International Climate Justice Frameworks

The textile sector's environmental implications face hurdles notwithstanding international climate justice requirements. The international legal community has weak tools of action, and corporate titans have destroyed any efforts to address them. International governance fragmentation makes it simpler for transnational firms to exploit loopholes across jurisdictions. The absence of developed-country pledges to support sustainable industrial transitions in the Global South is another issue. Technology transfer is limited by IPRs; therefore, commitments have always fallen short. The developing nations, particularly those countries that are in the business of textile production, have had challenges adapting to international clean production practices. To achieve climate justice, international legal instruments have to move beyond procedural ambitions to address structural inequality in international supply chains. To successfully manage the textile industries, international law needs to align itself with ideas of law and justice that are compatible with sustainable development.

9. CONCLUSION

The study was based on the effects of climate justice, constitutional environmental rights, and international law on the Global South. The study revealed that the environmental effects of textile production are disproportionate despite globalization. The Global South, which has a lower level of greenhouse gas emissions, bears a larger share of the environmental effects of the textile industry. The study revealed that the Global South's constitutional laws recognized environmental preservation as a constitutional right. The principles of sustainable development, polluter pays, and precautionary measures indicate that the Global South's constitutional laws protect the environment. Nevertheless, the judiciary cannot be relied on for climate justice.

On the international level, the UNFCCC, the Paris Agreement, and human rights conventions establish the principles of fair environmental governance. The principles of common but varied duties and intergenerational equity address the historical and structural climate injustices. However, these principles, including corporate responsibilities, are not enforceable, making them ineffective for the regulation of the textile industry. Although transnational business engages in the production of goods, the international legal system recognizes national governments. The research established that environmental degradation in the textile industry is related to exploitation, public health, and inequality. However, climate justice within the textile business must be restructured around sustainable activity, human dignity, and environmental sustainability.

In conclusion, the research established that it is a duty under the law and ethics to address climate justice within the textile industry. There is a need to incorporate all these aspects of constitutional rights, international obligations, and transnational business practices, as without them, the results will be counter to climate justice and sustainable development.

10. RECOMMENDATIONS

This study proposes recommendations to ensure equitable, enforceable, and sustainable outcomes.

- Strengthening Domestic Environmental Legislation and Enforcement - There is a need for countries in the Global South that have textile industries to strengthen their environmental legislation for the textile industry. This must be based on outcomes like reduced water pollution, reduced carbon emissions, etc. For this purpose, it is important that environmental regulatory bodies function properly.
- Inclusion and Diversity - For the transition of small and medium-sized textile industries towards sustainability, measures like delayed compliance rates, subsidies, etc., along with shared facilities like waste management plants, must be provided.
- Environmental Constitutional Rights - For the protection of the environment, proper legislation along with institutional efforts like judicial actions must be undertaken.
- Strengthening Environmental Impact Assessment - There must be improvement in methods like climate change vulnerability, cumulative impact, etc., especially for the textile industry.
- Developed countries must provide economic aid to countries that have textile industries for the purpose of mitigating climate change.

- Eliminate technology accessibility hurdles - International cooperation should address intellectual property challenges preventing clean and sustainable textile innovation in underdeveloped nations.
- Implement binding corporate accountability mechanisms; establish international and domestic regulations for environmental and climate change commitments throughout global supplier chains of multinationals. Environmental provisions in trade agreements must be accompanied by technical and financial assistance and cannot be a hidden trade barrier for developing nations.
- Encourage fair pricing and sustainable procurement rules globally, enabling suppliers to invest in sustainable production.
- Integrate labor rights and environmental standards with textile sustainability strategies to promote climate justice, including labor rights, health, and gender equity
- Promote community participation and responsible consumption. Governments and civil society may encourage sustainable fashion consumption and transparency to reduce environmental damage. Employees and communities must have access to environmental knowledge and decision-making power. They need an environmental and human rights grievance mechanism.

REFERENCES

Alqodsi, E. M., & Abdallah, A. K. (2024). Legal frameworks and educational strategies for sustainable development.

Alshawabkeh, I. K. (2025). Using the tax to reduce environmental pollution in the United Arab Emirates: Possibility and challenges. In *Legal frameworks and educational strategies for sustainable development* (pp. 1–12). IGI Global.

Baxi, U. (2010). The avatars of Indian judicial activism: Explorations in the geographies of (in)justice. In S. K. Verma & Kusum (Eds.), *Fifty years of the Supreme Court of India* (pp. 156–209). Oxford University Press.

Bhattarai, K. D. (2019). Environmental justice and the Supreme Court of Nepal: Public interest litigation and the right to a healthy environment. *Asia Pacific Journal of Environmental Law*, *22*(2), 195–214. DOI: 10.4337/apjel.2019.02.04

Bick, R., Halsey, E., & Ekenga, C. C. (2018). The global environmental injustice of fast fashion. *Environmental Health : A Global Access Science Source*, *17*(1), 92. DOI: 10.1186/s12940-018-0433-7

Bobaru, N. (2024). Reimagining Barbie: A feminist analysis of iconography and consumerism in the age of gender reevaluation. *Gender Studies*, *23*(1), 59–80. DOI: 10.2478/genst-2024-0005

Breggin, L. K., & Pendergrass, J. (2007). Addressing nanotechnology waste and product disposal: Can the Superfund safety net catch tiny particles? *Journal of Environmental Law*, *19*(3), 323–345. DOI: 10.1093/jel/eqm021

Bright, C., Marx, A., Pineau, N., & Wouters, J. (2020). Toward a corporate duty for lead companies to respect human rights in their global value chains? *Business and Politics*, *22*(4), 667–697. DOI: 10.1017/bap.2020.15

Chancel, L. (2022). Global carbon inequality over 1990–2019. *Nature Sustainability*, *5*(11), 931–938. DOI: 10.1038/s41893-022-00955-z

Chen, X., Cheng, X., Zhang, T., Chen, H., & Wang, Y. (2024). Decarbonization practices in the textile supply chain: Towards an integrated conceptual framework. *Journal of Cleaner Production*, *435*, 140452. DOI: 10.1016/j.jclepro.2023.140452

Divan, S., & Rosencranz, A. (2001). *Environmental Law and Policy in India: Cases, Materials and Statutes*. Oxford University Press.

Doan, T. M. H., & Vu, T. A. (2023). Sustainable supply chain in the textile and garment industry of Vietnam. *Economics and Business Quarterly Reviews*, *6*(3), 222–236. DOI: 10.31014/aior.1992.06.03.533

Dolšak, N., & Prakash, A. (2022). Three faces of climate justice. *Annual Review of Political Science*, *25*(1), 283–301. DOI: 10.1146/annurev-polisci-051120-125514

Exchange, T. (2023). *Materials market report 2023*. https://textileexchange.org/app/uploads/2023/11/Materials-Market-Report-2023.pdf

Fadly, D. (2020). Greening industry in Vietnam: Environmental management standards and resource efficiency in SMEs. *Sustainability (Basel)*, *12*(18), 7455. DOI: 10.3390/su12187455

Fang, B. (2023). Environmental and human impacts of fast fashion. *Communications in Humanities Research*, *15*(1), 166–171. DOI: 10.54254/2753-7064/15/20230660

Frankfurt, M. (n.d.). *Texpertise Econogy insights: Five central developments for a more sustainable textile industry* [Report]. https://marketdirectory.messefrankfurt.com/images/original/document_downloads/10000040202501/432571/1752696040146_1354155584.pdf

Füssel, H.-M. (2010). How inequitable is the global distribution of responsibility, capability, and vulnerability to climate change: A comprehensive indicator-based assessment. *Global Environmental Change*, *20*(4), 597–611. DOI: 10.1016/j.gloenvcha.2010.07.009

Heine, D., Faure, M., & Dominioni, G. (2020). The polluter-pays principle in climate change law: An economic appraisal. *Climate Law*, *10*(1), 94–115. DOI: 10.1163/18786561-01001004

Hossain, K. (1997). Public interest litigation in Bangladesh: A new horizon for the judiciary. *Bangladesh Journal of Law*, *1*(2), 1–20.

Intergovernmental Panel on Climate Change. (2012). Summary for policymakers. In Field, C. B., Barros, V., Stocker, T. F., Qin, D., Dokken, D. J., Ebi, K. L., Mastrandrea, M. D., Mach, K. J., Plattner, G.-K., Allen, S. K., Tignor, M., & Midgley, P. M. (Eds.), *Managing the risks of extreme events and disasters to advance climate change adaptation* (pp. 3–22). Cambridge University Press.

Jensen, F., & Whitfield, L. (2022). Leveraging participation in apparel global supply chains through green industrialization strategies: Implications for low-income countries. *Ecological Economics*, *194*, 107331. DOI: 10.1016/j.ecolecon.2021.107331

Joy, A., Sherry, J. F.Jr, Venkatesh, A., Wang, J., & Chan, R. (2012). Fast fashion, sustainability, and the ethical appeal of luxury brands. *Fashion Theory, 16*(3), 273–295. DOI: 10.2752/175174112X13340749707123

Khalequzzaman, M. (2019). Can the historic High Court judgement save our rivers? *The Daily Star*. https://www.thedailystar.net/environment/news/can-the-historic-high-court-judgement-save-our-rivers-1697419

Khan, M. A. (2025). The role of judiciary in environmental protection: A review of key cases from Pakistan judiciary. [IJMCP]. *International Journal of Multidisciplinary Conference Proceedings, 1*(1), 49. https://www.ijmcp.com/index.php/ijmcp/article/view/2024-12-31

Khan, M. R. (2015). Polluter-pays principle: The cardinal instrument for addressing climate change. *Laws, 4*(3), 638–653. DOI: 10.3390/laws4030638

Kingsford, R. T., Bino, G., Finlayson, C. M., Falster, D., Fitzsimons, J. A., Gawlik, D. E., Murray, N. J., Grillas, P., Gardner, R. C., Regan, T. J., Roux, D. J., & Thomas, R. F. (2021). Ramsar wetlands of international importance—Improving conservation outcomes. *Frontiers in Environmental Science, 9*, 643367. DOI: 10.3389/fenvs.2021.643367

Kozlowski, A., Searcy, C., & Bardecki, M. (2018). The reDesign canvas: Fashion design as a tool for sustainability. *Journal of Cleaner Production, 183*, 194–207. DOI: 10.1016/j.jclepro.2018.02.014

Lie, D., & Alam, M. (2025). Legal Framework and Practical Challenges of Bangladesh Environmental Legislation: A Critical Examination of Climate Change Adaptation, Environmental Judiciary, and Enforcement. *Environmental Claims Journal, 37*(3), 366–388. DOI: 10.1080/10406026.2025.2454662

Mahmud, F., Roy, H., Bhuiyan, M. M. K., & Islam, M. S. (2025). Critical analysis, characterization, and treatment of microplastics in the peripheral rivers of Dhaka city: Buriganga and Turag. *Environmental Science. Advances, 4*(11), 1810–1833. Advance online publication. DOI: 10.1039/D5VA00118H

Mohammad, N. (2012). Environment and sustainable development in Bangladesh: A legal study in the context of international trends. *Mediterranean Journal of Social Sciences, 3*(15), 188–196. DOI: 10.5901/mjss.2012.v3n15p188

Muthu, S. S. (2014). *Assessing the environmental impact of textiles and the clothing supply chain*. Woodhead Publishing

Niinimäki, K., Peters, G., Dahlbo, H., Perry, P., Rissanen, T., & Gwilt, A. (2020). The environmental price of fast fashion. *Nature Reviews. Earth & Environment*, *1*(4), 189–200. DOI: 10.1038/s43017-020-0039-9

Page, E. A. (2008). Distributing the burdens of climate change. *Environmental Politics*, *17*(4), 556–575. DOI: 10.1080/09644010802193419

Pauw, W. P., Castro, P., Pickering, J., & Bhasin, S. (2020). Conditional nationally determined contributions in the Paris Agreement: Foothold for equity or Achilles heel? *Climate Policy*, *20*(4), 468–484. DOI: 10.1080/14693062.2019.1635874

Price, K. A. (1985). Review of *International environmental policy: Emergence and dimensions*, by L. K. Caldwell. *Natural Resources Journal*, *25*(1), 254. https://digitalrepository.unm.edu/nrj/vol25/iss1/15

Rajamani, L. (2012). The changing fortunes of differential treatment in the evolution of international environmental law. *International Affairs*, *88*(3), 605–623. DOI: 10.1111/j.1468-2346.2012.01091.x

Razzaque, J. (2004). Public interest environmental litigation in Bangladesh: A study on BELA's experience. *Review of European Community & International Environmental Law*, *13*(1), 79–92. DOI: 10.1111/j.1467-9388.2004.00393.x

Roberts, J. T., & Parks, B. C. (2007). *A Climate of Injustice: Global Inequality, North–South Politics, and Climate Policy*. MIT Press.

Ruggie, J. G., & Sherman, J. F.III. (2017). The concept of "due diligence" in the UN Guiding Principles on business and human rights: A reply to Jonathan Bonnitcha and Robert McCorquodale. *European Journal of International Law*, *28*(3), 921–928. DOI: 10.1093/ejil/chx047

Sakamoto, M., Ahmed, T., Begum, S., & Huq, H. (2019). Water pollution and the textile industry in Bangladesh: Flawed corporate practices or restrictive opportunities? *Sustainability (Basel)*, *11*(7), 1951. DOI: 10.3390/su11071951

Sharma, S. (2025). Constitutional mandate for environmental protection in India. *International Journal of Environmental Sciences*, *11*(21S), 312–321.

Wang, D., & Fang, Y. (2024). Global climate governance inequality unveiled through dynamic influence assessment. *npj. Npj Climate Action*, *3*(1), 75. DOI: 10.1038/s44168-024-00159-5

Wang, T., & Gao, X. (2018). Reflection and operationalization of the common but differentiated responsibilities and respective capabilities principle in the transparency framework under the international climate change regime. *Advances in Climate Change Research*, *9*(4), 253–263. DOI: 10.1016/j.accre.2018.12.004

Yonzon, D. M., & K. C., O. (2021). Right to freedom of religion and right to clean environment: A case of Bagmati River, Pashupatinath Temple. *Kathmandu School of Law Review*, *7*(1), 112–129. DOI: 10.46985/kslr.v7i1.1205

Chapter 4
Climate Justice in Natural Resource Governance:
Legal Pathways for Equitable and Sustainable Resource Management in a Warming World

Lawal Olamilekan Abdulwahab
http://orcid.org/0000-0003-4955-244X
School of Business and Economics, Universiti Brunei Darussalam, Brunei

ABSTRACT

Climate change is reshaping natural resource governance, amplifying inequalities in access, control, and decision-making. This chapter examines how legal frameworks can embed climate justice principles into managing critical resources such as land, water, forests, and minerals. It situates climate justice within theories of distributive and procedural fairness, intergenerational equity, and human rights, while interrogating the legacies of colonial extraction and structural inequalities that persist in resource regimes. By analyzing international treaties, regional disparities, and national legal innovations, the chapter highlights the evolving role of climate justice in reshaping governance norms. It further explores litigation, community-based models, and policy reforms as pathways to address accountability gaps, empower marginalized populations, and advance sustainability. The chapter concludes with forward-looking recommendations for aligning ethics, law, and institutional reform to secure equitable and resilient futures.

DOI: 10.4018/979-8-3373-6635-7.ch004

1. INTRODUCTION

Climate change is no longer a distant threat; it is a lived reality with profound implications for ecosystems, societies, and legal systems. Among its most pressing consequences is the intensification of inequalities in access to and control over natural resources such as land, water, forests, and minerals. These resources essential for survival and development are increasingly strained by environmental degradation, extreme weather events, and shifting climate patterns. The result is a heightened vulnerability for already marginalized populations, including Indigenous peoples, women, and rural communities, especially in the Global South. The urgency of this crisis cannot be overstated (Grazini & Guarini, 2025; Hariharan & Mahees, 2024).

While traditional natural resource governance systems have often emphasized economic efficiency and state sovereignty, they have frequently failed to account for justice-oriented considerations such as equity, participation, and historical accountability (Larson et al., 2022). This omission is especially critical in the context of climate change, where the legacies of colonial extraction, unjust land tenure systems, and unequal carbon footprints persist. These issues highlight the need for a paradigm shift toward legal frameworks grounded in climate justice. The current legal frameworks are insufficient to address climate change challenges. A radical shift in our approach is needed to ensure a fair distribution of the burdens and benefits of climate change and the recognition of differentiated responsibilities across generations, nations, and communities (Puaschunder, 2019). It emphasizes not only distributive equity but also procedural fairness, ensuring that those most affected by environmental decisions have a voice in shaping them (Yazar et al., 2024).

Natural resource governance stands at the intersection of these urgent concerns. Responses to climate change from carbon markets and land use changes to renewable energy transitions reshape resource use and ownership patterns. As such, embedding climate justice into the legal architecture of natural resource management is not just desirable, but essential (Onifade, 2021). This chapter explores how domestic, regional, and international legal systems can operate climate justice principles in the governance of natural resources. It examines the potential of existing human rights instruments, constitutional environmental rights, litigation strategies, and participatory models to make natural resource governance more inclusive, accountable, and sustainable. The chapter also offers concrete policy and legal reform pathways to rebalance power asymmetries, correct historical injustices, and prepare societies for a climate-resilient future. The potential of legal systems to bring about these changes offers hope amid the climate crisis.

2. BACKGROUND AND CONCEPTUAL FRAMEWORK

2.1 Climate Justice in Resource Governance

Climate justice tackles the unequal causes and impacts of climate change, advocating for fairer sharing of burdens and benefits. It recognizes that communities, especially in the Global South, contribute least to emissions but suffer most from droughts, floods, and resource shortages (Abdelrahman, 2024; Ali et al., 2025). In natural resource governance, climate justice is a rights-based framework emphasizing equity, participation, and accountability. It critiques traditional state-centric or neoliberal approaches that marginalize vulnerable populations and seeks inclusive decision-making centering marginalized groups like Indigenous peoples, women, youth, and low-income populations (Abdulwahab & Azam, 2026; Uche & Ekine, 2025). This framework expands the discourse beyond environmental conservation to integrate human rights, development justice, and intergenerational equity. It demands attention to the resources used and how, as well as to who decides, how decisions are made, and who ultimately benefits (Ensor & Hoddy, 2020). For instance, resource sovereignty is being reconceptualized to encompass global egalitarian principles that ensure fair access to and governance of natural resources such as forests, water, and land (Angell, 2021).

Climate justice also challenges the structural roots of inequality embedded in both colonial histories and modern economic systems, calling for radical transformations rather than token reforms (Islam, 2024). It is increasingly viewed as a prerequisite for sustainable development, particularly in contexts where natural resources are central to both survival and sovereignty (Chang et al., 2021). Climate justice in resource governance holds that a just response to climate change must address environmental degradation and long-standing injustices in political power and resource access. This relationship is illustrated in the conceptual framework below (see Figure 1), which maps the interconnections between climate justice, legal systems, resource governance, and human rights. This figure illustrates how climate justice is supported by legal frameworks, human rights, and resource governance, grounded in interconnected principles of ethics, equity, law, and participation. It emphasizes that achieving climate justice requires an integrated approach where laws uphold rights, governance ensures equity, and all processes are ethically grounded and inclusive.

Figure 1. Conceptual framework linking climate justice, human rights, legal frameworks, and resource governance in a warming world

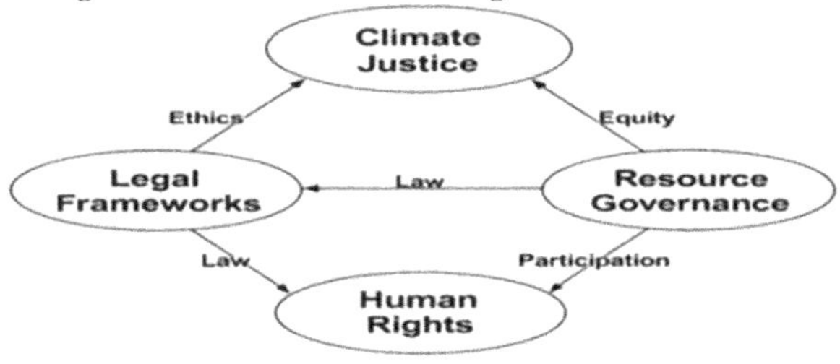

2.2 Theoretical Framework of Climate Justice

Climate justice draws from interdisciplinary moral, political, legal, and philosophical theories. It includes foundational frameworks like Rawls' justice as fairness, Sen and Nussbaum's capabilities approach, and intergenerational equity. These offer critical views on the equitable distribution of climate change burdens and benefits across people, regions, and generations. Intergenerational justice has gained prominence as climate impacts grow, with scholars emphasizing current generations' ethical duty to protect the environment for the future (Jat, 2025). Rawls' veil of ignorance remains a pivotal thought experiment in climate justice literature, proposing that climate policies should be designed without knowledge of one's societal position, generation, or geographic location, thus ensuring impartiality and fairness (Barsotti & Koçer, 2022; Marcellinus et al., 2021). When applied to climate governance, this framework compels policymakers to prioritize environmental sustainability and equal opportunity for all, including the unborn (Tirkey, 2023).

Simultaneously, the capabilities approach has emerged as a powerful supplement to distributive justice, arguing that climate justice should enhance people's fundamental freedoms, their abilities to live lives they have reason to value, rather than merely focusing on resources (Hwang, 2025). A central challenge in climate justice theory lies in negotiating the friction between national sovereignty and the concept of a shared global common. Increasingly, scholars argue that sovereign rights must be reinterpreted to encompass global responsibilities, particularly for high-emitting and resource-rich states (Puaschunder, 2020). These evolving theoretical frameworks push climate justice beyond a narrow focus on emission rights, urging a more holistic perspective that integrates procedural fairness, historical accountability, and long-term planetary stewardship.

2.3 Justice in Natural Resource Governance

Natural resource governance has historically been anchored in the principles of sovereignty and economic efficiency, often privileging state control and market-driven exploitation over equity and fairness. This traditional paradigm has been criticized for neglecting distributive justice, procedural inclusiveness, and the recognition of marginalized communities most affected by environmental degradation. Scholars emphasize that equitable access, fair distribution of burdens, and participatory decision-making must form the backbone of any governance framework for sustainable resource management (Sánchez-Simbaña, 2025). Contemporary models have moved toward integrating ecological, social, and economic dimensions through systems-thinking approaches.

For instance, participatory co-management of forests and fisheries has demonstrated how collaborative governance can enhance ecological resilience and social justice (Abdulwahab et al., 2025; Estévez et al., 2020). Similarly, community-based conservation initiatives have highlighted the value of empowering indigenous and local populations in shaping governance outcomes, promoting recognition and representation in line with environmental justice principles (Fariss et al., 2023). Benefit-sharing frameworks have also emerged as critical mechanisms for correcting historical injustices and reducing inequalities in natural resource use. By redistributing revenues from extractive projects or ecosystem services, such models ensure that local communities are not excluded from the benefits of conservation and climate finance (Santarlacci et al., 2024).

Evidence from Nepal's Asta-Ja governance framework illustrates how participatory and inclusive approaches can integrate justice considerations into multi-sectoral natural resource management strategies (Poudel, 2022). Climate justice perspectives have advanced hybrid governance models that merge top-down regulatory policies with bottom-up accountability. These models are particularly significant in transnational contexts where sovereignty claims intersect with global commons responsibilities, requiring frameworks that balance state authority with shared stewardship obligations (Caggiano & Landau, 2021). Such evolving governance structures highlight a critical transition: from extractive, exclusionary systems to equitable, participatory, and justice-centered approaches capable of addressing historical inequities and advancing sustainable development.

3. INTERNATIONAL AND REGIONAL LEGAL ETHICS

3.1 Evolution of International Legal Instruments for Climate Justice

The evolution of international climate governance demonstrates a gradual yet uneven shift from soft law declarations to binding treaty frameworks. The United Nations Framework Convention on Climate Change (UNFCCC), adopted in 1992, established the principle of common but differentiated responsibilities (CBDR), emphasizing that developed nations bear greater responsibility for historical emissions. While normatively significant, the UNFCCC lacked binding enforcement mechanisms, limiting its capacity to drive equitable action (Kaya & Kaya, 2025). The Kyoto Protocol (1997) marked a more decisive step toward binding commitments, imposing legally enforceable targets on industrialized nations. However, its rigid top-down structure and the withdrawal of major emitters undermined its effectiveness.

Scholars argue that the Protocol epitomized lex lata obligations but suffered from institutional design flaws, paving the way for the more flexible architecture of Paris (Turkamani, 2023). The Paris Agreement (2015) represents a paradigm shift by embracing a hybrid approach that combines nationally determined contributions (NDCs) with universal participation. Unlike Kyoto, Paris embedded justice considerations in its preamble, explicitly referencing human rights, gender equality, Indigenous rights, and workers' transitions. This expansion marked the first time climate justice became a substantive principle of international climate law (Ferreira, 2018; Goodier, 2018). Still, the reliance on voluntary commitments under Paris reveals its predominantly lex ferenda nature, raising doubts about long-term enforceability (Brunnée, 2021). One of the most significant justice-related developments was the recognition of loss and damage as a third pillar of international climate law.

Building on the Warsaw International Mechanism, Article 8 of the Paris Agreement, and subsequent negotiations, culminated in the 2022 COP27 decision to establish a Loss and Damage Fund. While hailed as a historic breakthrough, scholars emphasize the fund's ambiguous financing mechanisms and political contestations, questioning whether it can deliver real justice to vulnerable states (Toussaint, 2023). Small island developing states (SIDS), for example, continue to push for justice-based financing models that account for disproportionate risks, advocating levies on fossil fuels and aviation as equitable solutions (Lai et al., 2022). More summits, including COP28, reinforced the urgency of bridging the gap between scientific imperatives and political commitments.

Despite institutional advances, critics contend that international climate law remains constrained by sovereignty, voluntary pledges, and weak enforcement, raising the need for structural reform to embed equity into global governance (Desai & Desai, 2024). A comparative overview of these legal instruments and their justice-related provisions is provided in Table 1, which summarizes the progression of international climate agreements from 1992 to 2022 and highlights the evolving integration of justice principles. It reveals a trend toward stronger equity language and recognition of vulnerable populations, but also shows persistent enforcement gaps and reliance on non-binding mechanisms.

Table 1. Comparative analysis of international climate agreements and their integration of justice principles.

Agreement / Event	Justice-Related Clauses	Enforceability	Limitations
UNFCCC (1992)	Introduced 'common but differentiated responsibilities' (CBDR); recognized equity, but lacked enforceability	Non-binding; dependent on national action plans and reporting	No vigorous enforcement or financial commitments
Kyoto Protocol (1997)	Legally binding targets for developed countries; minimal direct justice references	Binding for Annex I countries; the compliance mechanism is weak in practice	Withdrawal of major emitters; inequitable burden-sharing
Paris Agreement (2015)	Embedded human rights, gender equality, indigenous rights, and 'just transition' in preamble	Non-binding NDCs; peer pressure and transparency are key tools	Relies on voluntary commitments; enforcement gaps
COP26 Glasgow (2021)	Renewed focus on just transition; emphasized equity in climate finance discussions	Non-binding political commitments; reviewing mechanisms under development	Justice language remains aspirational; limited follow-through
COP27 Sharm El-Sheikh (2022)	Established Loss and Damage Fund; equity for vulnerable states highlighted	Fund still under design; no binding obligations for contributions yet	Unclear financing structure; operational details under negotiation

Source: Compiled by the Author.

3.2 Human Rights and Climate Governance Convergence

An emerging global consensus is that climate justice and human rights are deeply interdependent. Climate change exacerbates violations of life, health, food, housing, and self-determination, making it essential for international environmental law to be intertwined with human rights law. Increasingly, scholars and courts recognize that safeguarding communities against climate-induced displacement, environmental degradation, and resource depletion requires robust human rights protections embedded within climate governance frameworks (Chotouras et al., 2024; Jane & Emma, 2021). Litigation has become a crucial mechanism for advancing this convergence. Human rights-based climate litigation has expanded significantly in recent years, with national, regional, and international courts redefining state obligations. Landmark cases before the Inter-American Court of Human Rights, the International Court of Justice, and the International Tribunal for the Law of the Sea have established precedents linking state responsibility for climate inaction with fundamental human rights. These legal developments are reshaping norms of accountability, transboundary harm, and intergenerational equity (Luhandjula, 2023). Intergenerational justice has become a key area where human rights discourse intersects climate governance.

Courts in Germany, Colombia, and other jurisdictions have recognized the rights of future generations to a safe and stable climate, with the German Federal Constitutional Court's 2021 ruling establishing that inadequate mitigation efforts threaten fundamental freedoms for youth and unborn generations (Minnerop, 2021). Similarly, youth-focused climate lawsuits globally, from Juliana v. United States to ENJEU in Canada, have asserted that governments must respect and fulfill the rights of children and young people disproportionately affected by climate risks (Parker et al., 2022). Normatively, human rights frameworks also provide leverage in international negotiations. By framing climate harms as rights violations, advocates strengthen the political and moral weight of claims for stronger mitigation and adaptation commitments. Scholars argue that this integration expands the concept of human dignity to encompass ecological sustainability, positioning climate change as not merely an environmental issue but a profound justice challenge (Musa et al., 2026; Rodríguez-Garavito, 2021).

Despite these advances, challenges remain. Structural barriers such as limited standing rules, weak enforcement of international human rights law, and North–South inequities in climate negotiations hinder the full realization of justice. Nevertheless, the growing body of rights-based climate jurisprudence and advocacy for explicit recognition of a right to a healthy environment indicate that the convergence of human rights and climate governance will play a defining role in shaping future legal and policy frameworks (Schroder, 2021). The intersection of human rights and climate justice is illustrated in Figure 2, which emphasizes their shared obligations and complementary roles in advancing equitable climate governance. This Venn diagram illustrates the conceptual overlap between human rights and climate justice. It highlights how both frameworks share key obligations such as adaptation, mitigation, and remedy, while retaining distinct focuses, rights-based protections on one side and principles of equity and participation on the other.

Figure 2. Intersectional framework of human rights and climate justice obligations

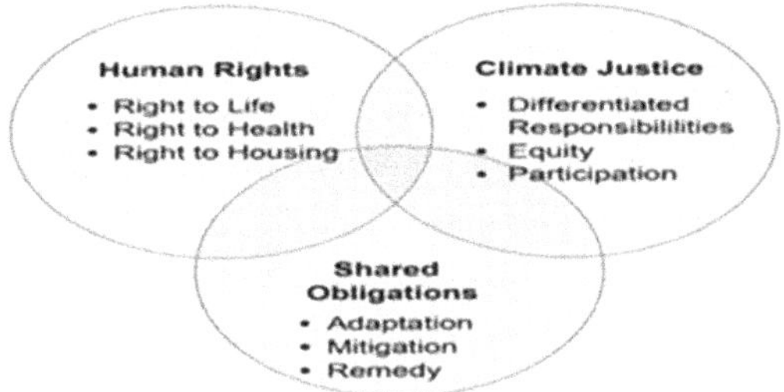

3.3 Regional Climate Law Frameworks and Disparities

While international climate law provides a global framework, much of the operational progress in climate justice occurs at the regional level, where legal traditions, institutional capacity, and political will vary significantly. The European Union (EU) remains the most advanced regional actor, pioneering binding emission reduction targets and climate directives that apply across its 27 member states. The EU's 2030 Climate and Energy Framework, with its legally binding goals on emissions, renewable energy, and efficiency, exemplifies its commitment to harmonized governance and reflects its leadership in integrating justice principles into environmental directives (Du, 2022). In contrast, Africa and South Asia rely more heavily on non-binding commitments and donor-driven frameworks.

Structural barriers, such as weak enforcement capacity, insufficient financing, and high vulnerability to climate risks, constrain the translation of justice principles into enforceable laws. Scholars argue that frameworks in these regions often prioritize adaptation and resilience, but struggle to ensure distributive and procedural justice for marginalized populations (Ehsan et al., 2022; Sorgho et al., 2022). A comparative analysis of legal frameworks across major economies reveals stark disparities. Countries like the USA, UK, Japan, China, Germany, and France have developed structured climate justice laws, often embedding principles of fairness, intergenerational responsibility, and just transition into domestic legislation. However, enforcement and accountability mechanisms remain uneven, with challenges including political resistance, limited judicial oversight, and competing economic interests (Nigam & Parikh, 2023).

The EU's expanding cooperation beyond its borders, such as its agreements with Korea, demonstrates the potential of regional legal instruments to link emissions trading systems and facilitate joint mitigation strategies. Despite the challenges of trade-related and sovereignty issues, such cooperation offers a hopeful path forward in the fight against climate change. (Lee, 2022). Similarly, regional climate finance disparities highlight the different capacities of Europe and Asia to fund just transitions. While the EU integrates justice considerations into recovery packages, China has relied more heavily on state-driven mechanisms without embedding distributive equity into its climate financing reforms (Minas, 2021). Other regions illustrate hybrid challenges. For instance, the Niger Delta in Nigeria epitomizes how local environmental degradation intersects with regional disparities, underscoring the need for justice-oriented regional governance frameworks (Ikporukpo & Ikporukpo, 2022). Meanwhile, Asia-Pacific frameworks reveal tensions between growth-oriented policies and ecological imperatives, reflecting uneven integration of justice within climate regimes.

As illustrated in Figure 3, the capacity and commitment to enforce climate justice principles vary considerably across regions, underscoring the global governance gap. This map visualizes the uneven global landscape of climate justice enforcement. It highlights how regions like Europe and North America have stronger enforcement mechanisms, while regions such as Africa, Southeast Asia, and parts of Oceania have weaker enforcement. The visual reinforces structural inequalities in climate governance and the need for capacity-building and policy reform in underrepresented regions.

Figure 3. Global disparities in climate justice enforcement mechanisms across regions

3.4 Challenges in Enforcing International Climate Law

The enforcement of international climate law remains one of the most pressing challenges for global governance. While treaties such as the UNFCCC, Kyoto Protocol, and Paris Agreement have established norms of cooperation, they lack strong enforcement mechanisms. The reliance on voluntary commitments and nationally determined contributions highlights the limitations of state consent and sovereignty as the backbone of international law, resulting in significant compliance gaps (Abdullahi & Musa, 2023). Legal scholars have proposed constructing a constitutional framework of lex specialis principles to give international climate law greater authority. However, institutional fragmentation and overlapping jurisdictions among international tribunals make this problematic. The proliferation of tribunals dealing with trade, human rights, and environmental disputes has created what some term a new medievalism in which legal pluralism undermines coherence and enforcement (Klerk, 2025).

Another central obstacle lies in the paradox of sovereignty. While states acknowledge climate obligations in principle, they often resist binding enforcement mechanisms that could infringe on domestic policy autonomy. This has led to a heavy reliance on soft law norms, with enforcement limited to reputational pressures, peer review, and diplomatic negotiation rather than sanctions or coercion (Liao, 2024). Developments in climate litigation highlight attempts to expand the jurisdiction of international tribunals to cover climate harms. Cases brought before the International Court of Justice and the International Tribunal for the Law of the Sea seek advisory opinions to clarify state obligations in preventing climate damage.

However, their outcomes remain non-binding (Segger & Gehring, 2025). Similarly, proposals to criminalize severe environmental destruction under the category of ecocide in international criminal law are still in their early stages, but signify a shift toward embedding climate accountability in binding legal frameworks (Shut et al., 2025). Efforts to create international liability schemes for climate-related loss and damage also face significant barriers. Victims of climate impacts struggle to bring claims against states or corporations due to jurisdictional limitations and the absence of a universal liability framework (Tigre & Wewerinke-Singh, 2023). Furthermore, political resistance from powerful states and competing interests in global trade and development undermines attempts to strengthen binding enforcement. The enforcement of international climate law is constrained by state sovereignty, institutional fragmentation, and the voluntary nature of compliance. While new pathways such as ecocide, climate litigation, and liability schemes represent promising avenues, they remain nascent. Without stronger global consensus and institutional reform, enforcement will remain the Achilles' heel of international climate governance.

4. NATIONAL LEGAL FRAMEWORKS AND IMPLEMENTATION PATHWAYS

National legal frameworks play a pivotal role in shaping the architecture for climate justice and natural resource governance. They are the operational backbone for translating international commitments into enforceable domestic policies, laws, and institutional structures. Effective national frameworks integrate constitutional principles, environmental statutes, judicial decisions, and administrative mechanisms to create equitable and sustainable pathways for natural resource management. Several countries have adopted constitutional and legislative provisions recognizing the right to a healthy environment, thereby anchoring environmental protection and climate justice within national legal systems.

For example, the South African Constitution explicitly guarantees the right to an environment that does not harm health or well-being. This provision is the foundation for litigation demanding climate accountability (Harvey et al., 2023). National judicial systems have increasingly been used to enforce climate obligations, as seen in the landmark Juliana v. United States and Urgenda v. Netherlands cases, highlighting how domestic courts can demand governmental compliance with emission reduction targets (Farah & Ibrahim, 2023). Legal pluralism in national contexts often results in both challenges and opportunities. On the one hand, centralized, top-down governance structures can marginalize local and indigenous communities from participating in decision-making. On the other hand, devolution of power can allow for locally grounded, context-specific legal responses, which are essential for climate justice (Husain et al., 2024).

In the Global South, for instance, there is increasing recognition of the need to harmonize statutory laws with customary legal systems to manage communal land, water, and forests equitably (Al-Hakim, 2023). Country-specific case studies illustrate how national frameworks implement climate justice differently. In Taiwan, a sophisticated spatial justice assessment model has been developed to identify regions facing environmental hazards and low adaptive capacity, informing the redistribution of public resources to promote climate equity (Chang et al., 2021). Similarly, Pakistan's polycentric governance model reveals the strengths and constraints of incorporating diverse justice framings within national policy processes, especially when confronted with centralized state power (Shawoo & McDermott, 2020). Furthermore, national legal frameworks increasingly rely on benefit-sharing to equitably distribute the gains from environmental conservation.

This principle, widely applied in access and benefit-sharing regimes under the Convention on Biological Diversity, is gaining traction in national climate policies to ensure that marginalized communities receive a fair share of benefits from resource use and conservation efforts (Laird et al., 2020). As natural resource pressures intensify under climate change, national governments are tasked with legal reform and fostering inclusive governance models. Integrating climate justice into national legal systems requires active participation by civil society, alignment with indigenous legal traditions, and the political will to implement reforms that reflect global responsibilities and local realities (Jones et al., 2024).

5. CLIMATE JUSTICE LITIGATION AND JURISPRUDENCE

The rise of climate justice litigation globally reflects a shift in the legal landscape, in which courts are increasingly recognized as critical arenas for addressing climate change and environmental injustices. Strategic litigation challenges state and corporate inaction, upholds human rights, and pushes for systemic legal reforms in climate governance. These legal efforts are reshaping jurisprudence and driving climate justice across jurisdictions. Table 2 presents an overview of landmark climate justice litigation cases and illustrates their broader jurisprudential impact. This table highlights major legal cases that have shaped the global climate justice landscape, showcasing how courts have responded to claims of human rights violations, youth-led activism, and corporate responsibility. These decisions have set legal precedents and influenced climate governance across jurisdictions.

Table 2. Notable climate justice litigation cases and their jurisprudential contributions

Case Name	Jurisdiction	Legal Focus	Outcome	Justice Implications
Urgenda Foundation v. State of the Netherlands	Netherlands	The government's duty to reduce emissions under human rights obligations	The court ordered the Dutch government to cut emissions by 25% by 2020	Set precedent for climate action as a human rights obligation
Leghari v. Federation of Pakistan	Pakistan	Failure to implement the national climate policy violates constitutional rights	The court compelled the government to implement the climate adaptation framework	Recognizing climate justice as part of the constitutional duty
Juliana v. United States	United States	Youth suing the federal government over inaction on climate change	Dismissed on procedural grounds, but sparked national discourse	Framed youth climate rights in U.S. constitutional discourse
Milieudefensie et al. v. Royal Dutch Shell	Netherlands	Corporate accountability for climate targets aligned with the Paris Agreement	The court ruled Shell must cut COâ,, emissions by 45% by 2030	Extended legal responsibility to private corporations
Neubauer et al. v. Germany	Germany	Inadequate climate targets violate the youth's constitutional rights	The court found the targets insufficient; Germany amended the climate law	Advanced intergenerational equity in climate policy

Source: Compiled by the Author.

Environmental Public Interest Litigation (EPIL) has emerged as a transformative legal strategy, particularly in jurisdictions like China, where it enables courts to hold entities accountable for ecological harm. This approach blends theories of public trust, human rights, and environmental ethics to enforce climate responsibilities and expand access to justice for vulnerable populations (Hui & Karim, 2025). Climate change litigation as a justice pathway is also evident in adaptation-focused lawsuits, which emphasize compensating communities affected by climate impacts. However, such litigation often reveals disparities in who truly benefits from judicial outcomes, raising questions about access to justice and inclusivity (O'Donnell, 2020). Rights-based climate litigation, especially in the post-Paris Agreement era, has effectively held states accountable for emissions targets and environmental protection. Landmark cases such as Urgenda v. The Netherlands and Juliana v. United States have set critical precedents, advancing intergenerational justice while revealing litigation's limitations in achieving distributive justice (Beauregard et al., 2021).

At the regional level, courts like the European Court of Human Rights (ECtHR) are emerging as influential venues for transnational climate litigation. Insights from the Global South, such as jurisprudence from South Africa and the Inter-American System, increasingly guide European decisions, contributing to a more unified global legal approach to climate justice (Murcott et al., 2023). A notable evolution in the field is the recognition of climate justice in labor law contexts. For example, the Quebec Court of Appeal's ruling in Environnement Jeunesse highlighted the growing role of labor jurisprudence in addressing climate injustice by enabling suits against corporate actors (Parent, 2023).

Beyond conventional litigation, innovative forms such as restorative justice have been proposed to address climate-related loss and damage. This approach emphasizes healing and reconciliation, especially in contexts where traditional litigation is inaccessible or ineffective, particularly for communities in the Global South (Robinson & Carlson, 2021). Additionally, legal critiques have identified inherent structural challenges in relying solely on litigation. The precedent-bound nature of judicial systems and the limits of procedural law often hamper radical legal transformations, requiring complementary policy and legislative reforms (Amato, 2025). While climate litigation is a powerful tool for climate justice, it is not a panacea. It must be embedded within a broader framework of legal pluralism, international cooperation, and structural reforms to effectively address the full scope of climate injustice.

6. COMMUNITY-BASED APPROACHES TO CLIMATE JUSTICE AND NATURAL RESOURCE GOVERNANCE

Community-based approaches are vital for climate justice and resource governance, emphasizing local knowledge, participatory decision-making, and the empowerment of vulnerable communities. Unlike top-down policies, they offer context-sensitive solutions reflecting social, cultural, and environmental realities. Procedural justice shows participation is not just moral but also pragmatic for better environmental outcomes. For example, Malawi's CB-CCD projects aimed to include local communities but faced power imbalances that threatened procedural justice. A six-step model was proposed to improve the recognition and participation of marginalized groups (Wood et al., 2018). Community-based organizations (CBOs) are vital in climate adaptation in urban settings. Research in San Francisco shows that CBOs address historical vulnerabilities and promote intersectional climate justice through coalitions and participatory planning mechanisms (Strange et al., 2024). Similarly, in northern Pakistan, community-based conservation programs helped vulnerable mountain communities become more resilient to climate impacts through improved governance, disaster preparedness, and ecosystem management (Zafar et al., 2022). Further supporting this, the study emphasized community science, highlighting how participatory knowledge production strengthens climate governance by democratizing environmental decision-making (Barry, 2022).

Moreover, the study showed that local climate narratives significantly influence policy in places like Miami, where historical environmental injustices intersect with present climate risks. Recognizing and incorporating these local narratives leads to more equitable and just policy outcomes (Donald et al., 2022). In Kenya, community-based resource management (CBRM) has fostered social capital, economic diversification, and the integration of indigenous knowledge, significantly contributing to climate security in Laikipia County (Keter & Mbali, 2025). Similarly, in Australia, community-based initiatives mainstreamed climate adaptation into disaster risk reduction by embedding environmental justice and adaptive planning into local governance strategies (Forino et al., 2018).

Lastly, collaborative strategies between public health nurses and community partners in the United States that addressed climate injustices related to pollution and colonial legacy. Their approach fostered resilience and built regenerative systems rooted in local justice (LeClair et al., 2024). These studies demonstrate that community-led and participatory governance models are essential for embedding equity and justice within climate and environmental frameworks. They empower local populations to actively shape and implement sustainable resource management strategies. Figure 4 illustrates a community-based governance model, emphasizing how participatory engagement leads to equitable outcomes in climate justice. This figure presents a governance model where climate justice emerges from community engagement, participatory decision-making, shared resource management, and climate adaptation/mitigation. It underscores how benefit-sharing flows from inclusive, bottom-up approaches.

Figure 4. Community-based governance process promoting participatory climate justice

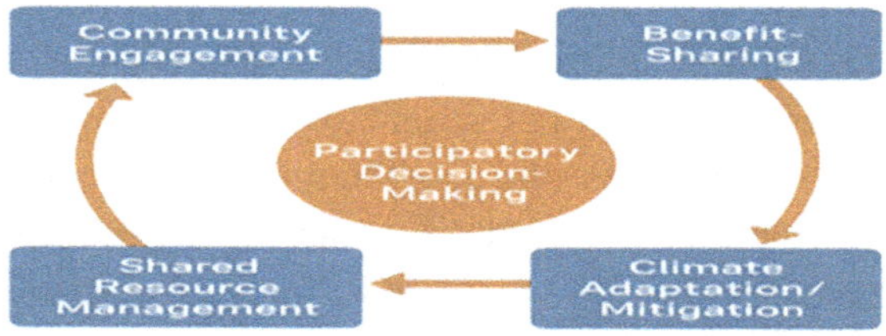

7. LEGAL INNOVATIONS AND POLICY REFORMS FOR CLIMATE JUSTICE AND SUSTAINABLE RESOURCE GOVERNANCE

As the global community faces growing environmental and climate injustices, legal innovations and policy reforms are crucial to achieving climate justice and effective resource management. The legal landscape is shifting to address climate change by ensuring fair access to resources, intergenerational equity, and adaptive legal frameworks for inclusive governance. A notable trend is the use of binding mechanisms to support climate justice, as India's experience shows that, despite political challenges, legal reforms can empower local communities and promote collective governance through judicial and legislative means (Diduck et al., 2021). Legal advocacy has proven instrumental in protecting rural livelihoods, especially when aligned with grassroots environmental activism and judicial remedies.

Contemporary environmental governance increasingly integrates social justice into policy, with legal tools such as the Paris Agreement and climate frameworks addressing the disproportionate impacts on vulnerable populations. Study shows legal advocacy, climate litigation, and participatory rights can hold governments accountable, close enforcement gaps, and support indigenous and marginalized groups' rights (Ahmed & Uddin, 2025). Innovative climate governance also requires aligning legal structures with emerging technologies. The rise of digital green laws underscores the need for regulatory frameworks that facilitate the deployment of climate technologies, such as AI for environmental monitoring and blockchain for carbon credit verification. There is emphasis on the value of adaptive, principle-based legal architectures to guide innovation while safeguarding transparency and equity (Vasile et al., 2025).

At the policy level, the integration of sustainability-focused judicial reforms is gaining traction globally. New judicial models, such as those aligned with the UN Sustainable Development Goals, incorporate ecological considerations and incorporate equity into traditional procedural fairness standards. This transformation in the judiciary's role reflects a broader conceptual shift toward multi-pillar justice that synthesizes environmental, economic, and social dimensions (Ieong, 2024). Additionally, the effectiveness of legal innovation in achieving sustainability is evident in the biofuel weed case study, which demonstrates how environmental risks necessitate novel legal and policy interventions that draw upon interdisciplinary insights from the social sciences, particularly in addressing wicked policy problems such as poverty and climate-induced displacement (Gal, 2016). Moreover, sustainable governance reforms must account for institutional capacity-building and equitable resource distribution.

Studies analyzing European environmental innovation confirm a moderate positive correlation between governance quality and sustainability outcomes, advocating for comprehensive policies that integrate market mechanisms with legal incentives to promote green technology uptake (Kruhlov et al., 2024). Finally, innovation in national climate reports often refers not to technological novelty but structural legal reforms that support sustainable development goals. These include regulations facilitating renewable energy deployment, conservation finance instruments, and mechanisms that embed sustainability into the administrative law frameworks of states party to the UNFCCC (He & Zahar, 2018). Table 3 compares selected national legal innovations and their contributions to advancing climate justice principles. This table highlights how countries have pioneered innovative legal and policy approaches to integrate justice into climate governance. From Ecuador's recognition of the Rights of Nature to the Philippines' adaptation fund, these reforms demonstrate varied pathways to embed equity, participation, and ecological protection within legal systems.

Table 3. Comparative overview of legal innovations and climate justice reforms in national frameworks

Country	Legal Innovation	Justice Element	Result
Ecuador	Recognition of Rights of Nature in the Constitution (2008)	Ecological justice; protection of ecosystems as rights holders	Enabled communities to litigate on behalf of ecosystems
Kenya	Climate Change Act (2016)	Mainstreaming equity in climate planning and policy	Strengthened accountability in climate governance
India	National Green Tribunal (2010)	Access to justice through specialized environmental court	Provided accessible legal remedies for climate victims
Germany	Amendment to Climate Protection Act (2021)	Intergenerational equity through stronger emission targets	Increased ambition in Germany's climate commitments
Philippines	People's Survival Fund (2012)	Equity in financing climate adaptation for vulnerable communities	Secured funding mechanisms for local adaptation projects

Source: Compiled by the Author.

8. FUTURE DIRECTIONS FOR CLIMATE JUSTICE AND SUSTAINABLE NATURAL RESOURCE GOVERNANCE

The future of climate justice in the context of natural resource governance is poised at a critical juncture, where law, policy, ethics, and innovation must intersect to forge paths that ensure environmental sustainability, equity, and global responsibility. The complexity of climate challenges necessitates integrative and anticipatory governance frameworks that address the long-term implications of natural resource exploitation and climate change, with a particular focus on intergenerational equity.

8.1 Intergenerational Justice and Governance Transformation

Intergenerational justice has transitioned from being a purely philosophical abstraction to a practical ethical framework guiding global climate governance. Contemporary scholars emphasize that climate decisions must ensure fairness between present and future generations, particularly with respect to ecological stability and sustainable development. Scholars argue that failure to integrate this perspective entrenches disproportionate burdens on youth and future societies (Cisco & Gatto, 2020). Building on this, frameworks for governance now call for future-oriented financial instruments, sustainable welfare systems, and institutional mechanisms dedicated to protecting unborn generations (Oliveira, 2018). Numerous studies further link intergenerational equity to climate tipping points, underscoring the urgency of adopting resilience-based systems to preserve fundamental human rights and environmental sustainability for future populations (Keizer, 2019).

8.2 Policy Innovation for Future-Oriented Sustainability

Policy innovation for future-oriented sustainability requires adaptive institutions that integrate scientific modeling, cross-sectoral governance, and technological advances. Emerging research emphasizes leveraging outputs from physical climate models and applying system-wide co-optimization strategies across energy, water, and food networks to strengthen resilience under climate unpredictability (Musetsho et al., 2024). Equally, open innovation strategies and green policies in urban planning have been identified as key tools for sustaining long-term ecological and social welfare, particularly in rapidly urbanizing contexts (Aboulnaga et al., 2021). As demonstrated in Australia's Murray-Darling Basin water governance reforms, institutional reforms that embed adaptive governance in environmental resource management are also critical for achieving sustainability outcomes.

8.3 Grassroots Mobilization and the Role of the State

Grassroots mobilization drives climate justice by challenging elite governance. Though states hold redistribution power, effective climate justice needs cooperation with community-led efforts. Research shows that grassroots movements, such as food sovereignty and energy remunicipalization, can reshape governance by integrating local knowledge and equity into climate policies (Routledge et al., 2018). Similarly, evidence from transnational mobilizations illustrates how bottom-up activism creates counter-hegemonic forces capable of influencing state agendas and shaping just transition policies (Almeida, 2019; Hoffmann, 2020). These dynamics highlight the need for hybrid governance models in which grassroots action and state authority co-produce pathways to sustainable and equitable climate futures.

8.4 Experimental Urban Governance and Socio-Ecological Justice

Experimental urban governance has increasingly been framed as a laboratory for advancing climate justice and socio-ecological equity. Study highlights how climate experiments can confront governance tensions while embedding distributive, procedural, recognition, and restorative justice principles (Manganelli, 2024). Complementing this, scholars emphasize the integration of socio-economic projections in adaptation planning to operationalize justice in urban resilience, ensuring that vulnerable groups are not excluded from benefits (Prall et al., 2023). Furthermore, comparative urban greening initiatives reveal that embedding justice principles into planning enhances equity and health outcomes, underscoring the transformative role of cities as drivers of just climate transitions (Breton-Carbonneau et al., 2025). These perspectives position cities not merely as sites of climate vulnerability but as incubators for innovative, justice-centered governance.

8.5 Global Norms and Institutional Reform

Global climate governance increasingly recognizes that safeguarding the rights of future generations requires deep institutional reform. Emphasis on the proposals, such as appointing UN special envoys, creating global citizens' assemblies, and embedding youth participation to overcome systemic short-termism (Caney, 2022). Complementing this, scholars on UN reform demonstrate how norm evolution, through mechanisms like Delivering as One, fosters system-wide coherence and strengthens the legitimacy of supranational institutions (Richter, 2018). Emerging work on climate assemblies highlights their potential to counter misinformation and democratize policy debates, embedding inclusivity and justice in governance structures (Suiter et al., 2025). These reforms reflect the growing momentum toward institutionalizing justice across time within global governance frameworks.

8.6 The Precautionary Principle and Ethical Reframing

The precautionary principle is a key ethical and legal tool for climate change, promoting proactive measures for future generations despite uncertainty. It relates to intergenerational justice and sustainability and requires integration into law. Building on Rawls's justice theory, recent views link precaution to obligations for environmental preservation, ensuring fairness across generations (Kim, 2019). Furthermore, ethical analyses underscore the principle's role in ensuring climate measures are adopted even under scientific uncertainty, framing precaution as a normative and institutional necessity for sustainable governance (Steele, 2006). These perspectives highlight how ethical reframing around precaution and justice provides a durable foundation for long-term climate policy.

The future of climate justice in resource governance depends on legal innovation, inclusive participation, and reforms. This involves embedding ethical principles, such as intergenerational equity and precaution, into systems, while ensuring that state mechanisms and grassroots mobilization amplify marginalized voices. Global institutions must be reimagined to address short-termism, improve accountability, and integrate justice across decision-making levels. Such a justice-centered approach is vital to equitable access to resources and resilience against climate risks. Protecting future generations' rights hinges on this shift toward fairness, sustainability, and shared responsibility.

9. POLICY RECOMMENDATIONS FOR CLIMATE JUSTICE AND EQUITABLE NATURAL RESOURCE GOVERNANCE

In crafting policy recommendations for climate justice and equitable natural resource governance, it is imperative to acknowledge the structural inequalities that shape climate vulnerabilities and environmental degradation. The disproportionate impacts of climate change on marginalized populations such as indigenous peoples, low-income communities, and Global South nations demand an intersectional and inclusive approach to policy design and implementation.

Institutionalizing Justice in Climate Governance

Climate policy must integrate mechanisms for distributive and procedural justice. This includes participatory frameworks that empower vulnerable communities in decision-making processes, ensuring transparency, fairness, and inclusiveness. Achieving just transitions in energy, land use, and governance mechanisms requires embedding equity into the legal and institutional fabric of environmental governance.

Designing Multi-Objective Policy Tools

Traditional climate policy models often prioritize economic efficiency, overlooking the trade-offs involved in equity and justice. This chapter proposes an innovative framework using Multi-Objective Multi-Agent Reinforcement Learning (MOMARL) to optimize climate policy for environmental outcomes and social equity.

Addressing Structural and Historical Injustice

Climate justice policy must also address historical injustices related to colonialism, unequal natural resource exploitation, and systemic underdevelopment. Integrating reparative justice mechanisms into international climate obligations, highlighting the role of colonial legacies in contemporary climate vulnerability.

Targeted Climate Financing and Loss & Damage Funds

The creation of accessible and non-discretionary financial mechanisms is essential. Critiques of existing financial instruments are insufficient and suggest that mandatory contributions to the Loss and Damage Fund can enhance justice for small island states and highly vulnerable nations.

Integration of Indigenous and Local Knowledge Systems

Policy frameworks must integrate indigenous legal traditions and local ecological knowledge. This supports climate resilience, cultural sovereignty, and sustainable resource stewardship. Equitable governance should merge formal systems with traditional models to ensure justice and sustainability.

Monitoring, Evaluation, and Adaptive Governance

Establishing robust accountability structures is critical. Recommend participatory monitoring and evaluation systems that allow iterative adjustments based on equity outcomes, thus ensuring long-term climate resilience.

Emphasizing Education and Awareness in Vulnerable Communities

Climate justice strategies must include robust public education, especially targeting communities that have historically been excluded from policy dialogue. There is a need for climate literacy, equitable access to infrastructure, and community-driven adaptation programs to counter systemic neglect.

Gender-Inclusive Policy Design

Justice-centered policies must be gender-responsive. Addressing the differentiated climate impacts on women and LGBTQ+ individuals requires deliberate inclusion in policy formulation and benefits distribution, as well as broader debate on climate justice narratives. Adequate and equitable climate governance requires legal and policy frameworks that go beyond mere emissions mitigation. They must proactively dismantle embedded structural inequalities and empower marginalized groups by expanding access to resources, ensuring participatory rights, and ensuring equitable distribution of benefits. The recommendations outlined here chart a practical, justice-oriented roadmap for policymakers, researchers, and civil society stakeholders committed to achieving climate justice and sustainable natural resource management.

Data Availability Statement

All relevant data are available online.

Conflict of Interest

The authors declare no conflict of interest.

Declaration of AI Use

While preparing this work, the authors used Grammarly and ChatGPT to improve English and enhance readability. After using it, the authors reviewed and edited the content as needed and took full responsibility for the publication's content.

REFERENCES

Abdelrahman, W. (2024). Climate Justice: Equitable Treatment for Future Generations and a Changing World. *Journal of Environmental Toxicology Research*. DOI: 10.59462/JETR.1.1.102

Abdullahi, A., & Musa, I. G. (2023). The legitimacy of international law: Challenges and the emerging issues. *Journal of Global Social Sciences*. DOI: 10.58934/jgss.v4i16.217

Abdulwahab, L. O., & Azam, H. (2026). Platform Work in Europe's Green and Care Economies: Regulatory Pathways for Sustainability, Social Protection, and Fair Work. In *Regulating the Gig Economy in the European Union: Challenges, Innovations, and Global Impacts* (pp. 99–124). IGI Global Scientific Publishing. https://www.igi-global.com/chapter/platform-work-in-europes-green-and-care-economies/395612

Abdulwahab, L. O., Raji, I. A., & Oladimeji, S. M. (2025). Impact of natural resource management on entrepreneurial activity and SME growth in emerging economies: Does the rule of law matter? *Resources Policy*, *108*, 105674. DOI: 10.1016/j.resourpol.2025.105674

Aboulnaga, M., Sala, M., & Trombadore, A. (2021). *Open Innovation Strategies, Green Policies, and Action Plans for Sustainable Cities—Challenges, Opportunities, and Approaches*. 49–68. DOI: 10.1007/978-3-030-57332-4_4

Ahmed, M. W., & Uddin, M. J. (2025). Environmental Justice and the Role of Legal Advocacy in Addressing Climate Change. *Journal of Humanities and Social Sciences Studies*. DOI: 10.32996/jhsss.2025.7.4.6

Al-Hakim, A. (2023). Navigating Legal Pluralism: A Socio-Anthropological Analysis of Governance and Law in Multicultural Societies. *Journal of Judikaltura*. DOI: 10.61963/jkt.v1i2.35

Ali, M., Elahi, M. M., & Farooqi, M. A. (2025). Coloniality of International Climate Litigation and Climate. In *Justice: Exploring Challenges and Perspectives from the Global South*. Social Science Review Archives., DOI: 10.70670/sra.v3i1.395

Almeida, P. (2019). Climate justice and sustained transnational mobilization. *Globalizations*, *16*(7), 973–979. DOI: 10.1080/14747731.2019.1651518

Amato, L. F. (2025). *Structural litigation, destabilization rights and trans-judicial cooperation networks: Lessons from comparative constitutional law*. Suprema – Revista de Estudos Constitucionais., DOI: 10.53798/suprema.2025.v5.nEspecial.a313

Angell, K. (2021). Chris Armstrong on Global Equality and Special Claims to Resources. *Global Justice : Theory Practice Rhetoric*. https://doi.org/DOI: 10.21248/GJN.13.01.184

Barry, L. (2022). Community science and the design of climate governance. *Climatic Change*, *171*(3-4), 24. Advance online publication. DOI: 10.1007/s10584-021-03295-7

Barsotti, F., & Koçer, R. G. (2022). MinMax fairness: From Rawlsian Theory of Justice to solution for algorithmic bias. *AI & Society*, *39*(3), 961–974. DOI: 10.1007/s00146-022-01577-x

Beauregard, C., Carlson, D., Robinson, S., Cobb, C., & Patton, M. (2021). Climate justice and rights-based litigation in a post-Paris world. *Climate Policy*, *21*(5), 652–665. DOI: 10.1080/14693062.2020.1867047

Breton-Carbonneau, A. C., Anguelovski, I., Triguero-Mas, M., & Cole, H. V. S. (2025). Just urban greening for climate adaptation & health equity planning: Lessons learned from 5 cities in the Global North. *Cities (London, England)*, *158*, 105677. Advance online publication. DOI: 10.1016/j.cities.2024.105677

Brunnée, J. (2021). *Legal Argumentation in the Evolving Climate Regime*. Talking International Law., DOI: 10.1093/oso/9780197588437.003.0012

Caggiano, H., & Landau, L. (2021). A new framework for imagining the climate commons? The case of a Green New Deal in the US. *Planning Theory*, *21*(4), 380–402. DOI: 10.1177/14730952211042765

Caney, S. (2022). Global Climate Governance, Short-Termism, and the Vulnerability of Future Generations. *Ethics & International Affairs*, *36*(2), 137–155. DOI: 10.1017/S0892679422000181

Chang, H.-S., Su, Q., & Chen, Y. (2021). Establish an assessment framework for risk and investment under climate change from the perspective of climate justice. *Environmental Science and Pollution Research International*, *28*(46), 66435–66447. DOI: 10.1007/s11356-021-15708-2 PMID: 34331639

Chotouras, D., Fragkou, R., & Tsadiras, A. (2024). *Environmentally Induced Human Displacement and Human Rights: Bridging Legal Gaps Through a Human Rights-Based Approach*. International Human Rights Law Review., DOI: 10.1163/22131035-13020001

Cisco, G., & Gatto, A. (2020). Climate Justice in an Intergenerational Sustainability Framework: A Stochastic OLG Model. *Economies, 9*. DOI: 10.20944/preprints202011.0664.v1

Desai, B. H., & Desai, J. B. (2024). The Climate Change Conundrum: A Case for Course Correction in the Global Regulatory Approach. *Environmental Policy and Law*, *54*(1), 3–14. DOI: 10.3233/EPL-239028

Diduck, A., Patel, K. A., & Malik, A. (2021). *Lessons for policy and institutional reform*. Advancing Environmental Justice for Marginalized Communities in India., DOI: 10.4324/9781003141228-15

Donald, R., Young, C., & Mach, K. (2022). The role of local narratives in emerging climate governance. *Environmental Research: Climate*, *1*(1), 015003. Advance online publication. DOI: 10.1088/2752-5295/ac7aca

Du, H. (2022). *Climate neutrality in the EU and China: An analysis of the stringency of targets and the adaptiveness of the relevant legal frameworks. Review of European, Comparative &amp*. International Environmental Law., DOI: 10.1111/reel.12453

Ehsan, S., Begum, R. A., & Maulud, K. N. A. (2022). Household External Vulnerability due to Climate Change in Selangor coast of Malaysia. *Climate Risk Management*, *35*, 100408. Advance online publication. DOI: 10.1016/j.crm.2022.100408

Ensor, J., & Hoddy, E. (2020). Securing the social foundation: A rights-based approach to planetary boundaries. *Earth System Governance*. Advance online publication. DOI: 10.1016/j.esg.2020.100086

Estévez, R. A., Veloso, C., Jerez, G., & Gelcich, S. (2020). A participatory decision making framework for artisanal fisheries collaborative governance: Insights from management committees in Chile. *Natural Resources Forum*, *44*(2), 144–160. DOI: 10.1111/1477-8947.12200

Farah, P., & Ibrahim, I. A. (2023). *Urgenda vs. Juliana: Lessons for Future Climate Change Litigation Cases*. University of Pittsburgh Law Review., DOI: 10.5195/lawreview.2023.908

Fariss, B., DeMello, N., Powlen, K. A., Latimer, C. E., Masuda, Y., & Kennedy, C. M. (2023). Catalyzing success in community-based conservation. *Conservation Biology, 37*(1). *Conservation Biology*, *37*(1), e13973. Advance online publication. DOI: 10.1111/cobi.13973 PMID: 35796041

Ferreira, P. (2018). The Paris Agreement on Climate Change: Analysis and Commentary. Edited by Daniel Klein, María Pía Carazo, Meinhard Doelle, Jane Bulmer, and Andrew Higham. Oxford: Oxford University Press, 2017. 447 + xxxii pages. *Canadian Yearbook of International Law/Annuaire Canadien de Droit International*, *55*, 666–672. DOI: 10.1017/cyl.2018.14

Forino, G., von Meding, J., & Brewer, G. (2018). Community based initiatives to mainstream climate change adaptation into disaster risk reduction: Evidence from the Hunter Valley (Australia). *Local Environment*, *24*(1), 52–67. DOI: 10.1080/13549839.2018.1548010

Gal, E. L. (2016). Exploring new research directions for achieving a sustainable future: What can be learned from the biofuel weed risk case study? *Asia Pacific Journal of Environmental Law*, *19*, 55–78. DOI: 10.4337/apjel.2016.01.03

Goodier, J. (2018). The Paris Agreement on Climate Change. *Reference Reviews*, *32*(4), 29–30. DOI: 10.1108/RR-12-2017-0250

Grazini, C., & Guarini, G. (2025). Climate Justice through the Capability Approach: Addressing Gender Inequality and Human Development in the Era of Climate Change. [JAES]. *Journal of Applied Economic Sciences*, 135. Advance online publication. DOI: 10.57017/jaes.v20.1(87).09

Hariharan, L., & Mahees, M. (2024). The Impact of Climate Change on the Environmental Rights of Dry-Zone Farmers. *Journal of Social Sciences - NISD*. DOI: 10.4038/jssnisd.v2i1.1

Harvey, N., Garmestani, A., Allen, C. R., Buijze, A., & van Rijswick, M. (2023). Identifying untapped legal capacity to promote multi-level and cross-sectoral coordination of natural resource governance. *Sustainability Science*, *19*(1), 325–346. DOI: 10.1007/s11625-023-01424-y PMID: 38362047

He, X., & Zahar, A. (2018). Innovation and the Law in State Reports on Climate Change Action. *Socially Responsible Investment eJournal*. DOI: 10.2139/ssrn.3210014

Hoffmann, P. J. (2020). *System change, not climate change – Analyzing the elements of counter-hegemonic climate justice mobilization*. https://consensus.app/papers/%E2%80%9C-system-change-not-climate-change-%E2%80%9D-%E2%80%93-analyzing-the-hoffmann/5b7474e5354f530caba3fffa1a634330/

Hui, Z., & Karim, M. R. (2025). Pursuing climate justice through public interest litigation: Theories, practices, and prospects. *International Journal of Innovative Research and Scientific Studies*. DOI: 10.53894/ijirss.v8i2.6385

Husain, S., Ayoub, N. P., & Hassmann, M. (2024). *Legal pluralism in contemporary societies: Dynamics of interaction between islamic law and secular civil law. SYARIAT: Akhwal Syaksiyah*. Jinayah, Siyasah and Muamalah., DOI: 10.35335/cfb3wk76

Hwang, J. (2025). The philosophical foundations of social justice: A critical analysis of Rawls, Nozick, and contemporary theories. *International Journal of Frontline Research and Reviews*. DOI: 10.56355/ijfrr.2025.4.1.0016

Ieong, M. C. A. (2024). Exploring New Judicial Models under the Sustainable Development Goals. *SHS Web of Conferences*. DOI: 10.1051/shsconf/202420002036

Ikporukpo, C., & Ikporukpo, N. N. (2022). Spatial Hierarchy of Climate Injustice. *Advances in Social Sciences Research Journal*, *9*(12), 354–370. Advance online publication. DOI: 10.14738/assrj.912.13648

Islam, M. S. (2024). Rethinking Climate Justice: Insights from Environmental Sociology. *Climate (Basel)*, *12*(12), 203. Advance online publication. DOI: 10.3390/cli12120203

Jane, M., & Emma, D. (2021). *Part 3 Protection, 12 Displacement related to the Impacts of Disasters and Climate Change*. The Refugee in International Law., DOI: 10.1093/law/9780198808565.003.0012

Jat, P. S. (2025). *A Sustainable Legacy: Navigating Climate Change through the Lens of Intergenerational Justice*. INTERNATIONAL JOURNAL OF MULTIDISCIPLINARY RESEARCH AND ANALYSIS., DOI: 10.47191/ijmra/v8-i03-02

Jones, R., Reid, P., & Macmillan, A. (2024). An Indigenous climate justice policy analysis tool. *Climate Policy*, *24*(8), 1080–1095. DOI: 10.1080/14693062.2024.2362845

Kaya, Y. Ç., & Kaya, H. (2025). Global Energy Policy: A Legal Perspective on Renewable Energy Initiatives. *Sustainability (Basel)*, *17*(9), 3991. Advance online publication. DOI: 10.3390/su17093991

Keizer, C. (2019). *Climate tipping points and intergenerational justice*. https://consensus.app/papers/climate-tipping-points-and-intergenerational-justice-keizer/c0f9d81c8fca55808812bf6fef7b9f5b/

Keter, J. C., & Mbali, K. (2025). The Impact of Community-Based Resource Management on Climate Security in Laikipia County, Kenya. *International Journal of Research and Innovation in Social Science*. DOI: 10.47772/IJRISS.2025.90300170

Kim, H. (2019). An extension of Rawls's theory of justice for climate change. *International Theory*, *11*(2), 160–181. DOI: 10.1017/S1752971918000271

Klerk, B. E. (2025). *The ITLOS advisory opinion on climate change: Revisiting the relationship between the United Nations convention on the law of the sea and the Paris agreement. Review of European, Comparative &amp*. International Environmental Law., DOI: 10.1111/reel.12588

Kruhlov, V. K. V., Moroz, V. M. V., & Tereshchenko, D. T. D. (2024). *Ecological Innovations in Shaping Sustainable Governance Policies*. Regional Formation and Development Studies., DOI: 10.15181/rfds.v42i1.2613

Lai, M., Robinson, S., Salas, E., Thao, W., & Shorb, A. (2022). Climate justice for small island developing states: Identifying appropriate international financing mechanisms for loss and damage. *Climate Policy*, *22*(9-10), 1213–1224. DOI: 10.1080/14693062.2022.2112017

Laird, S., Wynberg, R., Rourke, M., Humphries, F., Muller, M., & Lawson, C. (2020). Rethink the expansion of access and benefit sharing. *Science*, *367*(6483), 1200–1202. DOI: 10.1126/science.aba9609 PMID: 32165574

Larson, A., Barletti, J. P. S., & Vigil, N. H. (2022). A place at the table is not enough: Accountability for Indigenous Peoples and local communities in multi-stakeholder platforms. *World Development*, *155*, 105907. Advance online publication. DOI: 10.1016/j.worlddev.2022.105907

LeClair, J., Dudek, A., & Zahner, S. (2024). Climate Justice Strategies Implemented by Public Health Nurses and Their Community Partners. *Journal of Advanced Nursing*. Advance online publication. DOI: 10.1111/jan.16598 PMID: 39526565

Lee, S.-K. (2022). Will the EU Cbam Hurt Korean Manufacturers? An Empirical Analysis with Implications for Policy. SSRN *Electronic Journal*. DOI: 10.2139/ssrn.4315042

Liao, L. (2024). Synergy of soft and hard regulations in climate governance: The impact of state policies on local climate mitigation actions. *Environmental Policy and Governance*. Advance online publication. DOI: 10.1002/eet.2145

Luhandjula, Y. (2023). *Bridging the Gaps Between Human Rights and Climate Governance at the International and Regional Judicial Spheres*. Southern African Public Law., DOI: 10.25159/2522-6800/13679

Manganelli, A. (2024). Urban climate experiments: Governance tensions and opportunities for justice. *Environment and Planning C: Politics and Space*. DOI: 10.1177/23996544241306928

Marcellinus, B., Lucky, O., & Funke, A. C. (2021). *Rawlsian Veil of Ignorance and the Quest for Restructuring in Nigeria*. https://consensus.app/papers/rawlsian-veil-of-ignorance-and-the-quest-for-restructuring-funke-marcellinus/41af6bc266aa59f6b57139f628c60fda/

Minas, S. (2021). Financing climate justice in the European Union and China: Common mechanisms, different perspectives. *Asia Europe Journal*, *20*(4), 377–401. DOI: 10.1007/s10308-021-00644-0 PMID: 34899112

Minnerop, P. (2021). *The 'Advance Interference-Like Effect' of Climate Targets: Fundamental Rights, Intergenerational Equity and the German Federal Constitutional Court*. https://consensus.app/papers/the-%E2%80%98-advance-interferencelike-effect-%E2%80%99-of-climate-targets-minnerop/cacb9aa90bba5a9e98889289495afc68/

Murcott, M., Tigre, M. A., & Zimmermann, N. (2023). Transnational Insights for Climate Litigation at the European Court of Human Rights: A South-North Perspective in Pursuit of Climate Justice. *Verfassung und Recht in Übersee*, *56*(2), 299–320. Advance online publication. DOI: 10.5771/0506-7286-2023-2-299

Musa, S., Abdulwahab, L. O., & Beri, M. H. (2026). The Impact of AI Technologies in Transforming Financial Systems and Driving Innovation. In *Promoting Inclusivity and Accessibility With FinTech* (pp. 153–182). IGI Global Scientific Publishing., https://www.igi-global.com/chapter/the-impact-of-ai-technologies-in-transforming-financial-systems-and-driving-innovation/387218

Musetsho, K. D., Mwendera, E., Madzivhandila, T., Makungo, R., Volenzo, T., Mamphweli, N. S., & Nephawe, K. A. (2024). Assessing and mapping water-energy-food nexus smart innovations and practices in Vhembe District Municipality, Limpopo Province, South Africa. *Frontiers in Water*, *6*, 1253921. Advance online publication. DOI: 10.3389/frwa.2024.1253921

Nigam, A., & Parikh, M. (2023). Comparative Analysis of Climate Justice Framework in USA, UK, Japan, China, Germany and France. *Journal of Humanities and Social Sciences Studies*. DOI: 10.32996/jhsss.2023.5.9.6

O'Donnell, T. (2020). *Climate Change Adaptation Litigation: A Pathway to Justice, but for Whom?* 117–132. DOI: 10.1007/978-981-15-0466-2_6

Oliveira, R. V. (2018). Back to the Future: The Potential of Intergenerational Justice for the Achievement of the Sustainable Development Goals. *Sustainability (Basel)*, *10*(2), 1–16. DOI: 10.3390/su10020427

Onifade, T. T. (2021). *Climate Justice Under the Paris Agreement: Framework and Substance*. Carbon & Climate Law Review., DOI: 10.21552/cclr/2021/3/6

Parent, S. (2023). Hope for Climate Justice in Labour Law in Quebec. *Revue de Droit Comparé Du Travail et de La Sécurité Sociale*. DOI: 10.4000/rdctss.6746

Parker, L., Mestre, J., Jodoin, S., & Wewerinke-Singh, M. (2022). When the kids put climate change on trial: Youth-focused rights-based climate litigation around the world. *Journal of Qualitative Research in Tourism*. DOI: 10.4337/jhre.2022.01.03

Poudel, D. (2022). *Governance of Asta-Ja Resources in Nepal*. Nepal Public Policy Review., DOI: 10.3126/nppr.v2i1.48687

Prall, M., Olazabal, M., & Lehmann, M. (2023). Socio-economic projections in urban climate change adaptation planning: Practices and prospects for just adaptation. *Habitat International, 142*, 102946. Advance online publication. DOI: 10.1016/j.habitatint.2023.102946

Puaschunder, J. M. (2019). Governance and Climate Justice. *Socially Responsible Investment eJournal*. DOI: 10.2139/ssrn.3341494

Puaschunder, J. M. (2020). Global Leadership and Followership on Climate Justice. *Energy Policy & Economics eJournal*. DOI: 10.2139/ssrn.3691578

Richter, A. (2018). The Norm Life Cycle of UN Reform. *Delivering as One and UN System-Wide Coherence., 8*, 25–46.

Robinson, S., & Carlson, D. (2021). A just alternative to litigation: Applying restorative justice to climate-related loss and damage. *Third World Quarterly, 42*(6), 1384–1395. DOI: 10.1080/01436597.2021.1877128

Rodríguez-Garavito, C. A. (2021). Litigating the Climate Emergency: The Global Rise of Human Rights-Based Litigation for Climate Action. SSRN *Electronic Journal*. DOI: 10.2139/ssrn.3860420

Routledge, P., Cumbers, A., & Derickson, K. (2018). States of just transition: Realising climate justice through and against the state. *Geoforum, 88*, 78–86. DOI: 10.1016/j.geoforum.2017.11.015

Sánchez-Simbaña, S. E. (2025). *Una perspectiva emancipadora sobre la gobernanza de recursos naturales y soberanía en el contexto Ecuatoriano*. Revista Científica Zambos., DOI: 10.69484/rcz/v4/n1/82

Santarlacci, A. de S., Angelo, H., & de Souza, Á. N. (2024, March 23). Lima, M. de F. de B., Joaquim, M. S., Miguel, E. P., & Carneiro, J. de O. (2024). Benefit Sharing Governance Framework: Pathways for Financial Benefit Sharing in Traditional Communities. *Sustainability (Basel), 16*(7), 2650. Advance online publication. DOI: 10.3390/su16072650

Schroder, F. (2021). FUTURE GENERATIONS AND THE ENVIRONMENT: A RIGHT TO INTERGENERATIONAL EQUITY UNDER INTERNATIONAL LAW? *The Pretoria Student Law Review*. DOI: 10.29053/pslr.v15i1.3666

Segger, M.-C. C., & Gehring, M. W. (2025). *Climate Justice through International Courts and Tribunals: Advisory Opinions in the International Tribunal on the Law of the Sea (ITLOS), the Inter-American Court of Human Rights (IACtHR) and the International Court of Justice (ICJ). Region &amp*. Periphery., DOI: 10.12681/ri.40698

Shawoo, Z., & McDermott, C. (2020). Justice through polycentricity? A critical examination of climate justice framings in Pakistani climate policymaking. *Climate Policy*, *20*(2), 199–216. DOI: 10.1080/14693062.2019.1707640

Shut, V., Hanenko, I., & Yalanska, M. I. (2025). Geneva conventions and modern warfare technologies: Issues of effective legal application. *Uzhhorod National University Herald. Series: Law*. DOI: 10.24144/2307-3322.2025.87.4.35

Sorgho, R., Bhatt, M., Danquah, I., & Sauerborn, R. (2022). Institutional barriers to climate change and health adaptation in Burkina Faso. *Climate and Development*, *15*(7), 578–589. DOI: 10.1080/17565529.2022.2125786

Steele, K. (2006). The precautionary principle: A new approach to public decision-making? *Law Probability and Risk*, *5*(1), 19–31. DOI: 10.1093/lpr/mgl010

Strange, K. F., Satorras, M., & March, H. (2024). Intersectional climate action: The role of community-based organisations in urban climate justice. *Local Environment*, *29*(7), 865–885. DOI: 10.1080/13549839.2024.2315992

Suiter, J., Saude, K., & McNally, B. (2025). *Climate change assemblies as spaces for the potential mitigation of climate policy misperceptions: A survey experiment. Environmental Science &amp.* Policy., DOI: 10.1016/j.envsci.2025.103995

Tigre, M. A., & Wewerinke-Singh, M. (2023). *Beyond the North–South divide: Litigation's role in resolving climate change loss and damage claims. Review of European, Comparative &amp.* International Environmental Law., DOI: 10.1111/reel.12517

Tirkey, C. A. (2023). *Comparing Visions of Justice: Rawls' Idealism vs. Sen's Pragmatism.* Integrated Journal for Research in Arts and Humanities., DOI: 10.55544/ijrah.3.6.28

Toussaint, P. (2023). Loss and Damage, Climate Victims, and International Climate Law: Looking Back, Looking Forward. *Transnational Environmental Law*, *13*(1), 134–159. DOI: 10.1017/S2047102523000237

Turkamani, H. S. (2023). *The Legal Nature of the Climate Change Regime: Fluctuation between Lex Lata and Lex Ferenda.* Utrecht Journal of International and European Law., DOI: 10.5334/ujiel.603

Uche, O. L., & Ekine, S. B. (2025). Intersectional Approaches to Climate Justice: Examining Gendered Environmental Labor through the Lens of Equity and Sustainability. *Legal Research &amp. Analysis*, *3*(1). Advance online publication. DOI: 10.69971/lra.3.1.2025.62

Vasile, I., Mihaila, S., & Popa, R. (2025). Digital Green Laws: The Role of Legal Norms in Advancing Climate Technology for Sustainable Development. *Rechtsnormen: Journal of Law*. DOI: 10.70177/rjl.v3i2.2218

Wood, B. T., Dougill, A., Stringer, L., & Quinn, C. (2018). Implementing Climate-Compatible Development in the Context of Power: Lessons for Encouraging Procedural Justice through Community-Based Projects. *Resources*, *7*(2), 36. Advance online publication. DOI: 10.3390/resources7020036

Yazar, M., Haarstad, H., & Elfving, J. (2024). *Cities incorporate equity in their climate policies but overlook procedural justice in decision-making*. Nature Cities., DOI: 10.1038/s44284-024-00167-w

Zafar, S., Khan, M. Z., Mehmood, T., Begum, F., & Sadiq, M. (2022). Role of community-based conservation and natural resource management in building climate resilience among vulnerable mountain societies. *Climate and Development*, *15*(7), 608–621. DOI: 10.1080/17565529.2022.2135365

Chapter 5
Climate Vulnerability and Legal Equity:
Legal Pathways for Indigenous Peoples, Climate Migrants, and Vulnerable Groups

Sitesh Kumar Singh
http://orcid.org/0000-0002-7108-0808
College of Engineering, National University of Science and Technology, Oman

Jaishri Gothania
http://orcid.org/0000-0002-2656-642X
Lingaya's Vidyapeeth, India

Seema Bushra
Lingaya's Vidyapeeth, India

Abby Balkis Gorayah
http://orcid.org/0009-0006-7326-2249
College of Engineering, National University of Science and Technology, Oman

ABSTRACT

This chapter discusses climate vulnerability and justice of the law for the Indigenous people, climate migrants, and other disenfranchised people, such as women, children, and disabled individuals. Through the intersectional lens, it looks at the exacerbating effect of confluent social, cultural, and economic inequalities in increasing climate risks and analyzes deficiencies of existing legal systems. Community-based resilience and legal advocacy through case studies are complemented by pools of emerging responses in policy and best practices that help define the pathways to

DOI: 10.4018/979-8-3373-6635-7.ch005

inclusive governance. The recommendations call on the need to enhance human rights-based climate policies, institutionalize protection of climate migrants, integrate intersectionality into climate changes, and improve the involvement of the community to promote just and sustainable climate mitigation measures.

1. INTRODUCTION TO CLIMATE VULNERABILITY AND LEGAL EQUITY

Climate change has become one of the most burning and complicated problems of the twenty-first century. One can see its effects throughout all parts of the globe in an increase in the frequency and intensity of heatwaves, lengthy droughts, severe flooding, erratic rain patterns, and a rising sea level (IPCC, 2023). Even though no population is immunized, the climate affects it unequally. Socially, economically, politically, and geographically marginalized communities are the main victims of such shifts, and usually, lacking access to resources and resources creating defenses such as legal defense, they cannot adapt, recover, or find redress accordingly.

Such inequalities bring forth basic issues of legal equity, the idea that all people and groups must receive equal treatment in terms of case and availability of justice under the law. Equity in climate governance does not mean policymakers need to distribute resources used in adaptation equitably but acknowledge the needs, capacities and rights of various groups. The indigenous, climate migrants, and other vulnerable groups of people like women, children, and persons with disabilities also need specific legal means to protect their lives provisions, cultures, and human rights.

Nonetheless, existing national and international legal frameworks on climate change have not been able to assuage these realities adequately. A significant portion of policies take what is commonly referred to as the one-size-fits-all approach that presupposes homogeneous vulnerability and ability in all the populations and fails to consider accumulated disadvantages experienced by certain groups. The Paris Agreement and the Sendai Framework on Disaster Risk Reduction, as examples of international agreements, recognize the significance of human rights and the inclusive participation but do not contain adequate enforcement measures that would make states comply with the promises they give. Such mismatch between the ideal and the reality maintains inequalities with the most vulnerable affected groups taking the back seat of climate governance.

There is a strong correlation of law with historical and systematic injustices. Many Indigenous people of climate-sensitive homelands lack secure land tenure and formal jurisdiction to governance systems because of centuries of colonization, removal of people, and extraction of resources (Adger, 2006). In the same way, individuals dislocated by climate change, popularly termed as climate migrants, are mostly invisible in international refugee law. Climate effects are beyond these individuals because of gender roles, socio-economic marginalization, institutional discrimination, and sex and persons with disabilities. These nexuses help realize that fair climate governance cannot be restricted to environmental protection, as it is ideally necessary to prescribe systematic inequality at its origin. Global exposure to climate hazards is unevenly distributed, with hotspots in coastal, deltaic, and small island states are presented in Figure 1 (IPCC, 2023).

Figure 1. Global climate hazards and vulnerability hotspots

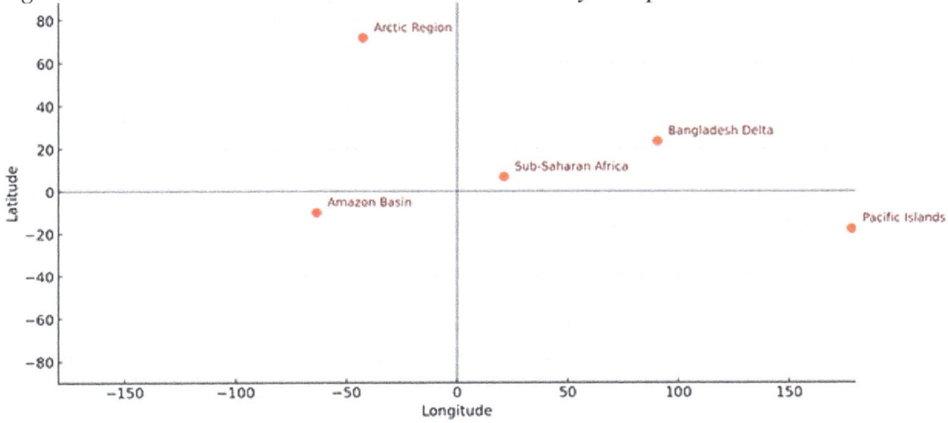

1.1. Overview of Climate Change Impacts on Marginalized Populations

The marginalized groups are disproportionately exposed to climate risks, since their residents tend to live in geographies that are environmentally vulnerable, as well as socio-politically marginalized. In the case of small island states, coastal communities of such states live in the constant threat of sea level rise and saltwater intrusion, which threatens food and water security in small island states. The lagging delta lands in Bangladesh, Nigeria and Vietnam are subject to flooding which destroys houses and farmlands and thus stifles the long-term development. These environmental risks are aggravated by the destruction of traditional lands by land grabbing, mining, deforestation, and other development initiatives of the state among the Indigenous peoples.

The association of Indigenous peoples with their lands and waters is not only economically oriented, but rather it is strong cultural and spiritual (United Nations, 2021). This deprives people of the land belonging to their ancestors which destabilizes governance structure, knowledge imparted between generations, and community unity. Moreover, it is not uncommon to find that the national legislative systems do not protect the Indigenous tenure rights even in cases where it is nominally accepted in the national constitutions and other international documents like the United Nations Declaration of the Rights of Indigenous Peoples (UNDRIP). These protections are often subverted by poor enforcement, conflicts of state interests, and the influence of the extractive industries.

Another issue is climate migration which is also increasing in size. Beneficiaries are displaced by sudden-onset disasters (including cyclones or flash floods), as well as slow-onset processes (including desertification, glacial retreat, and ocean acidification). However, these same studies fail to find clear support in the 1951 Refugee Convention or the 1967 Protocol to which the former is affixed, the provisions of which foresee the consequences of persecutions, but they do not entail environmental damage (Qil-Qdi, 2025). Consequently, climate migrants are likely to have uncertain legal status, not to have equal access to social services, and be exposed to the exploitation risk both in transit areas and in destinations.

These are aggravated by gendered and age-based inequalities. Women inhabitants in agrarian communities tend to be burdened with the role of providing food, water, and fuel supply which makes them highly vulnerable to resource insufficiency caused by a drought or flood (UN Women – Headquarters, 2025). Children lose access to schools, become at a higher risk of becoming malnourished, and expose themselves to exploitation when in displacement. People who have disabilities face inconvenience in the evacuation, entry to shelters, and involvement in decision making regarding the adaptation strategies (Twigg et al., 2011). These facts show that vulnerability can be not only the result of environmental exposure, but it also depends on inequalities that already exist in social aspects.

Unless legal action is considered with the consideration of the interest of the marginalized groups, there is a risk that climate action will end upholding the status quo of privilege and disadvantage. This suggests that equitable climate governance should consider not only environmental vulnerability factors but also structural factors that contribute to vulnerability and ensure that the most vulnerable populations have sufficient resources, empowerment and protection to live and flourish in a world characterized by shifting climate.

1.2. Conceptualizing Intersectionality in Climate Justice

The theory of intersectionality, initially developed by Crenshaw (1989) in research regarding the issue of anti-discrimination law, can be used as an effective tool to comprehend the influence of climate change on the social construct, resulting in varied consequences. Intersectionality is informed by the fact that individuals live with various, intersecting disadvantages (eg, racism, sexism, ableism, and economic marginalization) that cannot be either well conceptualized or effectively responded to in isolation.

In climate governance intersectionality reveals the insularity of legalisms. To cite an example, the environmental law can preserve the ecosystem, yet it overlooks gendered distributing labor that pre-determines the climate change effects experienced at household levels. On the same note, the disability rights laws can protect accessibility at urban infrastructural levels, and it can fail to clarify how people with disabilities will be addressed during rural evacuations. Such fragmentation does not allow legal institutions to respond fully to the ranges of vulnerabilities encountered by the marginalized communities.

An intersectional approach would compel the law and policymakers to incorporate several factors of identity and structural inequality into climate action. This translates to the existence of gender-sensitive, disable-inclusive and culturally appropriate laws protecting Indigenous territories. Climate migration policies should take both the relocation logistics and the maintenance of the cultural heritage, community cohesion and livelihoods. By doing so, intersectionality advances legal equity to the next level, which is real and situational, such as protections.

Intersectionality also creates an important burden in the tasks of governing. Considering the legitimacy and effectiveness of climate policies, increased inclusivity can be attained through the process of representation on the law-making level so that marginalized voices are given adequate representation. To illustrate, the Pacific Island countries have used traditional ecological knowledge in national adaptation plans, enhancing the level of community buy-in/ resiliency outcomes. Co-management between Indigenous governments with federal agencies has created more responsive and culture-tied climate adaptation planning in Canada. This is explained by the fact that intersectional governance is both massively, ethically necessary and practical.

Besides, intersectionality asks policymakers to understand that vulnerability is not invariable even though it is changing due to shifts in environmental conditions, migration patterns, and socio-political contexts. Law should thus be dynamic with provisions of continuous consultation and updating to keep itself responsive to the fact on the ground. In the absence of such flexibility, climate governance may be out of date and become inapplicable under the changes with changing times.

1.3. Literature Review and Research Gaps

The current literature on climate justice has extensively covered the theme of environmental vulnerability, human rights, and adaptation governance. The research has identified the unequal effects of climate change on Indigenous people, migrants, and marginalized groups (Adger, 2006; McAdam, 2012; IPCC, 2023). Gendered aspects of climate vulnerability and the impact of social inequalities in developing adaptive capacity have also been examined in the research (Sorensen et al., 2018; UN Women – Headquarters, 2025).

A lot of the literature however deals with such issues individually. The rights of indigenous land, climate migration, gender inequalities, and disability inclusion issues are usually studied in different disciplinary or policy models. Such fragmentation does not allow for gaining a holistic picture of how common social identities and structural disparities define climatic vulnerability.

In addition, the current systems of climate governance are characterized by a preference of environmental or technical solutions and the weak consideration of legal equity and intersecting justice in the design and implementation of policies. The ways of incorporating intersectionality into the climate governance systems and decision-making processes through participation in legal systems have received little attention.

This chapter attempts to fill these gaps by considering the vulnerability to climate under an intersectional legal framework; it discusses Indigenous peoples, climate migrants, and vulnerabilities based on gender. The chapter is synthesis of legal analysis and case studies and policy developments, which adds to the insights of the need of inclusive legal frameworks to enhance equitable climate governance.

2. INDIGENOUS PEOPLES AND CLIMATE CHANGE

2.1. The Unique Relationship between Indigenous Communities and Their Lands

The estimated indigenous population in the world is more than 476 million people who live their lives in various ecosystems such as tundra in Arctic regions and tropical rainforests. Their livelihoods, the way they govern themselves and their lands are inseparable (United Nations, 2021). These lands are not only economic properties but are also spiritual, cultural and ancestral land. Indigenous peoples also have access to adaptive strategies due to traditional ecological knowledge (TEK), which is formed because of generations-long direct interaction with local ecosystems. This is because TEK can offer viable solutions to resource management, conservation of diversity, and risk reduction in case of disasters.

An example is the Arctic Inuit, using centuries-old navigation technology, seasonal migration patterns and wildlife harvest technologies that are highly optimized to the weather patterns in the area. But Arctic warming is happening at almost four times the rate of the world, is not only affecting the formation of sea ice, changing animal migration patterns, but also exposes food safety (Ford et al., 2007). Elsewhere on the continent, there are the Indigenous peoples of the Amazon Basin, including the Kayapo and Ashaninka, who have advanced systems to manage the forest, and experience mining, agriculture, and infrastructure development-related deforestation. Tuvalu and Kiribati are Pacific Island countries where Indigenous peoples face the threat of rising waters, coastal erosion and loss of land, which signal not only displacement but also loss in culture and identity as a people (Kanungo, 2023).

Inclusion of these relationships weakens community integration as well as the carrying of knowledge across generations. Environmental degradation caused by climate change may undermine customary governance systems, decrease the sustainability of traditional livelihoods and make people more dependent on external aid, which is usually provided within structured approaches that do not necessarily reflect the needs of Indigenous populations. Unless it securitizes land tenure and respectfully adapts to temperature change, responses to climate change face the risk of exacerbating centuries of historical injustice, on top of marginalization. Indigenous knowledge systems provide adaptive strategies that strengthen biodiversity conservation and risk reduction as presented in Figure 2 as interconnectedness of traditional ecological knowledge and sustainability (Ford et al., 2007; United Nations, 2021).

Figure 2. Traditional Ecological Knowledge (TEK) and indigenous climate adaptation strategies

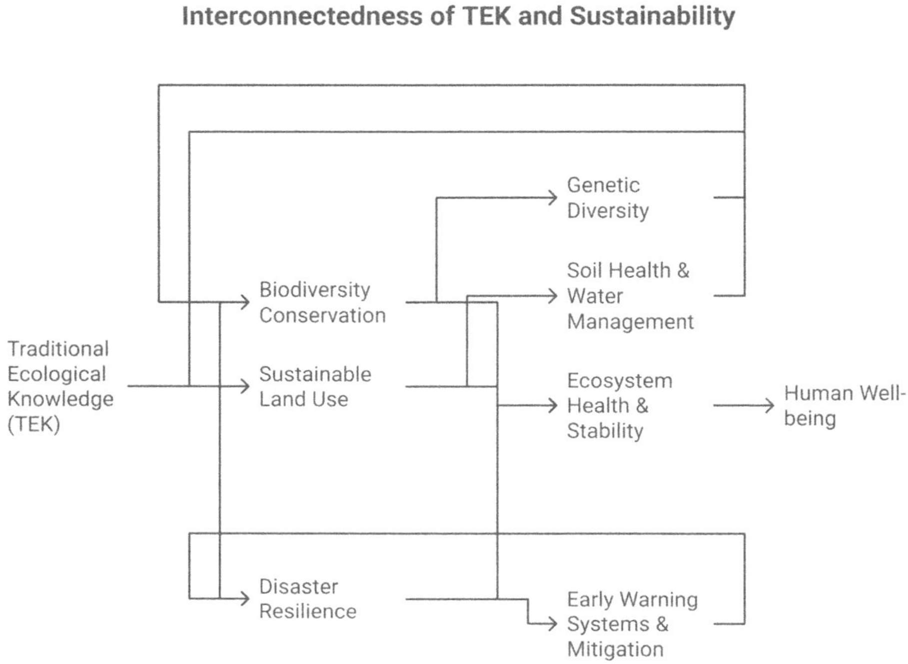

In terms of political economy, most of the Indigenous territories lie within the natural resource-rich regions in terms of minerals, forests, and hydrocarbons. In the past, colonization and current development policies have enabled the extraction of resources in such regions in large quantities, at the expense of national economic growth at the expense of Indigenous land ownership. Research based on anthropological and political economy demonstrates that such extractive projects are not only degrading the environment, but they also strengthen the structural inequalities by marginalization of Indigenous governance systems and limiting the ability of communities to control the land and other resources.

2.2. Legal Protections for Indigenous Land Rights and Traditional Knowledge

The Indigenous people have distinct rights to lands, territories, resources which are acknowledged by international law. The UN declaration of the rights of the indigenous people (UNDRIP) supports this right of the indigenous people to receiving free prior and informed consent before any project can receive approval on their territory or on their resources (United Nations, 2007). In the same breath, International Labor Organization (ILO) Convention No. 169 requires states to protect the territories of the Indigenous people and observe the use of customary land transfer (Convention C169 - Indigenous and Tribal Peoples Convention, 1989 (No. 169), 1989). Like other climate governance frameworks, the Paris Agreement does not ignore the importance of Indigenous rights in adaptation planning and capacity-building exercise.

States at the domestic level have some states that have included this with main legal cover. In Canada, the presence of constitutional rights of Aboriginals has enabled landmark cases in courts to support Aboriginal land claims through court decisions like in Tsilhqot'in Nation vs. British Columbia, 2014 (*Supreme Court of Canada Cases Involving Indigenous Peoples | SFU Library*, 2014). These settlement arrangements provided through the Treaty of Waitangi in New Zealand have helped the MÂori iwi to gain co-management agreements to control the natural resources in New Zealand, with MÂori worldviews centralized in environmental governance. In 2009, Bolivia became a pioneer in rights-based governance environmental management by introducing a constitution that incorporates Indigenous autonomy and rights of nature (*Bolivia (Plurinational State of) 2009 Constitution - Constitute*, 2009).

But implementation is patchy. Political opposition to legal protections is usually driven by the extractive industry press, low institutional capacity, and political opposition. The FPIC procedures can become hasty or badly executed, preventing the involvement of real participation. Also, traditional knowledge systems have been known to be used and misused without consent or mutual sharing, especially in such activities as climate resilience initiatives, and biodiversity protection initiatives. The absence of intellectual property protection makes Indigenous people particularly susceptible to exploitation, as well as being a loss of valuable expertise regarding the local ground to climate science.

Nevertheless, despite these identifications, most of the global climate and human rights are not practically legally binding as opposed to being normative. As an illustration, although the Paris Agreement recognizes Indigenous knowledge and involvement in climate action, it does not create any binding commitments to states to secure Indigenous land rights and provide free, prior, and informed consent to climate-related projects. Most of such provisions are heavily reliant on political will and national legislation. This is accompanied by economic incentives that are linked to extraction of natural resources, infrastructure projects and energy transitions which usually push political opposition against increased enforcement. This disparity between the international and national enforcement has persistently slowed down the efficacy of these legal frameworks in securing vulnerable communities.

Table 1. Key international legal instruments protecting indigenous peoples' land and knowledge rights

Instrument	Year	Key Provisions	Relevance to Climate Governance
UNDRIP (United Nations Declaration on the Rights of Indigenous Peoples)	2007	Free, prior, and informed consent (FPIC); land and resource rights	Ensures Indigenous participation in adaptation/mitigation
ILO Convention No. 169	1989	Customary land tenure recognition; self-determination	Protects Indigenous communities from displacement
Paris Agreement	2015	Recognizes Indigenous rights and traditional knowledge in adaptation planning	Encourages inclusive climate governance
Bolivia Constitution	2009	Rights of nature, Indigenous autonomy	Legal model for integrating rights into environmental law

As shown in Table 1, multiple international instruments acknowledge Indigenous peoples' land and knowledge rights, but enforcement and implementation remain weak (United Nations, 2007; Convention C169, 1989).

2.3. Case Studies of Indigenous-led Climate Resilience and Legal Advocacy

a) Oceania: Protection of the Pacific Islands

Native-driven mangrove rehabilitation programs in Vanuatu have integrated TEK and scientific practices in to increase resistance to change. These programs are backed by community bylaws, which control harvesting, and show how local politics can be co-joined with environmentalism (Veitayaki et al., 2017).

b) Canada – Consensus Agreements

The Inuvialuit Final Agreement in Northwest Territories of Canada gave the Inuvialuit and the federal authorities the burden of managing wildlife and environmental resources communally. This framework has made it possible to ensure that potential adaptation responses, like seasonal harvests and infrastructure design, incorporates Indigenous priorities along with best available scientific information (Executive and Indigenous Affairs, 1984).

c) Amazon Basin -Legal interventions to Deforestation

A historic legal success came in 2019 when the Waorani people of Ecuador won a six-year battle challenging an oil company in their ancestral territory following a courts-decision protecting their right to FPIC. This example highlighted the importance of a legal defense against environmental unfriendly projects on the territory of Indigenous people (Frontlines, 2023).

Throughout the examples, there are besides common denominators, such as community unity, assimilation of TEK and scientific knowledge, and resort to legal instruments, both domestic and international, to protect the territorial claims. These examples demonstrate that local climate governance in Indigenous hands not only has positive local impacts but also furthers the climate adaptation and mitigation agenda globally.

3. CLIMATE MIGRANTS: LEGAL RECOGNITION AND CHALLENGES

3.1. Defining Climate Migration and Displacement

Climate migration is the migration of people or communities who must abandon their regular homes because of the emergent or gradual environmental changes associated with climate change (IOM, 2019). This can be displaced by phenomena like cyclones, floods and wildfires as well as by long-run phenomena like desertification, sea level rise, and glacial retreat. Migration can either be internal, inside national boundaries or cross-border but the drivers behind the same are often multifaceted, where both the environmental, economic, social and political aspects are inextricably linked.

The difference between climate and environmental migrants is based on the anthropogenic climate change role. Climate migrants are those individuals whose movement occurs because of anthropogenic climate change in particular, whereas environmental migration may involve movements induced by natural variability or other non-climatic environmental calamities. In practice though the two types overlap with each other.

Available data on climate displacement across the globe accentuates the magnitude of the problem. The report of the weather-related hazards states that in 2022 alone, over 32 million people were displaced due to weather hazards (IDMC, 2022) as presented in Table 2.

Table 2. Global climate displacement trends

Year	Estimated Displaced Persons (millions)	Main Drivers	Source
2010	19.3	Conflicts, Floods, storms	IDMC (2012)
2015	24.0	Cyclones, droughts	IDMC (2016)
2020	30.7	Extreme weather events	IDMC (2021)
2022	32.6	Floods, storms, wildfires	IDMC (2022)

Most of these displacements are recurrent often communities are displaced during disaster and allowed back to damaged or degraded environment hence they remain perpetually vulnerable. Floods and storms remain the leading causes of displacement, followed by droughts and wildfires as presented in Figure 3 (IDMC, 2022).

Figure 3. Major drivers of climate migration

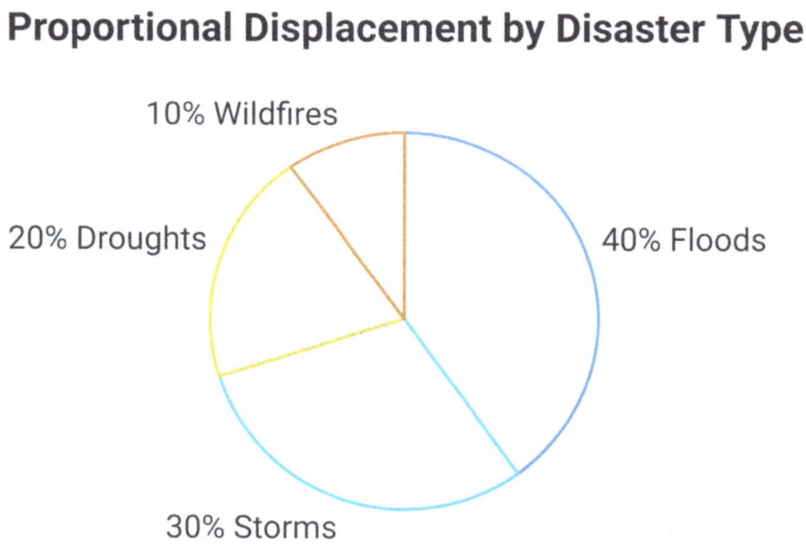

Displacement is becoming ever more widely accepted as a long-term adaptation approach in countries like Kiribati and the Maldives that are low-lying island states with planned relocation programs being discussed.

Social networks are another aspect of sociological research emphasizing the role of social networks in the migration decision and adaptation strategies. Relations within families, the community, and the diaspora networks tend to determine the migration of people, the place of migration, and the assimilation of migrants into the host communities. The networks may be able to offer financial assistance, information, housing, etc, therefore minimizing the risks of displacement and easing the resettlement processes.

3.2. Gaps in International and Domestic Law Concerning Climate Migrants

Among the main issues experienced by climate migrants is the lack of recognition in international law. The 1951 Refugee Convention and ensuing 1967 Protocol safeguard those who have fled persecution on grounds of race, religion, nationality, political opinion or membership of a certain social group. Nevertheless, they do not include individuals who are displaced just simply by environmental or climatic change (McAdam, 2012). This gap in the law means that the environmentally displaced are denied asylum and resettlement within the established, conventional framework of refugees.

Some countries have devised discretionary policies, e.g. Argentina special residency permits to environmental migrants and New Zealand humanitarian visas to the Pacific Islanders, but these are minimal and are not based on binding commitments. Natural disaster-induced displacement is recognized at the regional level in Africa in the Kampala Convention, although that recognition is unevenly applied.

Home legislation also comes in a big variance. Fragmented governance has been achieved in many states where internal displacement caused by climate impacts is managed under disaster management architectures instead of the law governing migration. Such partitioning may lead to housing, social, and livelihood provision blanks to displaced communities. Also, conflicts among host and displaced communities are inevitable following property rights that are never clear, especially in the relocation sites.

Lack of any coherent legal component not only hinders protection efforts but also makes data gathering, resource distribution and future adaptation plan difficult. Climate migrants have not been legally recognized, therefore, they form an invisible layer in policy discourse, reducing the possibility to seek justice and continuing the vulnerability cycle.

3.3. Emerging Legal Frameworks and Policy Responses

Such gaps notwithstanding, there are nascent efforts that might become the cornerstone of broader regimes of protection. Climate-induced migration Climate change has been recognized as a cause of migration through the adoption of the Global Compact on Safe and Orderly Migration (GCM) which is specifically addressing climate change as a cause of migration and urges the promotion of planned relocations policy and making migration policies part and parcel of climate change adaptation policies. Though not binding, the framework of cooperation and exchange of information can be considered with the GCM.

Nationally, Fiji has formulated a Planned Relocation Guidelines framework which provides procedural guidelines in relation to relocation of communities threatened by impacts of climate change that focuses more on consultations, human rights and preservation of culture (Link et al., 2025). Bangladesh has put into element in its Delta Plan 2100 considering climate migration and aims to make coastal high-risk areas resilient (Arfanuzzaman, 2024). Reforms to migration policy involving assessment of climate vulnerability are being investigated in Chile and Uruguay within Latin America (Fernández-Rodríguez & Freier, 2024).

Climate litigation is also forming a new face of legal innovation. Although majority of the climate cases have emphasized the cutting down of emission, an increment of cases with a note on displacement has been noted. An example is in Teitiota v. In New Zealand (2020), the UN Human Rights Committee acknowledged that the state might not deport persons to the countries where climate change threatens life due to climate-related emergency, in effect, expanding the non-refoulement rule to include some climate-related circumstances (Kent, 2025). Even though the ruling had no binding value, it can be regarded as a precursor of a future change in international human rights law in the focus on protection of climate migrants.

Moreover, policies are responding to advocacy by the community. The Pacific Island countries have also established regional groupings in a move to demand the legalization of climate displacement in international circles. The rights-based relocation can be found in the Native-progressed initiatives in Alaska and Louisiana, such as the development of relocation plans taking into consideration the cultural land attachments.

What is required now is to put these developing structures into firm products of law-statements that will be binding and enforceable, thereby ensuring the safety and humanity of climate migrants. To meet this, it would be necessary to tune international, regional, and domestic legal instruments, proper financing in relocation and adaptation, and making the affected communities the pivot of decision-making.

4. GENDER EQUALITY AND VULNERABLE GROUPS IN CLIMATE LEGISLATION

4.1. Gendered Impacts of Climate Change

Not all populations are affected in equal measure during climate change and gender plays a key role in determining vulnerability. There is common reason as to why women, especially those in rural and low-income environments, tend to lack economic resources, have lesser access to education, and engaged in less involvement in the decision-making processes. In farming economies, women are more reliant on natural resources because of their requirements in food, water and fuel supply, and these natural resources are constantly threatened by droughts, floods, and rain patterns (UN Women – Headquarters, 2025). During extreme weather events, women are unlikely to move freely because they have these responsibilities, which means that they are more likely to become exposed to harm.

Figure 4. Gendered vulnerability pathways under climate stress

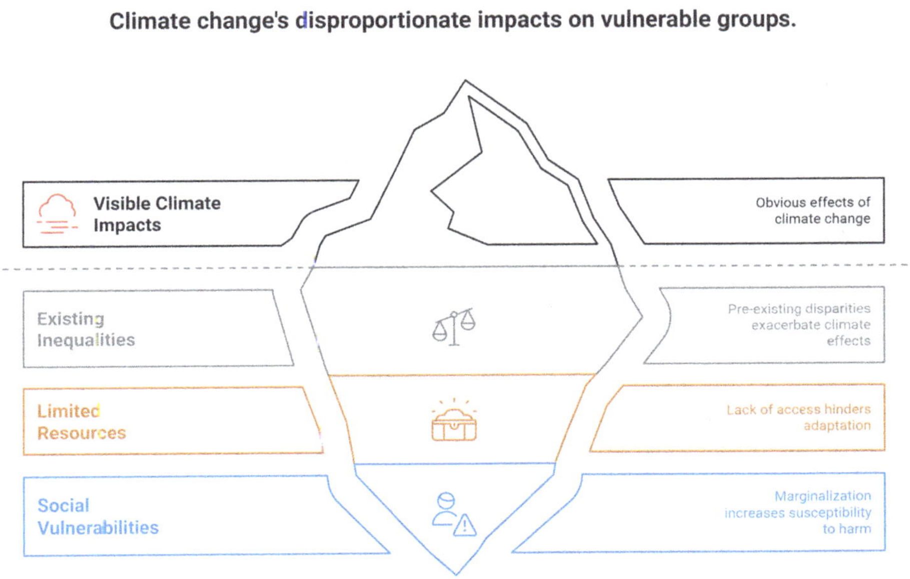

The gendered implications can also be noticed regarding health. As an example, women in predetermined areas are under a more profound threat of mortality in hot waves because of existing health inequality and involvement in society, which leads to enhanced exposure (Sorensen et al., 2018). Women face an elevated degree of gender-based brutality, trafficking and loss of livelihoods in post-disaster conditions, especially in temporary shelters or in a setting of displaced people. Gender roles and social inequalities shape how different groups experience climate shocks as shown in Figure 4 (UN Women – Headquarters, 2025; Sorensen et al., 2018).

Gender-specific vulnerabilities are possible in men and boys too. The physical danger various men face can be attributed to cultural norms likely to influence them to stay behind in the event of such a disaster to guard property at the expense of their lives. These disparate experiences highlight the need to have climate legislation be gender responsive, which means by meeting the needs, roles and capacity of all genders.

4.2. Inclusion of Women, Children, Persons with Disabilities, and Other Vulnerable Groups in Climate Laws

Vulnerability in climate governance does not apply only to gender, but also to children, people with disabilities, the aged, ethnic minority, or economically marginalized groups. Children are at risk of being malnourished, contracting waterborne diseases, and disrupting education in the event of a climate-related disaster (UNICEF, 2021). Individuals with disabilities often experience access impediments to evacuation, inaccessible infrastructure and lack of representation of their interests in adaptation planning, a condition that may eventually lead to high mortality and morbidity when a disaster strikes (Twigg et al., 2011).

The need to include these groups in climate legislation involves clear identification of these groups and action in their defense as permissible by the climate laws. There are some nations that started to combine these considerations with a legal system. The Climate Change Act (2009) of the Philippines requires that women and children, as well as marginalized sectors be brought into planning on climate change. The Climate Change Act (2016) in Kenya has incorporated gender and youth mainstreaming obligations with a view of ensuring that the responses to climate change are mainstreamed considering varied needs (REPUBLIC OF KENYA, 2016). The preamble to the Paris Agreement at the international level is a recognition of gender equality, women empowerment, and intergenerational equity but operational sections are rather few and have limited binding effect.

Nevertheless, there are challenges of the implementation. Most of the laws incorporate vulnerable groups on paper but fail to provide accountability or proper funding to the inclusive programs. Moreover, vulnerability across gender, disability, and environmental protection cannot be addressed comprehensively and need to be discussed in a siloed policymaking approach in view of climate governance.

4.3. Intersectional Analysis of Vulnerability and Legal Protections

An intersectional approach acknowledges the fact that vulnerabilities are not experienced alone. As an example, an Indigenous individual who has a disability might bear additional risks because of intersections of various modes of discrimination and marginalization. Anti-discrimination laws that cover only a single aspect of discrimination (gender or disability) might fail to help those at the intersection of more than one identity learn more about the cumulative experiences of depreciation at that intersection (Crenshaw, 1989).

Disaggregated data collection, inclusive consultative mechanism and multi-sectoral coordination is one of the requirements of intersectional analysis in climate legislation. As an example, the intersectional principles can be found in the National Climate Change Policy of Fiji where much gender, youth, and disability inclusion of all climate programs take place (Link et al., 2025). On the same note, the Gender-Based Analysis Plus (GBA+) system used in Canada has been utilized to understand the gap between the influences of climate policy on various demographics and dealing with them accordingly (Chyzzy et al., 2024).

Legally, intersectionality requires a cross-referencing of the environmental law and the human rights law to maintain consistency in the areas of protection. It also focuses on participatory governance where the vulnerable groups are well represented in the decision-making structures. This type of involvement does not only make the policies more legitimate, but it also makes them more effective as the various types of knowledge and experience can shape them.

5. INTERSECTIONALITY IN LEGAL RESPONSES TO CLIMATE VULNERABILITY

5.1. Applying Intersectional Frameworks in Climate Law and Policy

Intersectionality offers a framework through which one can learn to interpret how multiple social identities, and institutional inequalities exist in defining one to be vulnerable to climate change (Kimberlé W. Crenshaw, 1989). It adds that individuals are not just imposed on the effects of climate change due to their identity as Indigenous, female, disabled, and poor, but due to the intersection of these personal characteristics with the societal restrictions created in law, policy, and governance. When intersectional frameworks are applied in climate law, legal tools should serve the purpose of managing a variety of vulnerabilities at once.

To illustrate, such decisions as relocation policies concerning communities facing the rise in sea level should be made with taking into consideration the possibility of safe housing and preservation of the cultural heritage, female and male divisions of labor and needs of people with disabilities. In other jurisdictions, intersectional approaches are already appearing. Examples of the integration of gender-based analysis tools (like GBA+) into climate action planning have already been realised in Canada (Chyzzy et al., 2024), with establishment of Fiji Climate Relocation of Communities Trust Fund having encompassed both gender and disability inclusion in the relocation efforts (Link et al., 2025). These models indicate that an intersectional approach to policy designing can empower justice and efficacy in climate governance.

Nonetheless, there is poor adoption of existing national climate strategies because most of them do not come with concrete plans that locate and manage intersection vulnerability. This omission is usually followed by legal actions that are favorable to some groups of people more than to others thus increasing inequality instead of decreasing it. Table 3 illustrates the common legal issues, and new governance strategies of key vulnerable groups.

Table 3. Shared legal challenges and emerging legal responses across climate-vulnerable groups

Vulnerable Group	Key Legal Challenges	Structural Drivers	Emerging Legal and Policy Responses
Indigenous Peoples	Weak recognition of land rights; inadequate FPIC implementation; extractive industry pressures	Colonial land governance systems; resource extraction economies	Rights-based environmental governance; co-management agreements; Indigenous-led climate litigation
Climate Migrants	Lack of formal recognition in international refugee law; unclear protection mechanisms	Climate displacement, economic vulnerability, cross-border migration barriers	Global Compact on Migration; planned relocation frameworks; climate-related human rights litigation
Women and Gender-Marginalized Groups	Limited participation in climate decision-making; gendered economic inequalities	Gender norms, unequal resource access, labor burdens	Gender-responsive climate policies; gender mainstreaming in climate laws; community adaptation programs
Persons with Disabilities and Children	Limited accessibility in disaster response and adaptation planning	Institutional exclusion, infrastructure barriers	Disability-inclusive disaster risk reduction policies; child-centered climate resilience programs

As Table 3 indicates, there are similarities in the legal and institutional dilemmas within the various climate-sensitive populations and the new governance solutions to these problems.

5.2. Barriers to Participation and Representation in Legal Processes

The inclusion in legal and policy formulation is one of the most important pillars of climate justice but people who are marginalized lack access to meaningful participation because of the system. Such barriers are geographic isolation, language differences, a lack of available transportation, and a limited digital connection. In other instances, closed political or legal settings prevent such vulnerable communities, especially indigenous people researching against extractive industries.

The women, youth, and people with disabilities usually must encounter cultural and institutional barriers that make them silent in decision-making arenas (UN Women – Headquarters, 2025). In one case, an example is in some rural settings, the social behavior or etiquette does not allow women to go out and express themselves in a public gathering and the individuals living with disabilities are segregated whereby the facilities are not accessible to them or proper interpretation of sign language is not provided.

Intersectional analysis demonstrates that these obstacles are multiplied in the case of people with multiple marginalized identities. As an example, an Indigenous woman with a disability can face gendered and cultural discrimination and is, thus, almost completely isolated in policy consultations. The legal means to resolve them lie in the establishment of legal frameworks that dictate an inclusive process of participation, devote resources to accessibility, and should not allow representation to be tokenistic but rather substantial.

5.3. Best Practices for Inclusive and Equitable Climate Governance

The experience of a variety of countries in the world provides some best practices in terms of incorporating intersectionality into climate governance. Such practices are usually a blend of both stiff legal requirements and local community processes.

a. **Legal Requirements of Inclusion:** There must be legislations that compulsorily ask gender-responsive, disability-inclusive, and culturally neutral measures to all the climate related planning. As an example, under Kenya CHANGE Act, marginalized groups are obliged in the national climate change councils (REPUBLIC OF KENYA, 2016).
b. **Community-Led Governance:** Enabling affected communities to be at the helm of adaptation efforts keeps solutions close to the ground, and responsive to priorities and knowledge systems. Indigenous proposals and management projects established on the coast of Vanuatu deploy both the traditional ecological knowledge and scientific information that yield more culturally relevant practices and stronger sustainability (Veitayaki et al., 2017).
c. **Capacity Building and Education:** Training of government officials, as well as community leaders, can work towards getting mutual understanding and trust. Gender-sensitive training in disaster risk reduction has seen the engagement of women in disaster committees in Bangladesh at the local level (Arfanuzzaman, 2024).

d. **Cross-Sector Consultation and Liaising:** The collaboration that should occur in successful intersectional governance must be cross sector working between environmental agencies, human rights commissions, gender equality bodies and disability advocacy organizations. This assists in linking goals and prevention of competing mandates.

e. **Monitoring and Accountability:** To make the laws and policies effective, there should be measurable indicators of inclusion and equity, and public reporting on a regular basis, to increase accountability and ensure transparency.

The practices point to the fact that intersectionality is not a purely academic concept, but a working mechanism of creating more just, capable, and efficient climate solutions.

Figure 5. Intersectional climate governance framework

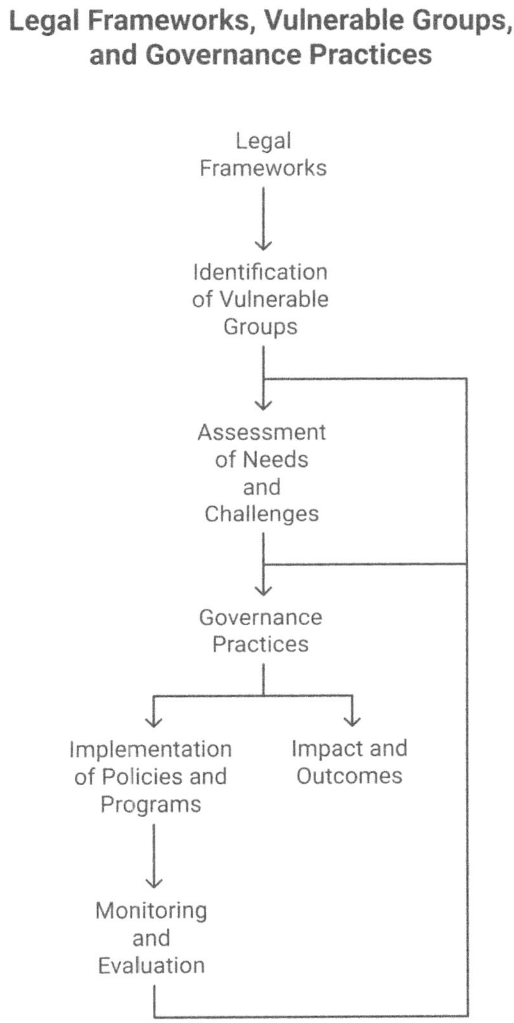

An intersectional governance framework integrates legal protections, equity principles, and inclusive participation is presented in Figure 5 (Crenshaw, 1989; Veitayaki et al., 2017).

6. INNOVATIVE LEGAL PATHWAYS AND POLICY RECOMMENDATIONS

6.1. Strengthening International Legal Instruments and Human Rights Frameworks

Most of these structures are based on soft-law and voluntary commitments, and this restricts their enforceability and leads to unequal application in various national settings. Fundamental and inherent weaknesses in existing laws to protect climate-vulnerable individuals, especially Indigenous peoples, climate migrants, and other discriminated cohorts necessitate the international law reforms. Current frameworks that include the Paris Agreement, the Sendai Framework for Disaster Risk Reduction, and the UN Guiding Principles on Internal Displacement are valuable normative frameworks without any enforceable actions. Such reforms are still difficult to implement politically. Particularly the states that fear having to bear greater migration responsibilities or face financial burdens because of climate displacement may not be willing to enter legal binding commitments. Big emitting states and countries that have high migration pressures would be especially wary of broadening international protection mechanisms. Consequently, the incremental strategies, i.e. regional agreements, soft-law instruments, and introducing climate displacement into current human rights mechanisms, might be more politically feasible in the short term.

The formal international law recognition of climate migrants is one such priority. This may be done by establishing another protocol to the 1951 Refugee Convention or establishing another treaty dealing specifically with climate-induced displacement (McAdam, 2012). The other alternative is to incorporate the legally enforceable obligations in the Global Compact on Migration where the states would agree to adhere to protection, relocation support, and sharing of resources.

International human rights law provides additional opportunities in enhancing protection. The Teitiota v. Prior legal precedent has been established by New Zealand (2020) at the UN Human Rights Committee, which connected recognition of the principles of non-refoulement to threats related to climate change (Kent, 2025). The multiplication of such jurisprudence might well force states to consider climate vulnerability when deciding on asylum and migration.

Moreover, the Indigenous rights framework such as the United Nations Declaration on the Rights of Indigenous People (UNDRIP) must be implemented via climate agreements with themes of free, prior, and informed consent (FPIC) applicable to all the climate-related initiatives on Indigenous lands. Convergence of environmental treaties and human rights obligations would establish compatible protection, which would support ecological and social justice.

6.2. Incorporating Intersectionality into Climate Action Plans

Intersectionality needs to be a part of the climate policy solution design. Although a few recognize gender or youth considerations in their national adaptation plans, there are few systematic initiatives to deal with the mixture of vulnerabilities including ethnicity, disability, and socio-economic condition. This negligence can lead to policy formulation where the goals end up discriminating against some groups of people as they are left out. Practically an intersectional impact review may serve the same role as environmental or human rights impact assessment already employed most jurisdictions. Governments would have to consider the impact of proposed measures on various demographic groups in terms of gender, indigeneity, disability, socio-economic status, etc. before implementing major climate policies or infrastructure projects. National human rights bodies, environmental control authorities, or independent review committees could offer oversight to make sure that there is transparency and adherence to equity principles.

The first strategy is to require intersectional impact review of every climate policy and program. Assessments of this sort would make policymakers assess the impact of the proposed actions on various population groups in different categories of identity. A model that can be adjusted to general intersectional standards such as disability, indigeneity, and migration status can be found in Canada and used in the Gender-Based Analysis Plus (GBA+) (Hankivsky, 2012).

Budgeting processes ought to incorporate intersectionality, as well. Investing in the marginalized populations in the climate programs will create a way of changing legal commitments into practical results. As an example, selective investments in Indigenous-controlled businesses in renewable sources of energy can benefit not only the ecological sustainability of the state but also the empowerment of communities.

Finally, monitoring and evaluation systems should include disaggregated data to observe the difference in the effect of climate interventions. Open reports on the beneficiaries and the ones left behind can spur accountability and consistent progress towards equity results.

The input of non-state actors is also significant to the further improvement of intersectional climate governance. Increased participation in climate negotiations, litigation, and policymaking is taken by International NGOs, Indigenous advocacy networks, and grassroots organizations. The legal advocacy groups have aided in the strategic litigation of climate and community-based rights claims and the community-based groups in society facilitate in the documentation of the effects of climate, mobilizing support and holding the institutions responsible for the implementation of the climate policy.

6.3. Enhancing Community Participation and Empowerment in Legal Decision-Making

The key principle of fair climate governance is community participation, which is sometimes no more than a symbolic representation, but not real control. The priority of the legal reforms must be the adoption of co-governance regimes contributing to running the climate-related projects and resources shared with the effect population.

To Indigenous peoples, this implies the acknowledgment of traditional systems of government in national laws and ensuring FPIC is an absolute condition of any project development or adaptation. In the case of climate migrants, planning of relocation as participatory (for example the Planned Relocation Guidelines in Fiji) are capable of undertaking resettlement locations that conform to cultures, economy and environment.

It is necessary to capacitate disadvantaged groups to participate in the legal procedures. This may involve legal literacy programs, grants to community-based organizations, and other means of technical assistance in taking part in international climate negotiations. The example is Bangladesh, where women are organizing community groups and their strategies on local adaptation plans have become successful once they were trained more specifically on advocacy and assessment of climate risks.

Digital solutions also have the capability of widening involvement, particularly among far-off or geographically isolated groups. Public comments online, reporting on climate risks via mobile and virtual stakeholder meetings, will create more inclusiveness when prepared in such a way that people with disabilities can use them and are available in local languages.

However, empowerment does not only entail access but also requires a change in power relations; the need to ensure that communities are not mere objects of policy but subjects of policy. The inclusion of legal requirements to achieve community consent, quota representation, as well as participatory budgeting in provisions of climatic legislation can codify power-sharing as a governance principle and not discretional power.

7. CONCLUSION AND FUTURE DIRECTION

7.1. Conclusion

Climate change is no longer a simple environmental issue; it is a great challenge to social justice/human rights and legal equity. It is unevenly distributed in its effects and marginalized groups, such as Indigenous peoples, climate migrants, women, children, and persons with disabilities are disproportionally affected. The intersecting socio-economic and cultural, as well as political, disparities, combined with exposures and the ability to render effective adaptations, will add to the vulnerabilities of these groups, as illustrated by the impacts of floods. An intersectional understanding of these layered risks is needed, understanding that the impacts of climate change are not universal, but are mediated through and by the many intersecting categories of marginalization and oppression.

The Indigenous communities are involved in one-of-a-kind challenges about climate change. Environmental degradation does have serious social and legal consequences since they are closely bound to land and natural resources: culturally, spiritually and economically. The traditional ecological knowledge may serve as a vital source of information about sustainable adaptations strategies, but legal instruments like the United Nations Declaration on the Rights of Indigenous Peoples (UNDRIP), the Convention No. 169 by the International Labour Organization, are sometimes not efficient in terms of implementation. This results into long-standing weaknesses especially when there are extractive industries and development projects which pose threats to the Indigenous land. Case studies show that Indigenous-led governance model that combine both traditional knowledge and modern legal and policy instruments offer promising culturally based approaches to resilience and climate adaptation.

There is also a high degree of legal invisibility of climate migrants. The displacement because of climate change is not expressly covered in the current international refugee law and, because of this, those affected by climate change have no rightful protection of asylum, relocation, or compensation. New tools such as the Global Compact on Migration as well as early-stage judicial claims such as Teitiota v. New Zealand, proposes some future measures that can be taken toward safeguarding climate migrants, yet such measures are still insufficient and non-unbinding. There is need to have stronger international legal processes that must guarantee equal treatment and the protection of human rights beyond climate-induced displacement.

Structural inequalities present in the conditions of vulnerable populations such as women, children and persons with disabilities result in heightened vulnerability risks. The climatic effects tend to further widen gaps that already exist socially since it affects everybody but people with vulnerabilities based on their age are usually more vulnerable, gendered division of labor is also prone to climatic effects as well as lack of democracy in decision making. Intersectional solutions within legal systems are necessary to address the intersection of vulnerability so that climate policies become inclusive and consider the different needs of various groups of people.

Intersectionality needs to be operationalized in climate governance using accessible, participatory and cross-sectoral legal means. As seen in some of the cases across different countries, a combination of intersectional principles would enhance justice and better policy effectiveness, therefore, contributing to fair play and building of resilience. Novel legal avenues and policy prescriptions are aimed at incorporating the idea of intersectionality into climate action plans, widening community engagement, and reinforcing international facilities in playing the role of guaranteeing the rights of marginalized communities.

Throughout all these dimensions, there is one common theme through which equity should be central to the climate law and policy. Climate action, however, will solidify structural injustices instead of fighting them unless there is a deliberate set of interventions that prioritize marginalized voices. The inclusion of justice and inclusiveness in climate governance preserves those who are most vulnerable and promotes more efficient, sustainable, and resilient climate initiatives in response to the climate crisis plaguing the world.

7.2. Future Directions for Research, Policy, and Legal Reform in Intersectional Climate Justice

In the future, there are some directions that demand immediate consideration to develop intersectional climate justice.

i. Creating International Climate Migrants Protections

Law practices of the future should be more obvious on the displacement that may be caused by climate changes. That may include a new treaty negotiation or through the amendment of existing conventions to cover climate migrants among the categories of formal protection. One responsibility should be carried out by research which ought to define clear eligibility standards, relocation standards and means of burden-sharing across states.

ii. How to operationalize Indigenous Rights in Climate Governance

The current issue is how to progress beyond the current stage of recognition of the Indigenous rights to the full operationalization of the same. This comprises the incorporation of free, prior, and informed consent (FPIC) on all climate-related undertakings and benefit-sharing relating to adaptation and mitigation projects utilizing Indigenous knowledge systems on equal bases. Comparative studies may be considered on what national models may be most effective: co-management in New Zealand, restoration of land in Canada, etc.

iii. Institutionalizing Intersectionality in Law/ Policy

Intersectionality is supposed to be institutionalized in the operations of legal writing, policymaking, and implementation. It will involve the need to come up with standardized intersectional assessment tools, the training of legal practitioners, and accountability mechanisms. The study needs to be conducted about how the idea of intersectionality can be unified within the legislation of environmental, human rights, and social protection to prevent the breaking up of the policy.

iv. Empowering Data and Evidence on equity-focused climate action

Aggregated data should be robust and disaggregated about who is being primarily impacted by climate and where the intervention is spread out. Future studies need to come up with strategies to gather and interpret intersectional climate data which should have privacy and community consent protection.

v. Modeling Community-Led and Co-Governance Expansion

Communities should have power to influence their adaptation measures so that the legal approach applied in situ is realistic and culture acceptable. Future research would require the assessment of the ways in which legal structures can streamline co-governance, especially in the scenarios involving political opposition or economic forces that jeopardize participatory practice.

vi. Promising Legal Solutions and Climate Lawsuits

Another area of increasing concern is litigation as a strategic method to gain rights to climate-vulnerable groups. Future research should assess how courts interpret the intersection of human rights, environmental law, and climate science, and explore opportunities for expanding legal standing and jurisdiction in climate cases.

vii. Breeding of International Unity and Interchange of Resources

Fair distribution of opportunities calls for the transfer of resources by high-income and high-emission states to the vulnerable nations and societies. The legal framework ought to consider the factors that promote climate finance that involves equity such as loss-and-damage funds, grant-based adaptation funding, and capacity-enhancement investment in the marginalized groups.

REFERENCES

Adger, W. N. (2006). Vulnerability. *Global Environmental Change*, *16*(3), 268–281. DOI: 10.1016/j.gloenvcha.2006.02.006

Affairs, E. I. (1984). Inuvialuit. Executive and Indigenous Affairs. https://www.eia.gov.nt.ca/en/priorities/concluding-and-implementing-land-and-resources-and-self-government-agreements-1#:~:text=What%20is%20the%20Inuvialuit%20Final,in%20the%20Inuvialuit%20Settlement%20Region

Arfanuzzaman, M. (2024). Bangladesh's pathways to climate-resilient development: A methodical review. *World Development Sustainability*, *4*, 100144. DOI: 10.1016/j.wds.2024.100144

Bolivia (Plurinational State of) 2009 Constitution - Constitute. (2009). https://www.constituteproject.org/constitution/Bolivia_2009

Chyzzy, B., Guruge, S., Schwan, K., Lee, J., & Stewart, S. (2024). Breaking Siloed Policies: Applying a Gender-Based Analysis Plus (GBA+) to Homelessness during Pregnancy in Canada. *Social Sciences (Basel, Switzerland)*, *13*(9), 486. DOI: 10.3390/socsci13090486

Convention C169 - Indigenous and Tribal Peoples Convention, 1989 (No. 169). (1989). https://normlex.ilo.org/dyn/nrmlx_en/f?p=NORMLEXPUB:12100:0:NO:P12100_ILO_CODE:C169

Crenshaw, K. (1989). Demarginalizing the intersection of race and sex: A Black feminist critique of antidiscrimination doctrine, feminist theory, and antiracist politics. *University of Chicago Legal Forum*, *1989*(1), 139–167. https://chicagounbound.uchicago.edu/cgi/viewcontent.cgi?article=1052&context=uclf

Fernández-Rodríguez, N., & Freier, L. F. (2024). Latin American immigration and refugee policies: A critical literature review. *Comparative Migration Studies*, *12*(1), 15. Advance online publication. DOI: 10.1186/s40878-024-00377-0

Ford, J. D., Smit, B., Wandel, J., Allurut, M., Shappa, K., Ittusarjuat, H., & Qrunnut, K. (2007). Climate change in the Arctic: Current and future vulnerability in two Inuit communities in Canada. *The Geographical Journal*, *174*(1), 45–62. DOI: 10.1111/j.1475-4959.2007.00249.x

Frontlines, A. (2023, December 22). Waorani people win landmark legal victory against Ecuadorian government - Amazon Frontlines. Amazon Frontlines. https://amazonfrontlines.org/chronicles/waorani-victory/

Hankivsky, O. (2012). Women's health, men's health, and gender and health: Implications of intersectionality. *Social Science & Medicine*, *74*(11), 1712–1720. DOI: 10.1016/j.socscimed.2011.11.029 PMID: 22361090

IDMC (Internal Displacement Monitoring Centre). (2012). *Global report on internal displacement 2012*. IDMC/NRC. https://www.internal-displacement.org/publications/global-overview-2012-people-internally-displaced-by-conflict-and-violence/

IDMC (Internal Displacement Monitoring Centre). (2016). *Global report on internal displacement 2016*. IDMC/NRC. https://www.internal-displacement.org/globalreport2016/

IDMC (Internal Displacement Monitoring Centre). (2021). *Global report on internal displacement 2021*. IDMC/NRC. https://www.internal-displacement.org/global-report/grid2021/

IDMC (Internal Displacement Monitoring Centre). (2022). *Global report on internal displacement 2022*. IDMC/NRC. https://www.internal-displacement.org/global-report/grid2022/

International Organization for Migration. (2019). *Glossary on migration* (2nd ed.). IOM. https://publications.iom.int/system/files/pdf/iml_34_glossary.pdf

IPCC. (2023). *Climate Change 2023: Synthesis Report*. Intergovernmental Panel on Climate Change. https://www.ipcc.ch/report/ar6/syr/

Kanungo, A. (2023, August 1). *How deforestation impacts indigenous communities | Earth.Org*. Earth.Org. https://earth.org/the-silent-cry-of-the-forest-how-deforestation-impacts-indigenous-communities/

Kent, S. B. A. (2025, June 13). The Teitiota Case and the limitations of the human rights framework - QIL QDI. QIL QDI. https://www.qil-qdi.org/the-teitiota-case-and-the-limitations-of-the-human-rights-framework/

Link, A., Piggott-McKellar, A., Nakoro, E., & Oakes, R. (2025). Climate-related partial relocation in Fiji impacts the wellbeing of those who relocated and those who stayed differently. *Communications Earth & Environment*, *6*(1), 394. Advance online publication. DOI: 10.1038/s43247-025-02357-3

McAdam, J. (2012). *Climate change, forced migration, and international law*. Oxford University Press., https://www.ohchr.org/sites/default/files/documents/issues/climatechange/cfi-hrc-53-session/submissions/2022-11-28/Jame-McAdam-5.pdf DOI: 10.1093/acprof:oso/9780199587087.001.0001

Qil-Qdi. (2025, July 31). Invisible in Law: Environmental and Climate Migrants and the Gap in International Protection - QIL QDI. QIL QDI. https://www.qil-qdi.org/invisible-in-law-environmental-and-climate-migrants-and-the-gap-in-international-protection/#_ftn13

REPUBLIC OF KENYA. (2016). THE CLIMATE CHANGE ACT, 2016. *Kenya Gazette Supplement*, 179. https://kenyalaw.org/kl/fileadmin/pdfdownloads/Acts/ClimateChangeActNo11of2016.pdf

Sorensen, C., Murray, V., Lemery, J., & Balbus, J. (2018). Climate change and women's health: Impacts and policy directions. *PLoS Medicine*, *15*(7), e1002603. DOI: 10.1371/journal.pmed.1002603 PMID: 29990343

Supreme Court of Canada cases involving Indigenous peoples | SFU Library. (2014). https://www.lib.sfu.ca/help/research-assistance/subject/criminology/legal-information/indigenous-scc-cases

Twigg, J., Kett, M., Bottomley, H., Tan, L. T., & Nasreddin, H. (2011). Disability and public shelter in emergencies. *Environmental Hazards*, *10*(3–4), 248–261. DOI: 10.1080/17477891.2011.594492

UN Women – Headquarters. (2025, April 21). How gender inequality and climate change are interconnected. UN Women – Headquarters. https://www.unwomen.org/en/articles/explainer/how-gender-inequality-and-climate-change-are-interconnected#:~:text=Women%20depend%20more%20on%2C%20yet,work%20harder%20and%20travel%20farther

UNICEF. (2021). *The climate crisis is a child rights crisis: Introducing the Children's Climate Risk Index*. UNICEF. https://www.unicef.org/reports/climate-crisis-child-rights-crisis

United Nations. (2007). United Nations Declaration on the Rights of Indigenous Peoples. https://www.un.org/development/desa/indigenouspeoples/wp-content/uploads/sites/19/2018/11/UNDRIP_E_web.pdf

United Nations. (2021). *State of the World's Indigenous Peoples: Rights to land, territories and resources*. UN Department of Economic and Social Affairs. https://www.un.org/development/desa/indigenouspeoples/news/2017/12/state-of-the-worlds-indigenous-peoples-iii-education/

Veitayaki, J., Waqalevu, V., Varea, R., & Rollings, N. (2017). Mangroves in small island Development States in the Pacific: An overview of a highly important and seriously threatened resource. In *Disaster risk reduction* (pp. 303–327). https://doi.org/DOI: 10.1007/978-4-431-56481-2_19

Chapter 6
Corporate Climate Adaptation Strategies and Their Implications for Human Rights Protection in Vulnerable Philippine Sectors:
An Analysis of Environmental Crime, Illegal Mining, and Deforestation

Froilan Delute Mobo
http://orcid.org/0000-0002-4531-8106
Philippine Merchant Marine Academy, Philippines

Jerielyn Reyes
Polytechnic University of the Philippines, Philippines

ABSTRACT

Research shows that the Philippines shows massive indications of extreme climate vulnerability (ND-GAIN Index, 2024; IPCC, 2022). Recent research shows (IMF, 2020) that the climate impact is exacerbated by resource-extracting environmental crimes like unregulated logging, illegal mining, and pillaging of natural resources. These crimes produce a collapse of ecosystems, weaken our natural defenses against the climate, and impact the climate-vulnerable more than the rest of the

DOI: 10.4018/979-8-3373-6635-7.ch006

world. These include the Indigenous, farming, and coastal communities and peoples (Global Witness, 2023). Corporations are still under-implementing climate adaptation actions, even though there is strategic alignment with Environmental, Social, and Governance (ESG) frameworks. These climate actions still neglect the systemic causes of vulnerability, stemming from environmental degradation and supply chain exploitation (UNEP, 2021). Effective adaptation research shows the need for policy on the inequities causing precarity.

INTRODUCTION

The Philippines is at an unstable crossroads of both geological and meteorological processes, which is why it can serve as one of the major locations to investigate the dangers of compound climate. It is a country that consists of over 7,600 islands, and it is also located within the Pacific Typhoon Belt and the Ring of Fire. It is confronted with a continuous flow of natural disasters, which are increasingly becoming frequent due to climate change brought about by human activity. According to the World Health Organization, roughly 60 percent of the landmass area of the Philippines is exposed to more than a single hazard. It makes it an extremely significant case study on global climate vulnerability (WHO, 2023). But this exposure is not a background risk anymore; it is an existential crisis, which is growing. Unstable tectonics coupled with unpredictable weather are the two factors that make the situation a high-stakes situation where there cannot be much room to make mistakes when it comes to environmental governance. Many islands comprise the Philippines; thus, there is no place to flee to the hinterland. Rather, each province is an upland or coastal border that is susceptible to the evolving weather conditions.

The Philippine Atmospheric, Geophysical and Astronomical Services Administration (PAGASA, 2023) estimates that the rate of sea level rise in the Philippine Sea is three times higher than the average on the planet. This rapidly expanding ocean is also threatening to drown coastal megacities and agricultural hubs that are already experiencing land subsidence due to excessive groundwater pumping and the burden of improvised city development. Also, the intensity of tropical cyclones has altered significantly. The long-term historical records can reveal that there has been no change in the total quantity of storms occurring on land annually, yet the category of storms that are categorized as Super Typhoons and which are of Category 4 and 5 has increased. They feature increased wind velocity and extreme precipitation incidences that have never occurred in the past and inevitably outperform the existing drainage systems in the city and infrastructure in rural regions (IPCC, 2022). This has been directly caused by the increase in the temperatures of the surface waters of the western Pacific. Such temperatures provide the storms with the thermal energy they require in order to rapidly accelerate in the days leading up to their landing on land, and this often results in early warning measures being missed.

We cannot simply observe the wind speeds and rainfall levels from a purely scientific perspective to justify the enormous human cost of the calamities in the archipelago. In order to get the whole picture of the crisis, you must refer to Climate Fragility, which is a socio-ecological condition where the environmental harm transforms controllable threats into life-threatening ones. The natural buffers in the Philippines have been destroyed by the systematic destruction of the primary rainforests and mountain watersheds in the country. In the absence of these buffers, they serve as biological infrastructure, regulating the flow of water and preventing gravity-oriented slopes. Once they are destroyed, usually by bypassing the law, illegal mining, land grabbing, etc., the impact of a typhoon is not merely the aggregate of its components. A degraded ecosystem turns a storm that would have resulted in a flood that would have been managed into a catastrophe in a healthy ecosystem in terms of landslides, siltation, and loss of lives. This vulnerability is artificial, and the ecological lack of naturalness of the tragedy is a convenient fiction that conceals the truth about the actual source of ecological pillaging (Lagon, 2022).

At least in reaction to these increasing threats, large business corporations in the Philippine mining, energy and infrastructure industries have included into their business plans under their Environmental, Social and Governance (ESG) systems, climate responsive activities. However, there is a worrying paradox of resilience in the world of business in the Philippines. The adaptation of corporations is largely concerned with the so-called gray infrastructure, involving the construction of seawalls, fortified processing facilities, and personal drainage systems to ensure that companies can continue to operate (Deloitte, 2023). These measures guard the company properties and shareholders without frequently considering or even aiding the destruction of so-called green infrastructure, such as forests and mangroves within their respective supply chains. This is the situation of adapting to the few that actually transfer the risk off the balance sheet of the company, and the people who are not able to afford to construct their own defenses, called the balance sheet. It is a form of spatial injustice since the fortified corporate enclosure is not affected by the storm, as the community surrounding it is filled with water and debris (Concha, 2020).

This situation is even more fragile due to the fact that the Philippine economy depends on natural resources that are prone to climate change. Since the activities of agriculture, fishing, and mining are the primary components of the economy, the environment suffers, and any harm to it poses an immediate danger to the stability of the country. The International Monetary Fund (IMF, 2020) has indicated that the geographical location of the country makes it one of the most exposed to natural disasters, but it is also because the country is also based on a resource base that is deteriorating rapidly. Risk multipliers, in this case, are environmental crimes such as illegal mining and logging without authorization. They render the watersheds, on which corporate infrastructure relies, less stable, thus forming a cycle of vulnerability that the established ESG frameworks cannot address. This creates so-called criminogenic spaces where the environmental rule of law has failed, and it can no longer be possible to obtain the so-called resilience that is often propagated in corporate glossy reports.

In order to overcome this dilemma, we must swiftly redefine the concept of adaptation by applying the concept of Environmental Human Rights Due Diligence (EHRDD). Climate change has been perceived as a technical rather than a human rights concern by many businesses today. The same can be said about corporations, making indigenous, farming, and coastal communities vulnerable since they fail to recognize the systemic vulnerability factors that arise as a result of using the supply chain (Global Witness, 2023). True adaptation must not be limited to the responses of incremental and responsive change; it must include revolutionary responses that can eradicate the causes of inequity that give rise to and support precarity (Taylor and Francis, 2024). This necessitates a radical re-thinking of corporate strength. The example of the Philippines, where a resilient entity cannot be found behind a seawall, and the resiliency should be measured by the ecological quality of the entire landscape where it is placed, and safeguarding the rights of the most vulnerable citizens to climatic extremes.

Notable about this crucible is that it shows that the current policy on climate of the whole world does not reverberate in the archipelagic areas that are at the high risk. The Philippines is one of the Global South, which proves that the absence of a rights-based approach to adaptation is the factor why even the so-called green transition and climate resilience can be another clarification to the back of the weak. The most significant moral issue of the decade becomes whether or not someone is strong enough to face increasing sea levels and more severe storms, and at what cost. We must address the environmental crimes and supply chain failures that render rain deadly to overcome climate fragility. It requires a change of orientation towards a justice-oriented adjustment whereby the idea of restoration of natural defenses is considered an essential prerequisite to the protection of human rights in the 21st century.

BEYOND THE NATURAL: WHAT DOES CLIMATE FRAGILITY MEAN?

An entirely scientific analysis of wind speeds and the rate of precipitation does not help to understand the colossal human and economic consequences of catastrophes in the archipelago. In order to understand the true nature of the crisis, we have to examine an example of a socio-ecological condition when environmental degradation transforms manageable natural hazards into carnal threats, namely, climate fragility. In the Philippines, the organized deforestation of primary rainforests and mountain watersheds has eliminated the natural buffers of the country, and the landscape has become weak, incapable of absorbing external shocks.

In good condition, these buffers form a biological infrastructure or a green infrastructure. Forests are considered sponges, which slow down the runoff process and allow the groundwater to refill. Root systems are the ones that bind the soil so that they prevent the slopes from succumbing to the power of gravity (Concha, 2020). When such systems are destroyed, as is frequently the case through the illegal extraction process, uncontrolled mining, or the process of land conversion, the effects of a typhoon are not added but rather multiplied. In a healthy ecosystem, flooding can be easily dealt with and can only be temporary due to a storm. In a poor ecosystem, however, landslides, massive siltation, and numerous deaths are the consequences of the storm (Lagon, 2022). According to the International Monetary Fund (IMF), the extreme risks of natural disasters in the country have been explained by the fact that the country heavily depends on climate-sensitive natural resources that are overused currently (IMF, 2020).

THE CORPORATE ADAPTATION PARADOX

To address the emerging threats, the mining, energy, and infrastructure sectors of the Philippines have introduced climate-responsive operations in their business models as Environmental, Social, and Governance (ESG) strategies. This appears to be one step in the right direction towards sustainable development. However, there has been a business ecosystem in the Philippines that has an alarming resilience paradox. According to Deloitte (2023), the corporate adaptation is dominated by so-called gray infrastructure, thereby consisting of the construction of seawalls, reinforced processing plants, high-rise warehouses, and personal drainage systems that serve to keep businesses in operation.

These measures secure company resources and maintain the value of shares at a high level, but are frequently in disregard of or even facilitative of that so-called green infrastructure (forests, mangroves, and wetlands) of which they themselves are part of their supply chains. This creates a scenario whereby only a few individuals will be capable of adjusting, and the strength of a business is pegged on the further intensification of the instability of a community in and around it. Once a company has established solid defenses around personal property, it transfers risk out of its own balance sheet and onto the shoulders of already impoverished individuals who lack the money and political strength to raise their own defenses (Taylor & Francis, 2024).

SPACES THAT CAUSE CRIME AND COMPLICITY IN THE SUPPLY CHAIN

Living with the paradox is made worse by the presence of what has come to be known as the criminogenic spaces, geographical areas in which crime thrives in the environment because of a lack of effective state regulation and the involvement of the corporate world. Unregulated logging and illegal gold mining are not only issues with the regulations, but also the constituents of numerous supply chains of the industrial world (UNEP, 2021). When a firm receives raw materials out of these degraded sections, they produce the so-called Climate Fragility; they say they are adjusting to bad.

The impacts on human rights are very poor. The extraction of the natural buffers illegally contravenes the right to life, health, and a balanced and healthy ecology guaranteed by the Philippine Constitution (Ramos, 2020). To the indigenous groups, the disappearance of forest cover that is caused by deforestation associated with corporations is an attack on their ancestral land and their traditional means of addressing climate change. Corporate adaptation projects, including large dams or renewable energy companies, in many cases proceed without actual Free, Prior, and Informed Consent (FPIC), which further marginalizes people who already face the problem of climate change (Global Witness, 2023).

THE RESILIENCE GAP AND THE NEED FOR EHRDD

The gap between the reported ESG practices by corporations and the reality of the environmental situation highlights the fact that there is a significant accountability gap. The modern research suggests that traditional Human Rights Due Diligence (HRDD) is insufficient; there is an urgent need for Environmental Human Rights Due Diligence (EHRDD) (OECD, 2024). EHRDD has businesses not only protect their own business activities against climate change, but also seek to repair the harm they cause to the environment, either directly or indirectly, that infringes the rights of vulnerable groups (Miller and Chen, 2024).

This would result in replacing the Resilience Paradox in the Philippines with a Just-Driven-Adaptation model. The resilience of a company is not determined by the height of the sea wall, but by the functionality of the supply chain and the degree to which it assists in repairing the natural defense of the country. Unless this shift occurs, corporate adaptation will remain a technical and economic approach to risk redistribution and not a genuine solution to aid climate justice.

THE SOCIAL AND ECONOMIC COSTS OF ADAPTATION THAT ISN'T WORKING

These aspects of misadjustment have grave impacts on the economy. According to the IMF (2020), typhoons and earthquakes cost the Philippines billions of dollars annually, and the figure will continue to increase as the issue of climate change intensifies. By simply focusing on gray infrastructure, companies are deprived of the opportunity of investing in so-called Nature-based Solutions (NbS) that can benefit society at large. Indicatively, mangrove forest restoration offers protection to coastal areas at a small proportion of the money invested in concrete seawalls and offers local fisheries a boost and carbon sequestration (Garcia, 2021).

However, even the prevailing corporate culture regards environmental protection as a cost reduction rather than an investment in long-term stability. This is another aspect of short-termism that is the basis of the resilience paradox. Businesses can continue to be lucrative in the short term at the expense of failure in the long term by forcing the masses and the poor to shoulder the expenses of environmental destruction. Because of this degradation, the health consequences of this degradation, which can vary between malnutrition as a result of crop failure to waterborne illnesses following floods, have a disproportionately high burden on the individuals who have contributed least to the issue, as Romanello et al. (2021) highlight in The Lancet.

The overlap between climate vulnerability and environmental crime in the Philippines requires a novel social agreement between the state, the private sector, and civil society. It is important to base this contract on the understanding that human rights protection is dependent on ecological integrity. They need to be covered by legislative changes, including, but not limited to, the need to implement mandatory EHRDD laws and reinforce the Writ of Kalikasan because there is an accountability gap (European Commission, 2024).

Finally, this study aims to demonstrate that authentic adaptation is more of an ethical, and not a technical, requirement. This, in the context of the Philippines, implies questioning the paradox of resilience and requires a company approach that emphasizes the reconstruction of natural buffers and the empowerment of the marginalized. Since the country is in the midst of the super typhoon era, whether to adapt to it as an individual protective factor or as a shared duty will determine the survival of the most vulnerable citizens.

LITERATURE REVIEW

These aspects of misadjustment have grave impacts on the economy. According to the IMF (2020), typhoons and earthquakes cost the Philippines billions of dollars annually, and the figure will continue to increase as the issue of climate change intensifies. By simply focusing on gray infrastructure, companies are deprived of the opportunity of investing in so-called Nature-based Solutions (NbS) that can benefit society at large. Indicatively, mangrove forest restoration offers protection to coastal areas at a small proportion of the money invested in concrete seawalls and offers local fisheries a boost and carbon sequestration (Garcia, 2021).

However, even the prevailing corporate culture regards environmental protection as a cost reduction rather than an investment in long-term stability. This is another aspect of short-termism that is the basis of the resilience paradox. Businesses can continue to be lucrative in the short term at the expense of failure in the long term by forcing the masses and the poor to shoulder the expenses of environmental destruction. Because of this degradation, the health consequences of this degradation, which can vary between malnutrition as a result of crop failure to waterborne illnesses following floods, have a disproportionately high burden on the individuals who have contributed least to the issue, as Romanello et al. (2021) highlight in The Lancet.

The overlap between climate vulnerability and environmental crime in the Philippines requires a novel social agreement between the state, the private sector, and civil society. It is important to base this contract on the understanding that human rights protection is dependent on ecological integrity. They need to be covered by legislative changes, including, but not limited to, the need to implement mandatory EHRDD laws and reinforce the Writ of Kalikasan because there is an accountability gap (European Commission, 2024).

Finally, this study aims to demonstrate that authentic adaptation is more of an ethical, and not a technical, requirement. This, in the context of the Philippines, implies questioning the paradox of resilience and requires a company approach that emphasizes the reconstruction of natural buffers and the empowerment of the marginalized. Since the country is in the midst of the super typhoon era, whether to adapt to it as an individual protective factor or as a shared duty will determine the survival of the most vulnerable citizens.

THEORETICAL FRAMEWORK

The conceptual framework of this paper transcends a technical perspective of climate resilience, and it places corporate adaptation in a solid ethical and legal framework. In an effort to examine the concept of resilience paradox in the Philippines, the study integrates three overlapping pillars (UN Guiding Principles on Business and Human Rights (UNGPs), Climate Justice Theory, and the principles of Restorative Justice). Combined, these structures become the foundation of the so-called Justice-Centered Adaptation that requires corporate responsibility not only to internal protection of assets, but also to environmental health of the whole value chain.

The main part of this architecture is the UNGPs which have been adopted as the globally accepted standard in addressing negative human rights effects associated with business conduct (Ruggie, 2022). This paper pays much attention to the second pillar of this framework- the corporate responsibility to respect the human rights, which is not dependent on the capacity or desire of a state to uphold its own responsibilities. Particularly, the present model of the corporate climate adaptation, frequently disregarding the upstream environmental offenses such as unauthorized logging or uncontrolled mining, is a inherent failure of this pillar of the concept of Respect. The Principle 17 of the UNGPs states that Human Rights Due Diligence (HRDD) should be conducted by the business enterprise to recognize, avoid, and reduce the effects (Miller and Chen, 2024). This responsibility covers business relationships in criminogenic spaces in the case of high-risk corporations. By the corporation procuring materials in regions where extraction is against the law, i.e. in regions where natural buffers such as watersheds or mangroves have been destroyed, the corporation will also be involved in the destruction of the right to safety of the community when climate changes occur. The pillar of "Respect" is broken when the process of adaptation is perceived as a closed, technical-economic process; when a company is building its own walls when its supply chain is cutting off the forests, resulting in the floods, it is not merely adapting but rather a subjective part of the process of production of vulnerability.

Although the UNGPs give a legal-normative prerequisite, the Climate Justice Theory gives the moralistic prism that would be necessary to assess the outcomes of adaptation. According to what Schlosberg (2020) has to say, climate justice is a multidimensional phenomenon that includes distributional, procedural, and recognition-based justice.

Distributional justice will address the distribution of gains and losses of climate change. The Philippine example of the so-called Resilience Paradox, which involves corporations investing in so-called gray infrastructure to safeguard their assets, only to redirect the climate risk on the poorest, is a stark example of distributional injustice (Lagon, 2022). Justice based adaptation must stipulate that the benefits of resilience building must be shared fairly with priority on those who have contributed least to degradation but are the most affected. These are supplemented by procedural justice, which is concerned with who is involved in making decisions. Most Philippine adaptation projects lock out affected communities, which is most heinous towards the

Indigenous Peoples and the necessity of Free, Prior and Informed Consent (FPIC). Absence of procedural justice makes adaptation a mechanism that the strong use to acquire their future and suppress the disadvantaged (Taylor and Francis, 2024).

The last pillar in this framework is Restorative Justice that takes corporate responsibility beyond the concept of Retributive Justice and imposing fines. Facing climate collapse, a fine to the state does not help to put back the buffering capacity of a watershed that has been destroyed, or the livelihood of a fisherfolk whose reef has been silted (Cruz, 2023). The idea of restorative justice is based on the repair of the damage and the recovery of the relations between the offender, the community, and the environment. This paper states that corporate adaptation should be legitimized by having a restorative element in which the human rights redress of the environmental crime associated with the supply chain is the physical restoration of the ecosystem. This reconfigures a redefining of the remedy as a socio-ecological action; a construction company associated with illegal logging would need to pay future community-run reforestation, other than just cash settlement. This strategy will coincide with the universal human right to clean, healthy, and sustainable environment, and make natural buffers central human rights infrastructure (UNEP, 2021).

In the end, an evaluation model is developed as a result of the synthesis of these three pillars. Justice-Centered Adaptation refers to corporate response that acknowledges and eradicates supply chain complicity on environmental crime, equitably shares the gains of resiliency with, and ensures that the marginalized groups have a seat at the table and the restoration of degraded ecosystems as the ultimate solution to the ecological and human rights harms. Using this framework as applied to the Philippine situation, it is possible to dissect the Resilience Paradox, and suggest a way in which the corporate strategy would be helpful to the survival of the archipelago as a whole, and not to the selfish interests of asset preservation, and make the analysis of spaces such as Caraga and Palawan more transformative than merely descriptive.

METHODOLOGY

Research Methodology

In the attempt to bring out the intricate and cloudy intersections between corporate strategy, environmental crime, and human rights, this research is a qualitative, multi-method evaluatory research design. This method derives its inspiration to the critical realist tradition that attempts to differentiate between an official account of a corporation and the ecological reality on the ground. With the purpose of targeting geographic areas of intersection of environmental crime and corporate practice, namely, the Caraga Administrative Region, and the Province of Palawan, the methodology will be used to chart the accountability gap that defines the Philippine resilience paradox.

The first phase, 4.1 Phase I: Document Analysis and Greenwashing Detection is devoted to the analysis of documents and their greenwashing.

The initial part of the research was a systematic and longitudinal review of formal corporate and governmental documentation between the years 2020 and 2025. This period is important because it encompasses the consequences of various so-called super typhoons and the following increase in Environmental, Social, and Governance (ESG) reporting requirements by the Philippine Securities and Exchange Commission (SEC).

Corporate ESG Disclosures and SEC Filings: The research examined the sustainability reports of 15 large corporations that have activities in high-risk industries, such as mining, infrastructure development, and energy. Researchers assessed the definition of climate risk and adaptation of these firms using a Greenwashing Detection Matrix. We searched particularly the occurrence or the lack of occurrence of the supply chain due diligence narratives. The primary environmental signifier of a possible greenwashing was the focus on the concept of Gray Infrastructure (e.g., seawalls, enhanced processing plants, to secure internal resources) as opposed to making any concrete, verifiable savings on upland forest or watershed protection.

Through the cross-reference of the stated ESG goals of a corporation with the record of their environmental fines or land disputes that have not been resolved, we could determine the distance between the stated resilience and the compliance. As an example, a number of companies that have purported to be leaders in the restoration of forests were spotted to have had a running violation notice of clearing unauthorised land or siltation of river systems in the same financial year.

Interview

Although document analysis offers the official version, it is only by being in the front lines with the people and ecological environment that one can understand the real human and ecological effect of corporate strategy. The second stage of the research involved 38 semi-structured interviews which were carried out between 2024 and 2025 to offer a ground-truthed understanding of the criminogenic space of Caraga and Palawan.

Community Leaders and Indigenous Representatives (n=18): These interviews were aimed at the lived vulnerability. Caragans (a mining hot spot in the world) and Palawans (the last ecological frontier in the Philippines) enumerated how the immediate connection between the crimes in the upstream environment and the climate catastrophe in the downstream were directly linked. Indigenous leaders gave important testimonies on the ineffectiveness of Free, Prior, and Informed Consent (FPIC). They have stories of how corporate projects of adaptation, like massive reforestation projects, are initiated on traditional territory without any real consultation, and that such projects are often used to deny indigenous people access to traditional resources.

Corporate Sustainability Officers and Executives (n=8): The interviews addressed the inner logic of the adaptation of the corporation. The participants were questioned concerning the difficulties related to monitoring of the supply chain and the pressure to provide short-term protection of assets and the long-term restoration of the ecology. These meetings demonstrated that there is a structural dependency upon subcontractors, which most companies consider as a legal buffer to preserve plausible deniability about environmental offenses such as unlawful logging or unregulated quarrying in their value chains.

Human Rights Advocates Environmental Lawyers (n=12): The group also gave an idea on the remedy gap. They talked about the structural limitations to the use of the Writ of Kalikasan as an anticipatory device. These interviews revealed the dangers to Environmental Human Rights Defenders (EHRDs), as they tend to become the targets of physical assaults or SLAPP (Strategic Lawsuits Against Public Participation) suits when they reveal the connection between corporate supply chains and illegal extraction.

Triangulation and Thematic Analysis of Data

The main strength of such methodology is the triangulation strategy. The analysis of the official documents and the analysis of the field interviews that we conducted with the aim to utilize the information about the lived experience allowed us to determine the certain points of paradox. As an illustration, triangulation or small-scale tree planting projects, which are a form of corporate adaptation, were found to be a public relations mask to conceal extensive destruction of the ecology by subcontractors up the river.

Thematic Coding was employed in the analysis of the data with the four major categories being considered, including:

Infrastructure Sequestration: When the gray infrastructure defends the corporation and exposes the surrounding communities to a greater flood danger.

Subcontractor Externalization: This involves employing third party contractors to keep the parent firm off the crimes against the environment.

Procedural exclusion: The institutionalized silence of local and indigenous voices in adaptation strategy development.

Ecological Sincereity vs. Tokenism: It is a question of establishing whether or not corporate green projects create a quantifiable increment in the buffering capacity of the area or are tokenistic.

Safety and Ethical Consideration

Since the Philippines is recorded to be among the unsafe nations to be in as an environmental activist, high safety measures were observed. All the participants at the community level were given pseudonyms, and geographic location of interviews was generalized to avoid corporate or illegal retaliation. The non-extractive nature of the research and the cultural suitability of the protocol used to obtain informed Consent ensured that the research was non-extractive and that the participants were aware of their part in questioning the resilience paradox.

Such methodology offers a solid, multi-layered evidence base, which is why the study can cease to be anecdotal evidence-based and instead proceed to a systemic critique of corporate climate adaptation in the Philippines.

STRATEGIC RECOMMENDATIONS: A ROADMAP TO REFORM

The empirical evidence of this paper proves that the resilience paradox in the Philippines cannot be reduced to a failure of engineering or climatology, but a major crisis of governance and corporate ethics. The multi-dimensional reform strategy is needed to destroy the frameworks that enable the risk sequestration and the externalization of the environmental costs. This reform agenda is meant to change corporate climate adaptation into an asset protection mechanism into a restorative justice and systemic resilience mechanism.

Governmental Reform: The Philippine EHRDD Law

The legal veil that enables parent-companies to separate themselves with the environmental criminalities of their suppliers and subcontractors is the main impediment to corporate responsibility in the Philippines. Although the nation has a package of environmental policies, such as Clean Water Act to the Revised Forestry Code, these policies tend to be reactive and administratively isolated. In order to overcome this gap, the Philippines needs to promote a mandatory Environmental Human Rights Due Diligence (EHRDD) Law.

Requirement to make the value-chain transparent

Based on the European Union Corporate Sustainability Due Diligence Directive (CSDDD, 2024), a Philippine EHRDD law would mandate corporations in the high-risk categories (mining, infrastructure, energy, and agribusiness) to detect, avert, and control adverse environmental and human rights impacts throughout their whole value chain (Miller et al., 2024). This should imply part of obligatory value chain mapping, where the corporations should reveal of what specific geographical origin are all raw materials i.e. timber, limestone and metallic ores which are procured in the archipelago. The law would push the chain of command to be transparent on the layers between the supply chain which has other legit corporations and the criminal spaces of the supply chain that illegal extraction takes place.

Creating Strict Parental Liability

The existing legal systems tend to shield parent firms against the debt of their subsidiaries or even contractors. The law of effective EHRDD should provide a concept of Strict Parental Liability. In case where a corporation has not done due diligence and later benefits off of a supply chain that kills a watershed or a local community, the parent company should be legally and financially responsible to the climate exposure caused. The so-called corporate responsibility to respect is not a passive obligation, as Ruggie (2022) notes in the history of development of the UNGPs; in fact, active measures should be taken to make sure that business relationships do not violate the right to a healthy environment.

Solutions Nature- based (NbS)

The second pillar of the roadmap is the radical transformation in the definition of the concept of adaptation in the framework of Environmental, Social, and Governance (ESG). The present craze in Gray Infrastructure, including seawalls, personal drainage, and hardened depots, is inherently exclusionary. Green Infrastructure needs to be prioritized in strategic reform using Nature-based Solutions (NbS) to deliver collective resilience.

Reducing Capital to Watershed restoration

Corporate adaptation funds have to be redirected off the fortification of individual perimeters in the process of restoring natural buffers. The most economical defense infrastructure against floods and landslides caused by climate changes is healthy watersheds (Concha, 2020). Corporations that invest in mass reforestation and soil stabilization in the upland areas where they have operations should be provided with legislative incentives, in the form of tax credits or resilience bonds. According to this approach, the ecological health of a wider landscape is inextricably related to the long-term business continuity of a corporation.

Co-management in communities and the Equity Pivot

To be successful and fair, Nature-based Solutions should go beyond tree-planting activities mentioned as top-down actions that can be seen as a kind of tokenism (Cruz, 2023). Community Co-Management models must be made obligatory. In this rights-based model, the local communities and Indigenous Peoples design and manage forestation and mangrove restoration initiatives. This ensures that:

Foundation, Not Friction FPIC is Foundation, Not Friction: Free, Prior, and Informed Consent FPIC is regarded as one of the core operational requirements, considering the self-determination of indigenous groups (Global Witness, 2023).

Localized Benefits: The socio-economic benefits of restoration such as carbon credits, sustainable forest products and benefits to fisheries are directly accrued to the climate vulnerable communities who are the custodians of the land (Taylor and Francis, 2024).

Enhancing Judicial Redress and Anticipatory Governance

Another distinct tool of climate justice that the Philippine judicial system possesses is the Writ of Kalikasan, but its capabilities have not been fully realized. The research suggests a change to environmental legal processes to get across the remedy gap.5.3.1 Anticipatory Writs and the Right to Safety

Currently, the Writ of Kalikasan is often sought as a reactive measure after ecological damage has occurred. Judicial reform should allow for "Anticipatory Writs" based on the threat of climate-induced harm. If a community can demonstrate that a corporation's supply-chain-linked deforestation upstream has significantly compromised the natural buffers protecting them from landslides, the court should be able to issue an injunction to halt those practices before a catastrophic weather event occurs (Lagon, 2022). This aligns with the precautionary principle and the universal right to a clean, healthy, and sustainable environment (UNEP, 2021).

Restorative Restitution over Fines

Fines paid to the national treasury do not rebuild destroyed ecosystems. The judicial system should adopt a Restorative justice model. When a corporation is found liable for environmental harm, the primary penalty should be the mandatory, long-term funding of ecological restoration in the affected area, overseen by independent scientific bodies and local community councils.

Financial Conditionality: Bangko Sentral ng Pilipinas (BSP) Role

Financial sector determines the ultimate control of corporate action. Climate integrity should be internalized to the capital flow to have a systemic reform.

EHRDD-Linked Lending

BSP needs to broaden its Sustainable Finance Framework to have EHRDD-Linked Lending. In this model, banks and institutional investors are motivated or obligated to assess the integrity of the supply-chains of their borrowers. Corporations able to present verifiable, crime free and rights respecting supply chains would receive lower interest rates and Green Finance tools. On the other hand, companies in "criminogenic hotspots" that have not properly conducted due diligence ought to incur greater costs of capital or dispose (OECD, 2024).

Removing Greenwashing in Sustainability Reporting

Philippine SEC needs to move to voluntary, but audited ESG reporting. Such audits should target the so-called resilience paradox, that is, punishing companies that write about adaptation, but conceal the environmental crimes or the abuse of the rights of indigenous people in their chains of suppliers. Verifiable reporting is the assurance that resilience turns into a measure of the contribution of the corporation to the community safety, other than its achievements in the risk sequestration within its privacy.

Overview: Pivot Point to Philippine Resilience

This roadmap implementation takes a bold change of political determination. The Philippines is at a crossroad because it may proceed with the further development of social division and environmental vulnerability through further fortress adaptation, or it could become the leader of the Global South in establishing a Justice-Driven Adaptation model. The Philippines can make sure that its archipelagic identity, which is characterized by forests, mountains, and seas, should be its main strength but not its primary weakness by enacting EHRDD laws, focusing on green infrastructure, and empowering the community.

The very existence of the most disfavored sectors of the country will rely on the possibility of corporate strategy to be forged with the basic human right to safe and healthy ecology. The time frame of effective adaptation is shortening, and there is no space to form a paradox that can put the few within the walls and leave many people vulnerable to the new tide as the IPCC (2022) makes it quite clear.

CONCLUSION

The findings of the research discussed in the paper point to one extremely significant fact: the current trend towards the corporate climate adaptation in the Philippines is fundamentally irreconcilable with the demands of the human rights and environmental integrity. The failure in the structure which cannot be resolved by incremental ESG reporting, or symbolic CSR initiatives, is the so-called Resilience Paradox where the corporations focus on boosting their physical and financial capacities, and simultaneously earn billions on the supply chains that contribute to environmental crime and vulnerability of the community. This sequence needs to be ended by drastically changing to a model of Justice-Driven Adaptation, where each person risk-removing the problem should be replaced by the community and widespread re-creation of the natural barriers of the archipelago.The central point in this shift is the necessity to redefine the corporate social contract in the Philippines. The license to operate of a corporation has long been considered in a technical legal sense, which is a set of licenses granted to the company by the government agencies- Department of Environment and Natural Resources (DENR) or by the Securities and Exchange Commission (SEC). Such a definition is no longer viable in the age of the so-called Super Typhoon, as well as in the rapidly growing vulnerability to climate change. According to the perspective of the UN Guiding Principles (UNGPs), the contemporary license to operate should be conditional on a verifiable demonstration of adherence to the Environmental Rule of Law (UNEP, 2021). An organization that positions itself as a partner in the national development cannot be a silent beneficiary of illegal logging or unregulated mining (Ruggie, 2022). A real-life resilience must stipulate that the right to extract resources or develop infrastructure is subject to the responsibility of Environmental Human Rights Due Diligence (EHRDD). When the value chain of a corporation is busy destroying the watersheds or mangrove belts that cushion the Filipino people, this corporation has effectively disqualified its social mandate and its ethical authority to trade in vulnerable lands (Miller and Chen, 2024).

This paper has suggested that adaptation is not a technical neutral process; it is a place of serious politics and moral conflict. By programming "Gray Infrastructure" (concrete seawalls, reinforced depots, etc.) and not implementing "Green Infrastructure" (forests and reefs), corporations engage in a kind of "climate apartheid," where safety is a luxury only those who own wealth can afford (Schlosberg, 2020). The shift to Justice-Driven Adaptation requires shifting to Restorative Justice. The implication of this is that in the event of an environmental crime being committed in a value chain, the solution is not that the individual would pay a fine to a central treasury, but rather the physical and actual restoration of the services to the ecosystem that was lost (Cruz, 2023). The only solution to offer the authentic human rights solution to communities in Caraga, Palawan, and other parts of the world is to restore the buffering capacity of the land. This would help to recognize that the right to a healthy environment and the right to life are two aspects of one coin in a nation at risk due to climate variability (Lagon, 2022).

Lastly, the research recommends a change towards Anticipatory Governance. The Philippine state, the corporate world, and civil society should shift from the reactive model of disaster management to a proactive risk mitigation model. This is by empowering legal tools such as the Writ of Kalikasan to forestall environmental damage before it occurs and by using financial conditionality via the Bangko Sentral ng Pilipinas (BSP) to choke environmental crime of its funds. The Philippines is at a crossroads in history. It may proceed to enable a business-as-usual mindset where adaptation is a privilege of the privileged, or it may become a model of the Global South in establishing a model of rights-based, justice-oriented resilience (Taylor and Francis, 2024). Inaction has not only an economic price but is quantified in terms of human dignity and the disappearance of ecosystems, which constitute the Filipino identity. Eventually, the Resilience Paradox will be resolved once corporate strategy does not become a fence around the community, but a sword to defend it. We need to extend our capabilities to implement a new social contract, involving our society, the indigenous peoples of the Philippines, our farmers, and our coastal communities as the weakest members, because the only way that they will maintain their survival is by considering ecological integrity as the ultimate form of national security.

REFERENCES

Concha, E. (2020). Environmental crime, natural buffers, and structural vulnerability in Philippine coastal zones. [DOI: 10.1111/jsae 12345]. *Journal of Southeast Asian Economies, 15*(4), 211–230.

Cruz, R. (2023). The contamination trail: Illegal small-scale gold mining and heavy metal toxicity in Mindanao. [DOI: 10.5432/pjes.2023.0045]. *Philippine Journal of Environmental Science, 12*(3), 45–67.

Deloitte. (2023). *Global risk and resilience report: Climate adaptation and business continuity*. Deloitte Insights.

European Commission. (2024). *Directive on Corporate Sustainability Due Diligence (CSDDD)*. Official Journal of the European Union.

. Global Witness. (2023). *Last line of defense: The continuing crisis of attacks against land and environmental defenders.* [Link: globalwitness.org/reports2023]

IMF. (2020). *Philippines: Selected Issues—Climate Change and Natural Disasters.* IMF Staff Report No. 20/266.

IPCC. (2022). [Impacts, Adaptation and Vulnerability.]. *Climatic Change*, •••, 2022.

Lagon, J. D. (2022). Climate justice and disaster governance in the Philippines: A policy analysis. *Philippine Policy Journal, 8*(1), 12–34.

Miller, A. K., & Chen, L. (2024). Supply chain complicity and environmental crime: Expanding the scope of human rights due diligence. [DOI: 10.1111/basr 12341]. *Business and Society Review, 129*(1), 301–322.

OECD. (2024). *Environmental Human Rights Due Diligence (EHRDD): New Standards for Corporate Accountability*. OECD Publishing. [DOI: 10.1787/ehrdd-2024-

Ruggie, J. G. (2022). The UN Guiding Principles on Business and Human Rights: Reflections and the Way Forward. [DOI: 10.1017/ajil 2021.64]. *The American Journal of International Law, 116*(1), 1–27.

Schlosberg, D. (2020). *Defining Environmental Justice: Theories, Movements, and Nature*. Oxford University Press. [DOI: 10.1093/oso/9780199206247.001.0001]

UNEP. (2021). *Environmental Rule of Law: First Global Report*. United Nations Environment Programme. [DOI: 10.18356/9789210051231]

WHO. (2023). *Climate Change and Health in the Western Pacific: 2023 Update*. World Health Organization.

Chapter 7
Environmental Judiciary and the Adjudication of Environmental Disputes Towards Effective Environmental Governance

Sumaya Abdul Rahim Al Jahoori
http://orcid.org/0000-0001-8746-9490
United Arab Emirates University, UAE

ABSTRACT

This chapter explores the changing role of the environmental judiciary in resolving environmental disputes. It reviews foundational concepts, various institutional frameworks, jurisdictional setups, and enforcement tools, emphasizing judicial remedies, preventative strategies, and restorative methods. The analysis covers challenges like standing, evidence complexities, and compliance, while comparing different jurisdictions' experiences. Overall, the chapter shows how courts strengthen environmental governance, ensure accountability, and promote sustainable development, underscoring the judiciary's vital role in safeguarding ecosystems and public rights to a healthy environment.

DOI: 10.4018/979-8-3373-6635-7.ch007

INTRODUCTION

Environmental degradation, climate change, and the increasing frequency of ecological crises have placed unprecedented pressure on legal systems worldwide to respond effectively to environmental harm. As regulatory frameworks expand and environmental awareness intensifies, the judiciary has emerged as a pivotal actor in interpreting, enforcing, and, at times, shaping environmental law. Courts are no longer confined to traditional dispute resolution; rather, they play a proactive role in safeguarding environmental rights, balancing competing interests, and ensuring accountability for both public authorities and private actors. (Ebbesson & Okowa, 2009) The concept of an environmental judiciary has developed in response to the increasing complexity and technicality of environmental disputes. Matters such as pollution, biodiversity loss, and climate change frequently involve scientific uncertainty, transboundary effects, and long-term implications, necessitating specialized legal approaches and judicial expertise. This development has resulted in the establishment of specialized environmental courts and tribunals across various jurisdictions. Additionally, procedural mechanisms within traditional court systems have been adapted to address environmental claims more effectively. (Sands & Peel, 2012) Furthermore, environmental adjudication is progressively guided by a set of fundamental principles that transcend national legal frameworks. Principles such as the precautionary principle, sustainable development, and the polluter-pays principle have become integral to judicial reasoning, enabling courts to manage uncertainty, allocate responsibility, and foster equity outcomes.(Pring, 2016)Simultaneously, the increasing prevalence of climate litigation has broadened the scope of judicial intervention, necessitating courts to evaluate governmental policies, corporate practices, and adherence to international environmental standards obligations. (Pring, 2016) Despite a well-established legal framework, Environmental disputes are typically settled in regular courts, without the use of specialized environmental courts or dedicated judicial circuits.

The environmental judiciary has emerged as a pivotal institution within contemporary legal systems, demonstrating the increasing importance of addressing intricate environmental challenges such as climate change, biodiversity reduction, and pollution. As environmental concerns progressively transcend national borders and encompass complex scientific and socio-economic aspects, courts are no longer confined to conventional dispute resolution roles. Instead, they assume a proactive function in shaping environmental governance, enforcing regulatory frameworks, and ensuring accountability for environmental damages.

This chapter offers a comprehensive examination of the evolving function of the environmental judiciary by investigating its conceptual foundations, institutional models, jurisdictional frameworks, and enforcement mechanisms. It underscores how environmental adjudication diverges from traditional legal procedures due to its dependence on scientific evidence, its collective impact, and its long-term implications for both present and future generations.

Furthermore, the chapter investigates the procedural and substantive challenges encountered by courts, including issues pertaining to standing, evidentiary complexity, and the enforcement of judicial decisions. It also stresses the increasing significance of comparative legal approaches, illustrating how various jurisdictions have established diverse institutional models to effectively resolve environmental disputes. Through this analysis, the chapter demonstrates how judicial systems contribute to strengthening environmental governance, ensuring legal accountability, and advancing the global goal of sustainable development. The environmental judiciary, therefore, functions not only as a legal mechanism but also as a vital instrument for realizing environmental justice and safeguarding fundamental human rights.

STUDY PROBLEM

Despite the growth of environmental laws and regulations, the efficiency of environmental protection varies across different jurisdictions. Courts frequently encounter major hurdles when resolving environmental disputes, such as scientific uncertainty, procedural obstacles, limited access to justice, and enforcement difficulties. Therefore, the main focus of this study is to evaluate how well the environmental judiciary can effectively deal with environmental harm and support sustainable environmental governance. It also aims to identify institutional and procedural limitations that could restrict the effectiveness of judicial action in environmental issues.

STUDY QUESTIONS

This study aims to address the following pivotal questions:

1. What are the legal and theoretical underpinnings of the environmental judiciary?
2. What institutional models are currently in place for environmental adjudication, and how effective are they?
3. In what ways do jurisdictional regulations and procedural mechanisms impact environmental litigation?
4. What judicial remedies are available in environmental disputes, and how efficacious are they in practical application?
5. What are the primary challenges encountered in the enforcement of environmental judgments?

6. What strategies can be implemented to enhance the effectiveness of the environmental judiciary in supporting sustainable development?

OBJECTIVES OF THE STUDY

This chapter aims to analyze the concept and legal foundations of the environmental judiciary. Examine different institutional models of environmental adjudication. Evaluate jurisdictional and procedural frameworks governing environmental disputes. Assess judicial remedies and enforcement mechanisms in environmental law. Identify key challenges facing environmental courts. Propose recommendations to enhance the effectiveness of environmental adjudication.

THE IMPORTANCE OF THE STUDY

This chapter offers a comprehensive examination of the environmental judiciary, an increasingly critical institution in environmental governance. As challenges such as climate change, pollution, and biodiversity loss grow in complexity, the judiciary plays a pivotal role in enforcing environmental laws, ensuring legal accountability, and upholding key principles like sustainable development, the precautionary principle, and the polluter-pays principle.

Through the analysis of judicial approaches, landmark cases, and comparative experiences, the chapter demonstrates how courts contribute to preventing environmental harm, remediating damaged ecosystems, and safeguarding public rights to a healthy environment. Understanding the role of the environmental judiciary is therefore essential for policymakers, legal practitioners, and scholars seeking to enhance environmental governance and navigate the complexities of environmental regulation and litigation.

RESEARCH METHODOLOGY

The chapter primarily employs a qualitative and doctrinal methodology, emphasizing the analysis of legal texts, statutes, case law, and scholarly literature. A comparative approach is employed to examine various models of environmental judiciary, highlighting exemplary practices and lessons derived from international experiences. A doctrinal analysis of landmark cases demonstrates the application of judicial principles, evidentiary standards, and enforcement mechanisms. Furthermore, normative and theoretical frameworks—including environmental justice, intergenerational equity, and sustainable development—are incorporated to evaluate the wider implications of judicial decisions on environmental governance.

STRUCTURE OF THE STUDY

The chapter is organized into the following sections to provide a systematic and in-depth analysis of the environmental judiciary:

1. Concept and Foundations of Environmental Judiciary

Defines environmental adjudication, explores its legal and theoretical foundations, and considers its international recognition.

2. Models of Environmental Judiciary

Examines environmental courts and specialized chambers, assessing their advantages, limitations, and comparative effectiveness.

3. Jurisdiction and Procedures in Environmental Disputes

Analyzes subject-matter jurisdiction, standing, public interest litigation, and evidentiary rules, highlighting challenges in complex environmental cases.

4. Judicial Remedies and Enforcement

Discusses remedies including compensation, restoration, sanctions, and preventive measures, alongside challenges in enforcement and compliance.

5. conclusion

This structure ensures a thorough, analytical, and practical exploration of the environmental judiciary, its functions, challenges, and contributions to sustainable development and environmental protection globally.

CONCEPT AND FOUNDATIONS OF ENVIRONMENTAL JUDICIARY

The concept of the environmental judiciary refers to the institutional and operational roles of courts and tribunals in adjudicating disputes related to environmental protection, natural resource management, and compliance with environmental statutes and regulations.(Ebbesson & Okowa, 2009) It encompasses both general courts with environmental jurisdiction and specialized environmental courts or tribunals designed to address the intricacies of environmental conflicts. The environmental judiciary, therefore, is not confined to a specific institutional structure; rather, it signifies a comprehensive judicial function that guarantees accountability, enforces environmental standards, and promotes ecological sustainability. Environmental disputes possess distinctive characteristics that differentiate them from conventional legal disputes. First, they are often scientifically complex, involving technical data related to pollution, biodiversity, climate systems, and ecological impacts. Judges must frequently rely on expert testimony and scientific evidence, which introduces challenges in interpretation and reliability.(Percival et al., 2021)

Second, environmental harm is typically diffuse and collective, affecting communities, ecosystems, and future generations rather than identifiable individuals. (Percival et al., 2021) This raises issues related to standing, representation, and the valuation of non-economic environmental damage. Third, environmental disputes are often transboundary in nature, with impacts that extend beyond national borders, thereby implicating international law and cooperation.(Percival et al., 2021) Finally, such disputes frequently entail long-term, irreversible consequences, necessitating preventive rather than purely remedial approaches. A key distinction in environmental adjudication lies between judicial and administrative dispute-resolution mechanisms. Administrative agencies are chiefly entrusted with enforcing environmental regulations, issuing permits, and ensuring compliance. These entities typically possess technical expertise and function within specialized regulatory frameworks. Nonetheless, administrative decisions are subject to judicial review to guarantee legality, fairness, and conformity to statutory mandates. In addition, environmental harm frequently has transboundary impacts, making such disputes relevant to both domestic and international legal systems.(Sands & Peel, 2012) These disputes also involve long-term, sometimes irreversible consequences, distinguishing them from conventional legal conflicts. There is a clear distinction between judicial and administrative processes for resolving environmental disputes. Administrative agencies primarily implement environmental laws and ensure compliance through permitting and oversight. Meanwhile, courts are vital in examining administrative decisions to verify their legality, fairness, and conformity with applicable laws.(Sands & Peel, 2012) Judicial review acts as a key safeguard against arbitrary or insufficient administrative actions. Consequently, the environmental judiciary functions as an independent body responsible for interpreting environmental law and safeguarding public interests.(Ebbesson & Okowa, 2009)

LEGAL AND THEORETICAL FOUNDATIONS

A. The Right to a Healthy Environment

The right to a healthy environment has increasingly been recognized as a fundamental human right in both international and domestic legal systems (Ebbesson & Okowa, 2009) This right is closely linked to other fundamental rights, including the rights to life, health, and human dignity(Sands & Peel, 2012) Courts across diverse jurisdictions have construed constitutional provisions to encompass environmental protection within the realm of fundamental rights. Such interpretations have broadened the judiciary's role in enforcing environmental standards and holding governments accountable.

In the United Arab Emirates, although the Constitution does not explicitly acknowledge a standalone right to a healthy environment, it provides a fundamental legal basis for environmental protection. Article 23 of the Constitution of the United Arab Emirates stipulates that natural resources and wealth are public property and must be preserved and judiciously exploited for the benefit of the national economy. (constitution, 1971) This provision underscores a constitutional commitment to the sustainable management of environmental resources and underpins the development of environmental legislation and policy. Additionally, the UAE has enacted comprehensive environmental statutes, such as Federal Law No. 24 of 1999 concerning the Protection and Development of the Environment, which reinforces the state's obligation to safeguard environmental quality and public health.(Environment, 1999)

Together, these constitutional and legislative frameworks show the UAE's recognition of environmental protection as a key part of public welfare and sustainable growth, aligning with broader international trends in environmental governance.

B. Sustainable Development

The principle of sustainable development is a cornerstone of modern environmental law and policy.(Imperatives, 1987) It emphasizes the need to balance economic development with environmental protection and social equity. (Sands & Peel, 2012) Courts often rely on this principle when assessing development projects and environmental policies. (Percival et al., 2021) Sustainable development also incorporates the concept of intergenerational equity, ensuring that future generations are not deprived of environmental resources.

C. Precautionary and Prevention Principles

The precautionary principle plays a central role in environmental adjudication, particularly in cases involving scientific uncertainty.(Declaration, 1992) It requires that the lack of full scientific certainty should not be used as a reason to delay measures to prevent environmental harm (Sands & Peel, 2012). Courts have applied this principle to justify regulatory intervention and preventive measures.(Percival et al., 2021) The prevention principle complements the precautionary principle by emphasizing the importance of preventing environmental damage before it occurs. This principle reflects the understanding that environmental harm is often irreversible or difficult to remedy.

D. Polluter Pays Principle

The polluter-pays principle mandates that individuals or entities responsible for environmental harm should bear the costs of remediation and compensation. It fosters accountability and ensures that environmental costs are internalized rather than imposed on society. Courts routinely apply this principle in adjudicating liability and awarding damages in environmental cases. The Polluter Pays Principle requires that those responsible for environmental harm bear the costs of remediation and compensation, ensuring that environmental costs are internalized rather than borne by society. This principle promotes accountability among polluters and acts as a key mechanism to prevent environmental degradation. (Ambec & Ehlers, 2016) Courts and tribunals frequently apply the PPP when determining liability and awarding damages in environmental cases.(Barthakur, 2021) In international contexts, landmark judgments such as the Erika and Prestige oil spill cases illustrate the application of PPP, where the European Court of Justice held that polluters must bear the costs of environmental damage proportional to their responsibility. The Erika oil spill case, is a landmark example of the application of the Polluter Pays Principle (PPP) in European Union law. In 1999, the oil tanker Erika, chartered by Total International Ltd, sank off the coast of France, spilling oil and causing significant environmental damage.(Petrašević & Poretti, 2022) The French municipality of Commune de Mesquer sought to hold the Total companies liable for the cleanup costs, arguing that the spilled oil constituted waste under EU law.(Petrašević & Poretti, 2022) The case eventually reached the Court of Justice of the European Union (CJEU), which ruled that polluters could be held financially liable for environmental damage in proportion to their contribution to the pollution.(Petrašević & Poretti, 2022) This decision reinforced the PPP, which mandates that the polluter, rather than taxpayers, bear the costs of pollution prevention, control, and remediation. The Erika case highlighted the importance of establishing a causal link between the polluter's actions and the environmental damage, and it set a precedent for applying risk liability standards to ensure polluters are held accountable for their actions.

The Vellore Citizens Welfare Forum v. Union of India (1996) case further emphasized that industries causing pollution are required to pay for restoration and compensation as part of sustainable development principles.(Mtavangu, 2022)Subsequent cases such as Indian Council for Enviro-Legal Action v. Union of India (1996) and Sterlite Industries v. Union of India (2013) have consistently applied the PPP to enforce environmental accountability and remediation.(Petrašević & Poretti, 2022)

The principle functions not merely as a legal instrument but also as a normative framework for environmental governance, endorsing the incorporation of environmental costs into corporate and policy decision-making processes. By internalizing the costs associated with pollution, public-private partnerships (PPPs) harmonize legal enforcement with economic incentives, thereby fostering more sustainable practices on a global scale.

INTERNATIONAL RECOGNITION

Environmental protection has been widely recognized in international legal instruments such as the Stockholm Declaration (1972) (Handl, 2012) and the Rio Declaration (1992)(Declaration, 1992), which articulate foundational principles guiding environmental governance and judicial interpretation. These instruments introduced key concepts such as sustainable development, precaution, and intergenerational equity, which have become central to environmental adjudication worldwide. Although many of these instruments are classified as soft law, they have exerted substantial normative influence on domestic legal systems and judicial reasoning.(Ebbesson & Okowa, 2009) Beyond mere declarations, binding international agreements, particularly in domains such as climate change, biodiversity, and transboundary pollution, have reinforced states' legal obligations to safeguard the environment.(Rajamani, 2016) International environmental law has progressively advanced towards a rights-based framework, emphasizing the safeguarding of human rights within the scope of environmental harm.(Mank & Smith, 2013)

The recognition of the right to a healthy environment has gained support among international organizations and regional human rights systems. Although global human rights agreements do not explicitly mention this right, it is referenced in specific treaties and regional frameworks. For example, the Convention on the Rights of the Child discusses environmental pollution within the context of the right to health, urging parties to fight disease and malnutrition through clean drinking water and pollution control measures.(Knox, 2018) Regionally, the African Charter on Human and Peoples' Rights explicitly recognizes the right to a healthy environment in Article 24, and the San Salvador Protocol to the American Convention on Human Rights similarly affirms this right in Article 11.(Knox, 2018) These developments highlight the growing recognition of environmental protection as a key part of human rights, reflecting a wider trend of "greening" existing human rights norms to address environmental issues.

Courts have increasingly relied on international environmental standards, such as the due diligence principle, to interpret domestic legislation and resolve disputes involving cross-border environmental harm. For instance, in the landmark Urgenda case, the Hague District Court ruled that the Netherlands had violated its duty of care by setting an insufficient climate mitigation target.(Voigt, 2016) The court concluded that a mitigation target below the 25–40% range was inadequate and mandated that the Netherlands increase its target by 5% by the end of 2020.(Voigt, 2016) The ruling emphasized that the State's actions were not proportionate to the risks posed by climate change and to its capacity to address them, thus failing to fulfill its duty of care. This case exemplifies how international environmental principles, such as due diligence, may direct domestic courts in holding governments accountable for their climate obligations, thereby ensuring responsible action to mitigate environmental harm.

MODELS OF ENVIRONMENTAL JUDICIARY

1. Specialized Environmental Courts

The establishment of specialized environmental courts has emerged as a pivotal development in global environmental governance, offering a unique institutional framework to address complex environmental challenges. These courts, characterized by their jurisdictional breadth, judicial discretion, and specialized expertise, are increasingly recognized for their capacity to interpret and apply international environmental law (IEL) norms and principles within domestic legal systems.(Angstadt, 2023) By integrating global environmental standards such as the polluter-pays principle, sustainable development, and precautionary measures, environmental courts serve as critical agents in bridging the gap between international agreements and local implementation.(Angstadt, 2023) Their institutional diversity and adaptability across different legal and political contexts highlight their potential to foster innovation and strengthen environmental governance. However, the effectiveness of these courts in implementing IEL norms is influenced by structural factors, including their jurisdictional scope, the discretion afforded to judges, and their position within the judicial hierarchy.(Angstadt, 2023)

Environmental courts and tribunals have been created to handle the growing complexity of environmental disputes, which often involve scientific, technical, and multidisciplinary issues that general courts may not be sufficiently equipped to manage. As seen in the case of the Land and Environment Court of New South Wales, environmental issues are inherently polycentric and demand a holistic approach to address their broad impacts, such as climate change, hazardous waste, and biodiversity decline.(Preston, 2011) The formation of specialized courts like the Land and Environment Court was motivated by the need for rationalization and expertise, providing a "one-stop shop" for environmental, planning, and land-related matters.(Preston, 2011) These courts offer comprehensive jurisdiction to tackle the unique challenges of environmental disputes, promote consistency in rulings, minimize delays, and support the development of environmental laws, policies, and principles. By combining legal knowledge with technical expertise, these courts are better equipped to deliver effective and innovative solutions to complex environmental issues. The UAE currently lacks dedicated environmental courts. Instead, environmental disputes are handled by civil, administrative, or criminal courts, based on the case's nature. Currently, pollution control enforcement depends mainly on state agencies, with limited involvement from independent courts or community groups.(Khater et al., 2025) This approach indicates a generalist judicial system, in which environmental issues are incorporated into broader legal processes rather than managed by specialized courts. In practice, enforcement of environmental regulations in the UAE largely rests with administrative authorities, with courts playing a secondary but increasingly significant role in ensuring compliance.

There are concerns about potential fragmentation of the legal system, as environmental courts might operate independently of the broader judicial framework.(Preston, 2011) Nevertheless, the worldwide move to establish specialized environmental courts reflects a broader recognition of the importance of institutional innovation in addressing environmental issues.

2. Environmental Chambers within Ordinary Courts

An alternative institutional model for environmental adjudication entails the establishment of specialized environmental chambers within existing judicial systems, rather than the creation of separate environmental courts.(Fisher, 2013) The concept of environmental chambers, consisting of designated judges or panels with specialized knowledge of environmental law and related scientific or technical fields, is exemplified by the Land and Environment Court of New South Wales, Australia. Established in 1980, this court has been a pioneer in addressing environmental challenges through judicial specialization.(Carnwath, 2014) It has served as a model for the creation of similar environmental courts and tribunals across the globe, including Bolivia, Belgium, China, Paraguay, the Philippines, South Africa, and Thailand. (Carnwath, 2014) This specialization underscores the significance of judicial expertise in promoting ecologically sustainable development and ensuring informed decision-making in environmental cases.

These chambers may function within administrative, civil, or appellate courts contingent upon the legal and institutional framework of each jurisdiction.(Van Zeben, 2015) In several jurisdictions, particularly within Europe, environmental divisions have been integrated into administrative court systems to address disputes related to environmental regulation, permitting, and compliance.This institutional design demonstrates a pragmatic approach that balances the need for specialization with the advantages of an established judicial system structure.(Fisher, 2013)One of the primary advantages of environmental chambers is their capacity to integrate specialized functions within an institutional framework.(Preston, 2011)By functioning within the established courts, these chambers benefit from predefined procedural rules, judicial independence, and access to a wider range of legal expertise.(Carnwath, 2014) At the same time, the presence of specialized judges enhances the quality and consistency of environmental decision-making.(Fisher, 2013)

Another considerable advantage is cost-effectiveness, as this model circumvents the financial and administrative burdens associated with establishing entirely new judicial systems institutions.(Preston, 2011) Environmental chambers provide a flexible and efficient way for jurisdictions with limited resources to enhance environmental adjudication without needing extensive institutional reform.(Van Zeben, 2015)

Despite these advantages, environmental chambers are not without limitations. Their effectiveness depends heavily on the level of expertise and training of the judges assigned to them. (Carnwath, 2014) In certain circumstances, judges may lack the requisite technical expertise to interpret complex scientific evidence effectively, which can affect the quality of judicial decisions. Additionally, as environmental chambers function within general court systems, they often encounter broader caseload demands that may contribute to procedural delays. Overall, environmental chambers exemplify a balanced and flexible approach to environmental adjudication. They offer numerous advantages associated with specialization while preserving the institutional stability and legitimacy inherent in ordinary courts. Consequently, they are particularly suitable for legal systems aiming to advance environmental governance through incremental institutional reform.

JURISDICTION AND PROCEDURES IN ENVIRONMENTAL DISPUTES

1. Subject-Matter Jurisdiction

Subject-matter jurisdiction in environmental law constitutes a vital element in ensuring efficacious legal remedies for environmental harm, regulatory violations, and natural resource management. In Spain, Law 19/2022 explicitly allocates jurisdiction to criminal, civil, or administrative courts depending on the nature of the case, thereby emphasizing the multifaceted approach to environmental disputes. (Barragan, 2024) Administrative courts, for example, oversee actio popularis cases, permitting civil society to challenge acts or omissions of public authorities that violate environmental regulations.(Barragan, 2024) These legal frameworks underscore the significance of subject-matter jurisdiction in comprehensively addressing environmental issues and safeguarding access to justice. (Barragan, 2024).

Environmental jurisdiction is generally categorized into civil, criminal, and administrative sectors, each addressing specific aspects of environmental harm. (Macrory & Woods, 2003) Civil jurisdiction often pertains to disputes between private entities, such as cases involving private nuisance or property damage linked to environmental factors.(Macrory & Woods, 2003) Criminal jurisdiction centers on sanctioning violations of environmental statutes, including unlawful waste disposal or pollution, which are typically prosecuted in criminal courts. (Macrory & Woods, 2003) Administrative jurisdiction concerns compliance with regulations, including appeals of permits, licenses, and enforcement notices issued by government authorities.(Macrory & Woods, 2003) This emphasizes the overlapping nature of these jurisdictions, highlighting the interconnectedness of regulatory appeals and criminal enforcement. It also recommends that a specialized Environmental Tribunal could improve the processing of administrative appeals and potentially contribute to civil and criminal enforcement, especially through administrative penalties and advisory opinions to lower courts. This highlights the complexity of environmental law and emphasizes the importance of a clear framework to tackle its various challenges effectively.

2. Standing and Access to Environmental Justice

Standing is an essential procedural prerequisite that serves as a "gatekeeper" in legal and administrative processes, determining which individual or entity has the statutory right to initiate a case before a court or administrative body.(Centre, 2014) It delineates whether hearings are to be convened, identifies who is to be heard, and specifies the issues and interests to be considered in the decision-making process.

(Centre, 2014) A person with standing is characterized as having the requisite legal status to initiate a hearing that would not otherwise take place or being granted full party status in an already initiated hearing. This concept is particularly pertinent in public law matters, as it ensures that only those with a legitimate stake in the outcome may participate, while also balancing the need for judicial efficiency with the safeguarding of third-party interests. (Centre, 2014)

To address these challenges, many jurisdictions have expanded standing rules to allow public-interest litigation, enabling individuals and organizations to bring cases on behalf of the public or the environment (Abbot & Lee, 2024).Public interest litigation plays a crucial role in environmental governance by allowing courts to address systemic issues and enforce environmental laws even in the absence of direct individual harm (Abbot & Lee, 2024) .

Environmental non-governmental organizations (NGOs) have played a crucial role in using litigation to challenge government decisions, enforce environmental laws, and encourage public participation.(Abbot & Lee, 2024) NGOs have actively used the Aarhus Convention, especially its third pillar on access to justice, to assert their rights and influence how environmental laws are interpreted and applied. (Abbot & Lee, 2024) While legal opportunity theory often highlights litigation as a key strategy for promoting social change, NGO legal mobilization goes beyond just courtroom actions(Abbot & Lee, 2024). Political advocacy and lobbying are also vital for making Aarhus rights effective domestically, enabling NGOs to shape policies and decision-making.(Abbot & Lee, 2024) This combined strategy of litigation and advocacy enables environmental NGOs to function as strategic actors, targeting both short-term victories and long-term changes to the rules and standards governing environmental management.

Access to environmental justice remains uneven due to various systemic and structural barriers that impede effective participation in environmental litigation. The significant challenges encountered by environmental justice movements, including persecution, criminalization, violence, and repression.(Scheidel et al., 2018) Grassroots activists, often from marginalized communities, face disproportionate risks, including social stratification, internal divisions, and targeted violence, which impede their ability to advocate for sustainability and justice.(Scheidel et al., 2018) These barriers not only obstruct the pursuit of environmental justice but also limit the transformative potential of environmental justice movements in driving sustainability transitions. Addressing these challenges is critical to ensuring equitable access to environmental justice and fostering inclusive sustainability processes.

Case studies that shed light on this matter include Urgenda Foundation v. State of the Netherlands (2019). The facts of the case are summarized as follows: The Dutch Supreme Court held that the government has a legal obligation to reduce greenhouse gas emissions to protect the right to life and well-being under the European Convention on Human Rights. (de la Cámara, 2022) This case is considered a landmark in climate litigation, as it established that courts can require governments to take concrete action to address climate change (Peel & Osofsky, 2015).

An additional significant case is Juliana v. United States (339 F. Supp. 3d 1062, D. Or. 2018), a pioneering climate litigation matter in which twenty-one young plaintiffs, alongside Earth Guardians and climatologist Dr. James Hansen, initiated legal proceedings against the federal government for infringing upon their constitutional rights to a healthy and livable environment. The plaintiffs contended that the government's endorsement and financial support of fossil fuels contributed to climate change, thereby endangering the survival and well-being of both current and future generations. (Brister, 2019) The United States District Court for the District of Oregon held that the plaintiffs may advance their claims, thereby denying most of the federal government's motions to dismiss.(Brister, 2019) The court asserted that the plaintiffs possessed the requisite standing to sue, having demonstrated injury-in-fact, causation, and the potential for redress. Furthermore, the court acknowledged the plaintiffs' contention that they hold a fundamental constitutional right to a climate system capable of sustaining human life.(Brister, 2019)

The Juliana v. United States ruling is important in climate litigation because it recognizes a potential constitutional right to a sustainable climate system. It establishes a precedent for courts to evaluate the government's role in climate change and its obligation to safeguard citizens from environmental damage. The Juliana v. United States case is a significant instance illustrating the considerations of standing and access to environmental justice within climate litigation. This case further promotes the discourse concerning legal standing and equitable access to environmental justice. It exemplifies how judiciary bodies can offer a forum for marginalized demographics, including youth, to pursue accountability for environmental damages and to advocate for their inherent rights to a sustainable future. Overall, establishing more inclusive standing rules is a vital move to improve environmental protection, increase accountability, and support sustainability. Allowing courts to tackle systemic environmental challenges expands access to justice and reinforces the judiciary's important role in advancing environmental governance and safeguarding the rights of both current and future generations.

3. Rules of Evidence in Environmental Litigation

Environmental litigation heavily relies on scientific and technical evidence to demonstrate environmental harm and establish causation. Courts are required to analyze complex data such as environmental impact assessments, pollution measurements, and expert testimony to determine the extent and cause of damage. This dependence on scientific evidence often poses challenges, as conflicting expert opinions can create uncertainty and disagreement. The burden of proof is a crucial issue; traditionally, the claimant must prove that environmental harm occurred and was caused by the defendant's actions. However, due to the complexity of environmental systems and causation difficulties, this can be hard to demonstrate. Some legal systems have adopted flexible approaches, such as shifting the burden to the defendant or lowering proof standards in environmental risk cases. In the landmark case Massachusetts v. Environmental Protection Agency (2007), the Supreme Court examined whether the Environmental Protection Agency (EPA) possesses the authority to regulate greenhouse gas emissions from new motor vehicles pursuant to the Clean Air Act (CAA).(Sugar, 2007) A coalition of private organizations petitioned the EPA to establish regulations for emissions of greenhouse gases, including carbon dioxide, citing scientific evidence that links these emissions to global warming and environmental degradation.(Sugar, 2007) The EPA dismissed the petition, asserting that the CAA did not explicitly authorize mandatory regulations for global climate change and that scientific uncertainty rendered regulation inadvisable at that juncture.(Sugar, 2007) The case was then brought before the Supreme Court, which ruled in favor of Massachusetts, finding that the state had standing to sue due to a specific and concrete injury such as rising sea levels threatening its coastal areas that could be linked to the EPA's failure to regulate greenhouse gases.(Sugar, 2007) The Court highlighted Massachusetts' sovereign interest in safeguarding its land and residents. It also found the EPA's decision not to regulate greenhouse gases to be arbitrary and capricious, criticizing the agency for lacking a well-reasoned explanation and for basing its decision on policy considerations outside the scope of the Clean Air Act (CAA).(Sugar, 2007) The Court overturned the D.C. Circuit's decision and remanded the case, directing the EPA to act in accordance with the CAA's statutory requirements.(Sugar, 2007) This case highlights the significance of scientific evidence in environmental litigation. The Court placed strong emphasis on the "objective assessment of the relevant science" and the consensus among experts about the connection between greenhouse gas emissions and climate change. It clarified that regulatory decisions should be based on scientific findings and statutory requirements, not on policy preferences or uncertain predictions. This highlights the crucial role of evidence in proving causation, injury, and the need for regulatory intervention in environmental disputes. Courts are frequently called upon to scrutinize conflicting expert testimonies, thereby prompting essential inquiries regarding judicial competence and the criteria employed to gauge scientific credibility in environmental litigation.

JUDICIAL REMEDIES AND ENFORCEMENT

1. Judicial Remedies in Environmental Disputes

Judicial remedies are vital for turning environmental principles into enforceable outcomes, bridging the gap between legal standards and real-world environmental protection. Courts play a key role in reviewing administrative actions for legal errors and resolving disputes over contentious environmental issues. This role is especially important in environmental law, where complex and multifaceted problems often demand careful legal interpretation. EU environmental law has traditionally offered not only standards and strategies for protection but also strong enforcement and accountability mechanisms.(Fisher, 2020) Through judicial review and case law, courts enforce these laws, interpret their provisions, and give substance to environmental obligations.(Fisher, 2020) This highlights the crucial role of judicial remedies in thatensuring sure environmental laws are more than just goals but are effectively implemented to address real-world challenges. Environmental litigation differs from traditional legal disputes because it demands remedies for both individual harm and broader ecological and collective damages, making judicial intervention inherently more complex.(Birnie et al., 2009) Compensation remains a frequently used remedy designed to offer financial compensation for environmental damage impacting individuals and their property.(Faure & Partain, 2019)

In recent years, restoration and remediation have emerged as pivotal components in environmental adjudication, signifying a substantial shift from purely compensatory justice to ecological restoration. This evolution underscores a growing recognition of the importance of addressing environmental damage caused by human activities, not solely through monetary compensation but also through proactive measures to restore ecosystems and prevent future deterioration. This trend is particularly evident in climate change litigation, wherein courts are increasingly tasked with evaluating the consequences of climate-related damage, including extreme meteorological events, rising sea levels, and declines in biodiversity.(Peel & Osofsky, 2020) For example, cases such as Urgenda Foundation v. State of the Netherlands underscore the significance of governmental proactive measures to reduce greenhouse gas emissions and safeguard citizens from the impacts of climate change.(Peel & Osofsky, 2020) Furthermore, adaptation litigation, which addresses these effects, has gained prominence in jurisdictions such as Australia, where courts have adjudicated cases involving coastal flooding and wildfires.(Peel & Osofsky, 2020) This strategy not only addresses immediate environmental harm but also promotes long-term sustainability and resilience. Such approaches are consistent with the objectives of international agreements and mirror a broader societal recognition of the intrinsic value of ecosystems and the necessity of their preservation for future generations.

The emergence of restoration and remediation as central elements of environmental adjudication signifies a considerable advancement in the ongoing efforts to combat climate change and environmental deterioration. By emphasizing ecological repair, judicial bodies are progressing beyond merely awarding compensatory justice to addressing the fundamental causes of environmental damage and fostering sustainable solutions. This transition highlights the capacity of environmental litigation to function as a formidable instrument in promoting climate governance and ecological sustainability. As the effects of climate change persist and intensify, the importance of restoration and remediation within environmental adjudication is anticipated to become increasingly pivotal in shaping the future landscape of climate justice.

2. Preventive and Restorative Environmental Justice

The role of the judiciary in environmental law extends beyond resolving disputes after harm has occurred; it also encompasses a preventive function aimed at avoiding environmental damage before it arises. Preventive environmental justice is grounded in principles such as precaution and prevention, which emphasize acting in the face of uncertainty to protect the environment. Courts play a crucial role in this preventive framework by issuing injunctions and interim measures that halt potentially harmful activities.(Peel & Osofsky, 2020) Such measures are particularly important in cases involving irreversible environmental damage, where waiting for harm to occur would render judicial remedies ineffective.(Birnie et al., 2009)

Preventive judicial action in climate litigation is increasingly emerging as a significant strategy, as courts are being called upon to assess the adequacy of governmental policies and corporate practices before the full manifestation of environmental damage. This methodology exemplifies a heightened sense of urgency and scope in addressing climate change. For instance, the youth-led #CancelCoal case in South Africa challenges a government energy strategy that includes plans to procure new coal-fired power, arguing that such policies entrench high-emission development trajectories and hinder climate objectives.(Setzer & Higham, 2022) By focusing on sectoral-level decisions rather than isolated projects, these litigation cases aim to proactively tackle systemic issues and avert extensive environmental damage.(Setzer & Higham, 2022) This developing trend highlights the expanding role of judiciary authorities in shaping climate governance and ensuring accountability of governments and corporations for their contributions to climate change.

3. Enforcement Challenges

Enforcing environmental judgments remains a considerable challenge in many jurisdictions, particularly when influential actors such as corporations or state entities are involved. Despite the availability of judicial remedies, non-compliance with environmental legislation frequently arises from rational political and economic decisions made by these entities.(Efthymiou et al., 2025) The punitive measures adopted by the European Union, which impose substantial fines on non-compliant member states, have largely proven ineffective in ensuring compliance with environmental laws.(Efthymiou et al., 2025) Numerous member states choose to pay penalties rather than implement the necessary measures to comply with the European acquis, thereby delaying significant progress in environmental protection.(Efthymiou et al., 2025) This can undermine the effectiveness of judicial decisions and weaken environmental governance. One major challenge is the technical complexity of environmental remedies, particularly those involving restoration and remediation. (Faure & Partain, 2019) Implementing such remedies requires specialized knowledge, long-term monitoring, and significant financial resources (Birnie et al., 2009). In addition, legal and procedural barriers such as limited access to enforcement mechanisms, delays in judicial processes, and insufficient penalties can further complicate enforcement efforts. These challenges are particularly pronounced in developing countries, where institutional capacity and resources may be limited. (Birnie et al., 2009) This reflects a broader political economy dynamic in which regulated entities may rationally choose non-compliance when the cost of penalties is lower than the cost of environmental compliance, thereby undermining the deterrent effect of environmental law.

Environmental enforcement effectiveness varies greatly across regions, especially between developed and developing nations. Environmental courts and tribunals in the Global South face major challenges that limit their effectiveness in tackling environmental issues. Common barriers include limited capacity, insufficient funding, and a lack of technical expertise, which hinder optimal functioning.(Yanti Sulistiawati et al., 2022) Many nations grapple with inadequate government support, corruption, and competing priorities, leading to underfunded institutions with limited resources for enforcement and adjudication.(Yanti Sulistiawati et al., 2022) For example, in Africa, even countries with established environmental courts often lack the necessary funding and technical skills to handle environmental cases properly. Likewise, in South-East Asia, environmental law and science training for judges is usually not mandatory, and administrative actions are often favored over litigation because of the high costs and complexity of legal proceedings.(Yanti Sulistiawati et al., 2022) These issues underline the importance of investing in capacity-building, judicial training, and resource allocation to strengthen environmental courts and improve access to justice in environmental matters.These structural issues often impede the enforcement of judicial decisions and diminish the effectiveness of environmental justice. Therefore, evaluating the success of environmental judicial systems requires considering their broader socio-economic and governance contexts.

This resistance underscores the systemic challenges in enforcing environmental judgments and emphasizes the need for tailored policies that account for the unique characteristics of each jurisdiction. Enhanced collaboration between governing authorities and non-compliant entities, along with the provision of expertise, financial support, and targeted initiatives, may constitute a more effective approach for addressing non-compliance and advancing sustainable environmental governance. Addressing these enforcement challenges requires strengthening institutional frameworks, enhancing judicial capacity, and promoting cooperation between courts and administrative agencies. Effective enforcement is essential for ensuring that environmental laws are not only declared but also implemented in practice. Effective environmental enforcement necessitates a structured interaction between judiciary entities and administrative agencies, wherein courts deliver binding rulings, and regulatory authorities execute technical measures under judicial oversight.

CONCLUSION

This chapter has explored the evolving role of the environmental judiciary as a key component of modern environmental governance. It has shown that environmental adjudication is no longer limited to resolving isolated disputes but now serves as a dynamic tool for enforcing environmental laws, promoting sustainable development, and protecting collective ecological interests. By analyzing conceptual foundations, institutional models, jurisdictional frameworks, and judicial remedies, the chapter has emphasized the growing scope and importance of judicial intervention in environmental issues.

One of the main findings of this study is that environmental disputes have unique characteristics that set them apart from traditional legal conflicts, especially regarding their scientific complexity, collective impact, and long-term effects. These features require specialized institutional responses, such as environmental courts, specialized chambers, and administrative agencies, each offering different benefits and limitations. While specialized courts improve expertise and consistency, environmental chambers promote institutional integration, and administrative agencies provide efficiency and technical skills. However, no single model is universally best, and their success largely depends on the legal, institutional, and socio-economic context of each jurisdiction.

The chapter further underscores the paramount significance of procedural mechanisms, notably standing and access to justice, in assessing the efficacy of environmental adjudication. The broadening of standing regulations and the increasing prevalence of public interest litigation have facilitated more extensive participation in environmental governance, thereby empowering civil society and environmental organizations to actively enforce environmental legislation. Nonetheless, enduring obstacles such as elevated litigation expenses, procedural intricacies, and imbalances of power persist, thereby restricting equitable access to environmental justice.

Regarding judicial remedies, the analysis demonstrates that effective environmental adjudication necessitates an integration of compensatory, restorative, and preventative strategies. Courts are progressively transitioning from conventional compensation frameworks towards more comprehensive solutions that emphasize environmental restoration, risk mitigation, and sustainable practices. This transition signifies an increasing acknowledgment that environmental harm cannot invariably be remedied solely through monetary compensation. Concurrently, enforcement persists as a notable challenge, especially in cases involving influential entities or intricate remediation procedures, thereby underscoring the disparity between judicial determinations and their practical enforcement.

From a broader analytical perspective, the environmental judiciary functions as both a legal and a governance body. Its responsibilities go beyond legal interpretation and application, encompassing the formulation of environmental policies, influencing regulatory practices, and fostering accountability at national and international levels. The heightened reliance on courts, especially in climate change lawsuits, underscores the judiciary's growing significance in addressing global environmental issues. Nevertheless, this broader role prompts concerns about judicial capacity, legitimacy, and the balance between legal judgments and political decisions.

In light of these findings, several recommendations can be proposed to enhance the effectiveness of the environmental judiciary:

1. Strengthening Judicial Expertise

Enhanced investment in judicial training and specialization is imperative to empower judges to proficiently manage the scientific and technical intricacies inherent in environmental disputes.

2. Enhancing Access to Justice

Legal reforms should prioritize expanding standing rules, decreasing litigation costs, and promoting public interest litigation in order to ensure wider and more equitable access to environmental justice.

3. Improving Evidentiary Frameworks

Courts ought to implement more adaptable methods for handling scientific evidence, such as involving expert panels and applying precautionary reasoning, to more effectively manage uncertainty in environmental cases.

4. Promoting Institutional Coordination

Effective enforcement of environmental judgments necessitates enhanced co-ordination among courts, administrative agencies, and regulatory bodies to ensure compliance and proper implementation.

5. Strengthening Enforcement Mechanisms

Legal systems should improve enforcement tools, such as monitoring mechanisms and proportional sanctions, to tackle non-compliance and increase the effectiveness of judicial decisions.

6. Encouraging Preventive and Restorative Approaches

Courts should persist in emphasizing preventive measures and restorative justice, ensuring the avoidance of environmental harm whenever feasible and its effective remediation when it occurs.

This can be accomplished through the implementation of structured judicial oversight mechanisms, which encompass mandatory reporting systems and court-appointed technical monitors. Ultimately, the environmental judiciary assumes a crucial role in bridging the divide between environmental legislation and practical environmental issues. Its effectiveness is contingent upon the robustness of legal frameworks and remedies, the capacity of institutions, the accessibility of justice, and the willingness of courts to embrace innovative and proactive approaches. As environmental challenges persist and escalate, the judiciary will continue to serve as a vital actor in shaping sustainable futures, ensuring accountability, and safeguarding the rights of both current and future generations. Although the judiciary's role in environmental governance is growing, concerns persist about its capacity, the risk of overreach, and the limitations of courts in managing complex policy issues like climate change.

REFERENCES

Abbot, C., & Lee, M. (2024). NGOs shaping public participation through law: The Aarhus Convention and legal mobilisation. *Journal of Environmental Law, 36*(1), 85–106. DOI: 10.1093/jel/eqad030

Ambec, S., & Ehlers, L. (2016). Regulation via the Polluter-pays Principle. *Economic Journal (London), 126*(593), 884–906. DOI: 10.1111/ecoj.12184

Angstadt, J. M. (2023). Can domestic environmental courts implement international environmental law? A framework for institutional analysis. *Transnational Environmental Law, 12*(2), 318–342. DOI: 10.1017/S2047102523000092

Barragan, G. S. (2024). Civil Litigation for Environmental Damages: Are the So-Called Rights of Nature an Alternative? *International Journal of Procedural Law, 14*(1), 73–98. DOI: 10.1163/30504856-14010004

Barthakur, A. (2021). Polluter pays principle as the key element to environmental law. *International Journal of Scientific and Research Publications, 11*(3), 274–277. DOI: 10.29322/IJSRP.11.03.2021.p11137

Birnie, P. W., Boyle, A. E., & Redgwell, C. (2009). *International law and the environment*. Oxford University Press.

Brister, D. (2019). Juliana v. United States. *Public Land & Resources Law Review*(9), 27.

Carnwath, L. (2014). Judges and the Common Laws of the Environment—At Home and Abroad. *Journal of Environmental Law, 26*(2), 177–187. DOI: 10.1093/jel/equ009

Centre, E. L. (2014). *Standing in environmental matters*. chrome-extension://efaidnbmnnnibpcajpcglclefindmkaj/https://elc.ab.ca/media/98894/Report-on-standing-Final.pdf

constitution, u. (1971). *UAE Contitution* Retrieved from https://uaelegislation.gov.ae/en/constitution

de la Cámara, M. A. (2022). Urgenda Foundation v. State of the Netherlands. *Revista Española de Derecho Constitucional*, (126), 299–330.

Declaration, R. (1992). Rio declaration on environment and development. In.

Ebbesson, J., & Okowa, P. N. (2009). *Environmental law and justice in context*. Cambridge University Press. DOI: 10.1017/CBO9780511576027

Efthymiou, L., Tsiamis, N., & Tsagarakis, K. P. (2025). Compliance with the European environmental law: Leaders and Laggards. *Euro-Mediterranean Journal for Environmental Integration, 10*(5), 4251–4264. DOI: 10.1007/s41207-025-00781-7

Environment, F. L. N. o. c. t. P. a. D. o. t. (1999). *Federal Law No. 24 of 1999 concerning the Protection and Development of the Environment.* Retrieved from chrome-extension://efaidnbmnnnibpcajpcglclefindmkaj/https://uaelegislation.gov.ae/en/legislations/1146/download

Faure, M. G., & Partain, R. A. (2019). *Environmental law and economics: theory and practice.* Cambridge University Press. DOI: 10.1017/9781108554916

Fisher, E. (2013). Environmental Law as 'Hot' Law. *Journal of Environmental Law, 25*(3), 347–358. DOI: 10.1093/jel/eqt025

Fisher, E. (2020). *Executive environmental law.* Wiley Online Library. DOI: 10.1111/1468-2230.12456

Handl, G. (2012). Declaration of the United Nations conference on the human environment (Stockholm Declaration), 1972 and the Rio Declaration on Environment and Development, 1992. *United nations audiovisual library of international law, 11*(6), 1-11.

Imperatives, S. (1987). Report of the World Commission on Environment and Development: Our common future. *Accessed Feb, 10*(42,427), 1-223.

Khater, M., Chami, Y., & Albakjaji, M. (2025). Legal Environmental Protection and Sustainable Development in the United Arab Emirates. *Journal of Human Rights. Culture and Legal System, 5*(2), 378–411. DOI: 10.53955/jhcls.v5i2.469

Knox, J. H. (2018). The Paris Agreement as a human rights treaty. *Human Rights and 21st Century Challenges: Poverty, Conflict, and the Environment (Akande et al. eds., Oxford University Press).*

Macrory, R., & Woods, M. (2003). Modernizing environmental justice: Regulation and the role of an environmental tribunal. In: UCL Laws.

Mank, B. C., & Smith, S. (2013). Book Review, David R. Boyd, The Environmental Rights Revolution: A Global Study of Constitutions, Human Rights, and the Environment (UBC Press, W. Wesley Pue General Ed., 2012). *The Environmental Rights Revolution: A Global Study of Constitutions, Human Rights, and the Environment*, 14-09.

Mtavangu, V. (2022). USING POLLUTER PAYS PRINCIPLE TO ENFORCE ENVIRONMENTAL REMEDIATION AND COMPENSATION OF POLLUTION VICTIMS IN TANZANIA. *The Eastern African Law Review*, *49*(2), 133–170.

Peel, J., & Osofsky, H. M. (2020). Climate change litigation. *Annual Review of Law and Social Science*, *16*(1), 21–38. DOI: 10.1146/annurev-lawsocsci-022420-122936

Percival, R., Schroeder, C. H., Miller, A. S., & Leape, J. P. (2021). *Environmental regulation: Law, science and policy*. Wolters Kluwer.

Petrašević, T., & Poretti, P. (2022). The 'polluter PAYS'PRINCIPLE: The Croatian experience. *EU and comparative law issues and challenges series (ECLIC)*, *6*, 3-13.

Preston, B. J. (2011). Benefits of judicial specialization in environmental law: The land and environment court of New South Wales as a case study. *Pace Environmental Law Review*, *29*(2), 396. DOI: 10.58948/0738-6206.1692

Pring, G. (2016). Environmental Courts & Tribunals-A Guide for Policy Makers.

Rajamani, L. (2016). The 2015 Paris Agreement: Interplay between hard, soft and non-obligations. *Journal of Environmental Law*, *28*(2), 337–358. DOI: 10.1093/jel/eqw015

Sands, P., & Peel, J. (2012). *Principles of international environmental law*. Cambridge University Press. DOI: 10.1017/CBO9781139019842

Scheidel, A., Temper, L., Demaria, F., & Martínez-Alier, J. (2018). Ecological distribution conflicts as forces for sustainability: An overview and conceptual framework. *Sustainability Science*, *13*(3), 585–598. DOI: 10.1007/s11625-017-0519-0 PMID: 30147788

Setzer, J., & Higham, C. (2022). *Global trends in climate change litigation. Snapshot Policy Report. Grantham Research Institute on Climate Change and the Environment*. LSE.

Sugar, M. (2007). Massachusetts v. Environmental Protection Agency. *Harv. Envtl. L. Rev.*, *31*, 531.

Van Zeben, J. (2015). Establishing a governmental duty of care for climate change mitigation: Will Urgenda turn the tide? *Transnational Environmental Law*, *4*(2), 339–357. DOI: 10.1017/S2047102515000199

Voigt, C. (2016). The Paris Agreement: What is the standard of conduct for parties? *QIL. Zoom-in*, *26*, 17–28.

Yanti Sulistiawati, L., Bouquelle, F., Lin, J., Lavrysen, L., Ortega, M., Pereira, R., & Tseng, S. (2022). *Environmental courts and tribunals: A guide for policymakers*. United Nations Environment Programme.

Chapter 8
Environmental Litigation as a Mechanism for Achieving Environmental Sustainability:
Principles and Frameworks

Mariam Ahmed Alsandal
http://orcid.org/0000-0003-0869-1208
University of Dhaid, UAE

ABSTRACT

Many natural and human factors impact the environment. Since the right to live in a healthy environment is one of the fundamental human rights, Achieving this sustainability requires the involvement of the legislative, judicial and executive authorities. judiciary, is considered one of the most important tools for enforcing legal provisions. Therefore, some countries have established specialized environmental courts, often referred to as "green courts." Others have opted to create specialized circles for environmental disputes within existing civil or criminal courts, some countries have chosen to train specific judges to specialize in such disputes. Beyond the judiciary, or alongside it. In all cases, This highlights the need to develop an advanced legal and judicial framework as an effective tool for environmental protection. Therefore, this article seeks to compare the rules of environmental litigation across legal systems in different geographic regions to identify the best proposals for achieving environmental sustainability.

DOI: 10.4018/979-8-3373-6635-7.ch008

INTRODUCTION

One of the primary purposes of environmental legislations is to safeguard the right to live in a safe and healthy environment, which represents a universal right that extends beyond human beings to include all living organisms that share life on Earth. The interconnection between environmental protection, human rights, and the rights of other living creatures has become increasingly clear (Mendes et al., 2025). Accordingly, the deterioration of the environment in any of its forms directly affects its integrity and health, which in turn impacts the well-being and health of humans and all other living organisms.

In this context, If climate change is one of the factors that negatively affect the environment and represents a global crisis that knows no geographical boundaries, with environmental impacts reaching every nation, then addressing climate issues has become an urgent global necessity that requires collective efforts at the international level (Karim, 2025).

Therefore, nations have sought to ensure a safe and healthy environment through various environmental legislations at both international and national levels, with the goal of achieving environmental sustainability that serves future generations. At the international level, the United Nations established, in 1972, the United Nations Environment Programme (UNEP). Over the years, several international agreements related to environmental protection have emerged. Among the earliest is the Convention on International Trade in Endangered Species of Wild Fauna and Flora (CITES), which entered into force in 1975 and regulates global trade in endangered animal and plant species to ensure their survival. Another significant milestone is the Montreal Protocol on Substances that Deplete the Ozone Layer, adopted in 1987, which seeks to phase out chemicals that harm the ozone layer. Moreover, in 1992, the United Nations organized the Rio Earth Summit in Rio de Janeiro, Brazil, with the aim of promoting global environmental protection and sustainable development. That same year witnessed the adoption of the United Nations Framework Convention on Climate Change (UNFCCC), which recognized climate change and its adverse effects as a common concern for all humanity. The same year also saw the adoption of the Convention on Biological Diversity (CBD), which aims to conserve biodiversity and ensure its fair and equitable use, thereby supporting the sustainability of biological diversity. Another key agreement is the Stockholm Convention on Persistent Organic Pollutants (POPs), which entered into force in 2004. This treaty focuses on protecting human health and the environment from harmful chemical substances that persist in the environment and accumulate in the food chain, by eliminating or restricting their use as much as possible.

As for national environmental legislation, in the United States, for example, there exists a range of federal laws aimed at protecting the environment, overseen by the Environmental Protection Agency (EPA). Among these is the Clean Air Act (CAA), which seeks to regulate air emissions from both stationary and mobile sources. Through the establishment of the National Ambient Air Quality Standards (NAAQS), the EPA sets limits designed to protect public health and promote general welfare by controlling emissions of air pollutants that pose risks to human health and the environment. Another key piece of legislation is the Resource Conservation and Recovery Act (RCRA), which grants the EPA full authority over the management of hazardous waste from its origin to its final disposal. This includes regulation of its generation, transportation, treatment, storage and disposal, ensuring comprehensive oversight throughout the waste's entire lifecycle.

In the Philippines, there are several environmental laws, including the Clean Air Act, the Clean Water Act, the Solid Waste Management Act, and others that collectively aim to protect the environment. India likewise has national legislation dedicated to environmental protection, such as the Environment Protection Act, the Water Act, and the Air Act, among other laws. In Indonesia, Law No. 32 of 2009 serves as the cornerstone of the country's environmental protection framework. It sets out the fundamental principles of environmental protection and the concept of sustainable development, alongside other related regulations such as the Environmental Impact Assessment (AMDAL), which evaluates the environmental impact of major projects, waste management policies, and additional environmental legislation.

In the United Arab Emirates, there are also various environmental laws, including Federal Law No. (24) of 1999 Concerning the Protection and Development of the Environment; Federal Law No. (23) of 1999 Concerning the Exploration, Protection, and Development of the Live Aquatic Resources; Federal Law No. (17) of 2009 Concerning the Protection of New Plant Species; and Federal Decree-Law No. (25) of 2022 Regarding the Regulation and Development of the Industry, among other related legislations.

It is certain that the mere existence of international and national environmental legislation is not sufficient to protect the environment or achieve sustainable development. These laws must be effectively implemented and translated into action (Amirante, 2011a) through comprehensive development plans, in which all state institutions, including the judiciary, play an active role in bringing environmental goals into practical reality.

The role of courts and judicial bodies in handling environmental cases varies from one country to another. In some nations, a specialized system of environmental courts and judicial bodies has been established specifically to address environmental disputes. This development stems, first, from widespread dissatisfaction with the performance of ordinary, non-specialized courts, and second, from the evolution of environmental laws and principles, which have led to recognition of the strong link between environmental protection, including efforts to combat climate change, and human rights (Smith, 2018). The establishment of specialized environmental courts and judicial bodies necessarily requires the development of substantive and procedural rules suited to their jurisdiction. Such frameworks are essential to promote sustainable development and ensure environmental justice.

In this study, we examine environmental litigation as one of the key mechanisms for addressing the various factors that negatively affect the environment and for mitigating their consequences; ultimately aiming to achieve environmental sustainability and justice. This will be explored through two main themes: specialized environmental courts: between reality and aspiration, and the challenges of environmental litigation: issues and solutions for achieving environmental justice.

Specialized Environmental Courts: Between Reality and Aspiration

Climate change in itself is an issue that requires an environmental legal framework to mitigate and address its negative impacts. This applies both at the national level and at the international level, as local environmental regulation alone is insufficient to tackle climate change. The strength of environmental legal frameworks varies from strong to moderate and weak across different countries. This, in turn, naturally reflects on the effectiveness of environmental litigation, thereby impacting the achievement of environmental justice and the fulfillment of societal responsibility (Sharma et al., 2025).

Climate change represents an accelerating threat to human health and to the environment (Negandhi et al., 2026). There is no doubt that humans themselves contribute to climate change, with both natural and human factors driving these changes. Since the impacts of such changes are not confined to the environments of the countries causing them but extend beyond, and because living in a healthy environment falls within the scope of fundamental human rights, addressing these environmental changes requires both local and international intervention. It also necessitates legislative, executive and judicial actions to achieve a sustainable environment.

While the absence of environmental legislation in some countries poses an obstacle to addressing environmental problems through the judiciary, many countries have local environmental laws. The main challenge, however, lies in the implementation of these laws by the executive and judicial authorities. This indicates that the issue of environmental protection is multifaceted; the problem may stem from deficiencies in environmental legislation or from challenges in executive and judicial procedures even where such legislation exists.

The lack of appropriate legislative and executive measures to address the climate crisis has led to the emergence of climate litigation as a tool to achieve the desired environmental protection (Alobaidi et al., 2025) It is based on constitutional rights (Maboudi and D'Amico, 2026). Accordingly, environmental litigation represents a powerful strategy with significant influence on legislative authorities to ensure legal accountability, ultimately leading to climate justice (Rydberg, 2024).

The judiciary is considered a key factor in achieving the Sustainable Development Goals (SDGs). For example, Goal 16, concerning "Peace, Justice, and Strong Institutions," motivates countries to enhance their environmental law frameworks, which fall within the scope of the United Nations' 2023 agenda for sustainable development and the resolution of environmental disputes (Yanti Sulistiawati et al., 2022).

Moreover, environmental prosperity and economic development in any country give rise to environmental disputes. This naturally necessitates the establishment of specialized and efficient judicial systems characterized by speed and flexibility, which can accommodate such developments, thereby promoting investment and supporting the economy (Abdel Shahid, 2001). Economic growth requires the development of a judicial system, and a primitive judicial framework may hinder economic progress, especially given the intense competition in the Middle East (Albaheth, 2020). Countries adopt different approaches to structuring their courts; some maintain a general division by creating circles for specific types of disputes, such as civil, commercial, or personal status matters, while others establish fully specialized courts. In some cases, the division may rely on assigning specific judges to particular dispute types, or a combination of these arrangements may be employed.

A court differs from a circle. A court is the judicial body whose establishment, organization, composition, circuits, jurisdiction, procedures, and appeal mechanisms are regulated by law. Specialized circles, however, are an internal organizational system within a court, coordinated with the Minister of Justice, designed to distribute work within the court. Specialized circles may be established for civil disputes, others for criminal disputes, and others for commercial disputes, and so on for other types of cases.

Accordingly, the level of national environmental litigation varies depending on the judicial system adopted in each country, whether the country has specialized environmental courts, specialized circles, or judges trained to handle these types of disputes. Each system has its own advantages and environmental disputes may fall under the jurisdiction of ordinary courts (the general scope of courts).

Figure 1. Types of national environmental lawsuits

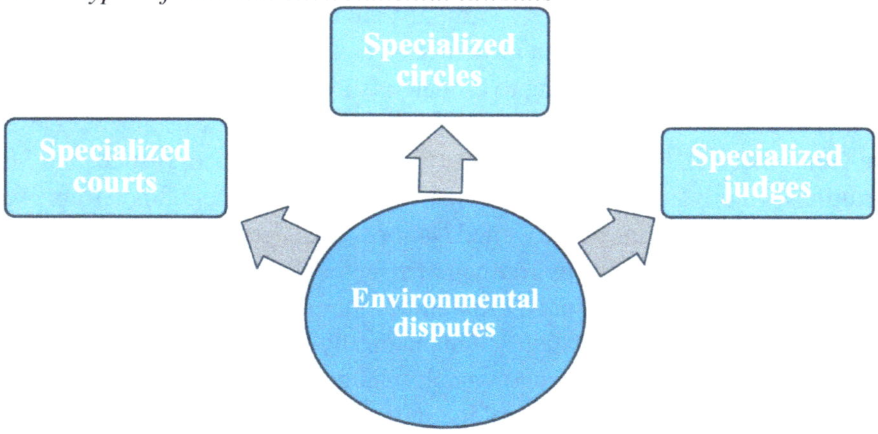

Specialized Judicial Courts

Specialized courts are: judicial bodies established either by a specific law or under the authority of a general legal framework. They typically operate at the level of a court of first instance and fall within the broader system of ordinary courts, yet their jurisdiction is limited to defined types of disputes. The judges serving in these courts possess advanced expertise. In some cases, these courts also apply procedural rules tailored to their nature. It has been argued that specialized environmental courts are better positioned than ordinary courts, as they are more capable of developing effective and innovative solutions to address complex environmental issues (Yanti Sulistiawati et al., 2022).

Specialized courts are sometimes referred to as "problem-oriented courts" meaning courts that focus on a specific type of issue such as drug courts, family courts, juvenile courts, and others. This type of courts emerged out of the need to address particular kinds of cases through nontraditional approaches, to reduce the burden on ordinary courts, and to provide litigants with more effective and responsive judicial services (Freiberg, 2001).

The establishment of specialized courts results in the withdrawal of jurisdiction from ordinary courts over matters that fall within the competence of these specialized judicial bodies whether such bodies are specialized courts or special judicial committees.

The establishment of specialized courts, at the first instance, appellate, and supreme levels, dedicated to a specific type of dispute and competent to hear all related cases, whether civil, administrative, or criminal, leads to a specialized judicial system enhances efficiency by saving time and effort for both litigants and the judiciary, while also ensuring stronger judicial protection and the realization of justice (Jiming, 2020). One of the best examples of specialized courts is the establishment of economic courts in Egypt in 2008, which marked significant progress and positioned Egypt as one of the leading countries in this field within the Middle East and North Africa region (Albaheth, 2020).

The same principle applies to environmental disputes. Establishing specialized environmental courts serves two purposes simultaneously; namely, resolving disputes and protecting the environment (Zhang, 2019). Creating such courts to handle all types of environmental cases, administrative, civil and criminal, would lead to the development of an integrated judicial system characterized by efficiency, flexibility, and fairness in adjudicating environmental matters. This is particularly important because delays in resolving environmental disputes often exacerbate their harmful effects over time, making swift judicial action essential. Moreover, having a single specialized court familiar with all aspects of environmental disputes would ensure a comprehensive understanding of cases that may involve overlapping administrative, civil and criminal dimensions.

Figure 2. Jurisdiction of specialized environmental courts

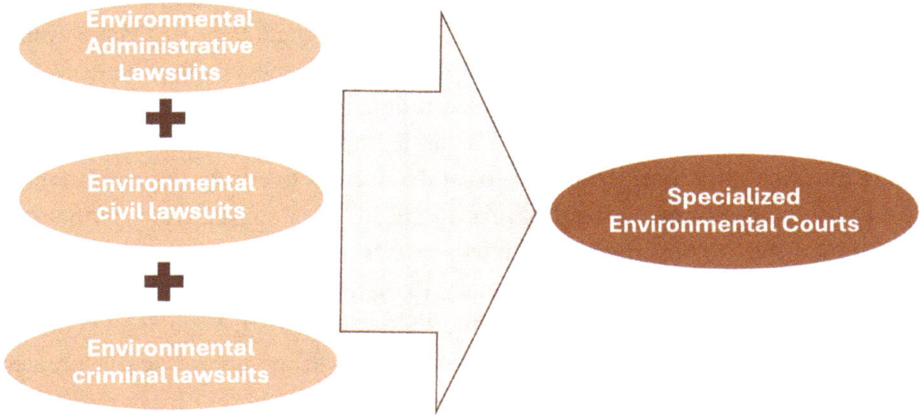

The system of specialized courts requires the establishment of dedicated legislation or the inclusion of specific provisions within procedural laws. These should contain procedural rules that enhance the effectiveness of such courts in applying substantive laws. This means that the framework for specialized courts must be formally incorporated either through separate legislation or within existing procedural codes, with clear procedural rules governing their operation. Such rules should define the procedures followed before these courts, ensure the swift resolution of cases within their jurisdiction, and facilitate the enforcement of their judgments.

Although there is firm belief that environmental litigation serves as a powerful tool for driving policies toward establishing a legal framework for cleaner energy (Popielarski, 2016), the international community still lacks a specialized global court dedicated to environmental disputes. This is despite the existence of international judicial bodies such as the International Court of Justice and the International Criminal Court, which on occasion address environmental issues. It is therefore essential to establish an international environmental court (Heyvaert, 2024) with jurisdiction over all types of environmental disputes, whether the parties responsible for environmental harm are individuals, corporations, or even states. Such a court should also retain the authority to impose reparations and to issue penalties, including imprisonment or fines, as appropriate to the nature and severity of the violation (Sayem, 2025).

As an example of applying the concept of specialized courts at the national level to address the challenges of environmental litigation, the Philippines has adopted within its judicial system the concept of environmental courts, known as “Green Courts” (Minchun and Bao, 2012). These courts operate under specific litigation procedures designed to implement environmental laws and other related regulations. The procedures cover all aspects of trial, including filing fees and evidence submission, ensuring the swift adjudication of environmental cases (Mayo-Anda, 2025). The legal ground for establishing these courts is derived from general constitutional authority. The 1987 Philippine Constitution, in its provisions concerning the judiciary, mandates the establishment of a single Supreme Court and lower courts, granting the Supreme Court the power to establish rules to ensure the enforcement of constitutional rights and to regulate procedures in lower courts. It also authorizes the Supreme Court to designate certain trial courts and define their jurisdiction in specific matters. Furthermore, the Judiciary Reorganization Act allows the establishment of multiple lower courts across the Philippines. As for specialized courts established by special laws, a relevant example is the Philippine statute that created a specialized court to combat corruption.

One of the most notable judicial rulings from the Supreme Court of the Philippines is its decision in the case Metropolitan Manila Development Authority v. Concerned Residents of Manila Bay, in which it emphasized that environmental protection is not optional but a duty of the government. To enforce this, the court established an important principle granting courts the authority to continuously monitor the government's compliance (such as cleaning and rehabilitating Manila Bay) and its commitment to environmental protection, requiring regular reports on the matter. The court also rejected the argument of insufficient budget from any government agency as an excuse to avoid its legal responsibility regarding implementation(2008).

Environmental courts in China date back to 2007 and were established with the aim of achieving environmental justice. The early emergence of these courts, along with the development of environmental jurisprudence under the leadership of the Supreme People's Court, has played a key role in advancing China's environmental judicial system (Gao et al., 2023).

Similarly, India established a specialized environmental court in 2010 known as the National Green Tribunal (NGT) Its jurisdiction is specifically limited to civil cases concerning the protection of the environment, forests, and other natural resources (Rengarajan et al., 2018). Moreover, the decisions of India's Supreme Court in environmental matters have had a profound influence on the development and shaping of various environmental laws in the country (Desai and Sidhu, 2010). For example, the Supreme Court of India in the case of Vellore Citizens Welfare Forum v. Union of India established important judicial principles, namely the polluter pays principle and the precautionary principle, which later contributed to the formation of subsequent environmental legislation (1996).

Although some environmental courts exist in Indonesia, such as the Fisheries Courts, of which there are around ten across different provinces and which function as criminal courts handling cases related to fisheries resources, demands for the establishment of additional specialized environmental courts continue to grow. These calls are grounded in the Indonesian Constitution of 1945, which, under Article 28, recognizes the right to a good and healthy environment as a fundamental right of every Indonesian citizen. Furthermore, the Indonesian Judicial Authority Act stipulates the need for specific legislation when creating specialized courts, thus legally permitting their establishment. The Environmental Protection and Management Act No. 32 of 2009 reinforces this framework by setting out key principles, including state responsibility and the pursuit of sustainability. Undoubtedly, the establishment of a specialized environmental court in Indonesia would strengthen the country's capacity to fulfill its role in advancing the goals of sustainable development at both the national and international levels (Wahid and Magassing, 2025).

From the foregoing, we conclude that the role of the judiciary is not limited to merely resolving disputes and applying laws, but it also plays a creative role in establishing substantive and procedural judicial principles to fill legislative gaps in environmental laws. This is evident in the rulings and judicial principles established by specialized environmental courts, the best example of which is the Indian Supreme Court ruling mentioned earlier.

Given that the United Arab Emirates is a federal state composed of seven emirates, its judicial system is divided between federal courts, which serve some emirates, and local courts, retained by other emirates that have chosen not to join the federal judiciary, as permitted by the Constitution. Some emirates have also opted to withdraw from the federal system after previously joining it. In general, both the federal and local court systems in the UAE are structured similarly to those of other nations, comprising courts of first instance, courts of appeal, and a Federal Supreme Court, in accordance with the UAE Constitution of 1971. Federal Decree-Law No. 42 of 2022 on Civil Procedure (as well as the repealed Federal Law No. 11 of 1992 on the same subject) provides in Article 2 that any emirate which has not transferred its judiciary to the federal system may establish special courts or judicial committees to hear and decide on specific disputes or matters of civil rights, in accordance with the laws of that emirate. From this, two points can be inferred. First, the authority to establish a special court or judicial body must derive from local legislation issued by the emirate itself, not from federal law. Second, the federal judiciary is excluded from this exception and therefore cannot establish special courts under the current framework. Consequently, under existing legislation, it is not possible to establish a specialized environmental court in the UAE. Doing so would require amending the Civil Procedure Law to expressly allow the establishment of specialized courts in specific areas, including environmental matters.

It should be noted that the previous Civil Procedure Law, Federal Law No. 11 of 1992 and its amendments, permitted the establishment of a special court with mandatory jurisdiction over civil, commercial and inheritance disputes only, excluding personal status cases. It also had optional jurisdiction over cases in which the claim value met or exceeded the minimum threshold for cassation appeals, set at AED 500,000, as well as cases in which the parties mutually agreed to refer their dispute to such a court. This provision was later amended by Federal Decree-Law No. 42 of 2022, which replaced the earlier law. Under the new legislation, the legislator converted the concept of a "special court" into a "judicial circle" and revised its scope of jurisdiction accordingly as to be discussed below.

In general, however, it is worth noting that the increasing number of environmental disputes alone is not the primary reason for establishing specialized environmental courts. Rather, it is the complex and technical nature of such disputes, and the need for expertise and specialization in handling them, that provides the strongest justification for establishing these courts in the present era (Minchun and Bao, 2012).

Looking at various comparative judicial experiences, we find that the success of the environmental judicial system depends on the role of the environmental judge, who goes beyond merely adjudicating disputes to actively and effectively developing environmental principles, as is the case in the Indian experience. In contrast, the Philippine experience embodied the continuous supervisory role of environmental judiciary, while a certain degree of judicial environmental restraint appeared in other judicial systems, where the judge's role was limited to applying legal rules without contributing to their development or monitoring their implementation. This does not achieve the goal of establishing a specialized environmental judiciary. Fundamentally, it is the role of the judge, not the structure of the courts, that matters, which explains the clear differences regarding environmental judiciary in comparative legal systems.

Specialized Judicial Circles

To begin with, it is important to distinguish between the concepts of "court" and "judicial circle" on one hand, and the concept of "circuit court" on the other. The first two concepts, court and circle, are the ones relevant to our discussion here and in the following sections. However, the term "circuit court" refers to a judicial system adopted in some countries, including the United States, where the term "District Court" denotes the trial-level courts distributed across the country's 94 districts. Meanwhile, the term "Circuit Court" refers to the Courts of Appeal, of which there are 13, that hear appeals from the District Courts within its circuit.

As for the distinction between "court" and "judicial circle," legal scholars note that a court represents a legal division, whereas a judicial circle is an internal administrative division within a court. To the best of our knowledge, there are currently no specialized judicial circles for environmental matters in the courts of the Arab Gulf states, with the sole exception of one existing within the local judiciary of the Emirate of Abu Dhabi.

If the distinction between courts and judicial circles, as discussed earlier, is traditionally based on the notion that the division of courts is a legal classification while that of judicial circles is an administrative one, this distinction does not fully apply in the case of the United Arab Emirates. According to Article 15 of Federal Decree-Law No. (32) of 2022 on the Federal Judicial Authority, the Federal Court of First Instance is composed of one or more judicial circles that handle criminal, civil, commercial, administrative, and personal status matters. The legislator further expanded the scope of such federal primary judicial circles by permitting the establishment of additional specialized circles in other fields, which could potentially include a circle dedicated to environmental matters, even though such a circle does not currently exist. The composition of these circles and the assignment of judges to them are determined by the General Assembly, based on a proposal from the President of the Court and the approval of the Head of the Judicial Inspection Department. The President of the Court or one of its judges presides over each judicial circle.

Figure 3. An illustrative example of the court circuits

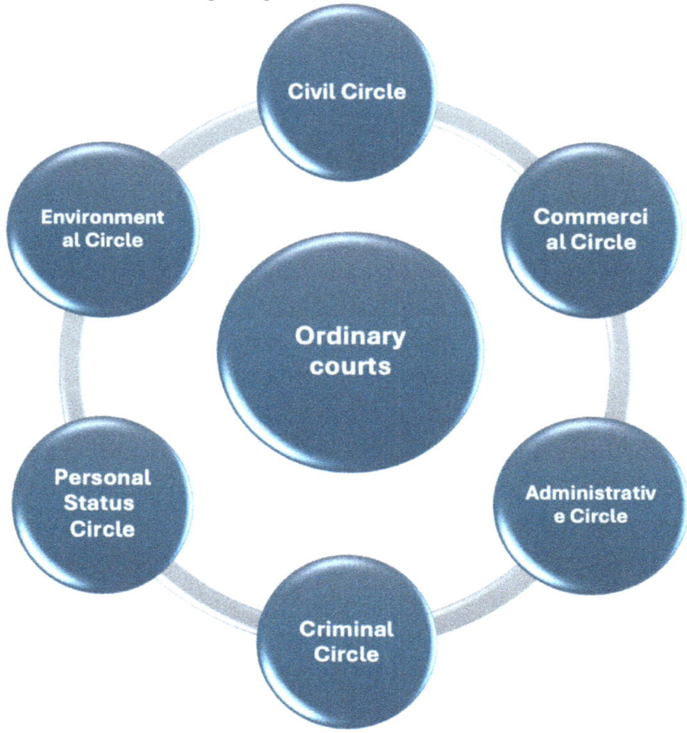

Based on the foregoing, the UAE Federal Civil Procedure Code authorizes the Minister of Justice or the head of the judicial authority of any emirate with a judiciary independent from the federal system to establish special judicial circles dedicated to inheritance cases, while also defining their jurisdiction, procedures, and the methods of appeal against their judgments. This represents the judicial circle whose legal framework was amended by the UAE legislator, transforming it from a special court into a specialized judicial circle, as previously discussed.

Moreover, the Abu Dhabi Courts established in 2013 a set of environmental judicial circles known as the "Green Criminal Circles" in Abu Dhabi City, Al Ain, and Al Dhafra. As their names suggest, these circles are dedicated exclusively to criminal environmental disputes, rather than civil or administrative ones. In practical application, one of these circles ruled in favor of imposing a fine of AED 200,000 on a defendant company for committing two offenses: causing damage to the soil and affecting its natural properties, and conducting activities with environmental impact without obtaining a license from the Environment Authority. This judgment was upheld by the Abu Dhabi Court of Cassation in Appeal No. 1022 of 2024. Similarly, in Appeal No. 10 of 2024, the Abu Dhabi Court of Cassation affirmed a judgment imposing a fine on an individual for using and possessing prohibited fishing tools, as well as engaging in activities harmful to human health and environmental safety without the required authorization from the competent authorities.

In view of all the foregoing, the traditional division of courts of first instance, into civil, commercial, personal status and other specialized judicial circles can be regarded as a form of adopting a system of specialized judicial divisions. However, environmental cases remain largely outside this structure in the Arab Gulf states, with the exception mentioned earlier. This absence may be attributed either to the limited number of environmental lawsuits or to the lack of sufficient judicial expertise in this relatively specialized and technical field.

Specialized Judges

Due to the unequal distribution of the effects of climate change among nations and populations, as being the primary global factor threatening both the environment and humanity, there has been a growing international demand to achieve environmental justice. This concept refers to ensuring fair and non-discriminatory treatment in the formulation of policies and projects that affect the environment. However, this ideal has not yet been fully realized, as developing countries, the poor, women, and children continue to suffer disproportionately from environmental harm. This reality necessitates reforming existing legislation and legal procedures related to environmental protection (Yanti Sulistiawati et al., 2022).

According to the Sixth Assessment Report of the Intergovernmental Panel on Climate Change (IPCC), this situation has led to a rapid increase in the number of environmental lawsuits worldwide. As a result, a substantial body of environmental case law has emerged, helping to define this legal field more clearly. The report provides a comprehensive overview of global judicial trends concerning environmental issues and serves as a vital database for policy-makers, judges, and environmental and human rights advocates (Burger and Tigre, 2023).

In recent years, many judicial systems have undergone fundamental transformations (Cid Teles and Casaleiro, 2025), accompanied by significant developments in legal education (Kang and Hong, 2025). These changes have extended across all areas of law, including those pertaining to environmental protection.

Accordingly, the role of judges has evolved significantly in the context of climate governance, contributing positively to environmental protection. This evolution is reflected in the exercise of judicial oversight over the legislative and executive branches, as well as over public and private institutions and corporations, to ensure their compliance with both national and international environmental legislation (Hussein, 2024). This judicial evolution has also extended to assessing the degree of commitment among environmental judges to the proper application of environmental laws, in order to prevent these laws from remaining merely nominal provisions written but not effectively enforced (Amirante, 2011b).

There is no doubt that judges with expertise in environmental matters are better equipped to interpret and apply both national and international environmental laws effectively. Their specialized knowledge directly contributes to enhancing environmental protection and promoting sustainability through their judicial decisions (Preston, 2012).

It should be noted that the system of specialized judges cannot be regarded as entirely separate from that of specialized courts or judicial circles. Environmental judges often acquire their expertise through practical experience, which makes them the most suitable candidates for appointment to specialized courts or divisions dealing with environmental matters.

Some scholars argue that rotating judges between different courts helps in broadening their professional development, as it allows them to exchange experiences and gain new practical insights. After serving for a period, typically up to five years, in one court, a judge carries the knowledge, tools, and practices acquired from previous work into the new judicial environment (Ackermann and Bastard, 1995).

However, given that legal knowledge alone is insufficient to resolve environmental disputes due to the inherently scientific and technical nature of such cases, judicial training should not be limited solely to legal aspects (Zhang, 2019). To address this challenge, some experts believe that the frequent exposure of non-specialized judges to environmental cases within specialized environmental courts will, over time, help them develop the necessary expertise and improve their ability to issue more effective judicial decisions in environmental matters (Stern, 2014). Nevertheless, the standards and methodologies by which judges should evaluate and handle scientific environmental information remain largely undefined (Doherty et al., 2025).

The presence of specialized judges in various fields does not eliminate the need for expert involvement. The continuous advancement of science and technology increases the judiciary's reliance on expert knowledge to support well-founded and scientifically accurate judicial decisions (Frankel, 2001). This need is even more pronounced in complex environmental matters, which often requires the participation of specialized experts (Kaplan, 2006).

In all cases, whether before environmental courts, specialized circles, or ordinary courts, environmental disputes must be heard by judges who are knowledgeable in environmental matters, to avoid the negative consequences of judicial decisions that lack a sound scientific and environmental basis (Preston, 2014).

We conclude this section by emphasizing that each country's choice of an environmental judicial system, whether to adopt specialized courts, specialized circles, or specialized committees, and how to assign judges, should be determined in accordance with its own economic, political and social conditions, as well as its prevailing legal culture (Hussein, 2024).

Challenges of Environmental Litigation: Issues and Solutions for Environmental Justice

The legal standards governing environmental lawsuits are largely similar at both the national and international levels. In both contexts, they aim primarily to uphold fundamental human rights, including the right of all individuals to access the judiciary to claim their rights or seek judicial protection, the right to non-discrimination, and the guarantee against the arbitrary exercise of governmental power. Collectively, these principles compel policymakers to safeguard the rights of future generations (Sulyok, 2024).

Nevertheless, environmental law remains one of the greatest challenges faced by judiciaries worldwide. It is also one of the least familiar fields for judges, both in terms of basic concepts and procedural mechanisms. This highlights the urgent need to activate the discretionary powers of judges in resolving environmental disputes presented before them, given the relative novelty of environmental law and, in some countries, the absence or limited scope of dedicated environmental legislation (Shelton and Kiss, 2005).

In addition, environmental laws often contain many substantive provisions but few procedural ones. In some cases, only the general procedural rules found in national civil or criminal procedure laws are applied, which are not always suitable for the unique nature of environmental disputes.

It cannot be denied that environmental legislation has developed significantly worldwide due to the increasing number of environmental problems and growing public awareness. However, many countries, particularly developing ones, have failed to put these laws into effective practice (Pring and Pring, 2009). The frequency of environmental disasters highlights the need for strict environmental legislation to address such crises (Tao et al., 2025). Simply having environmental laws does not necessarily mean achieving their intended goal, which is the protection of the environment, unless effective mechanisms for their implementation exist (Wang et al., 2025). While climate change poses a serious threat to both the environment and humanity, litigation serves as a vital tool for achieving environmental justice in its various dimensions (Tigre et al., 2025).

Under the current judicial system, the environmental goals of sustainable development may not be achieved for several reasons. These include the traditional classification of lawsuits, difficulties in proving and presenting evidence, the high costs of litigation, the limited expertise of environmental decision-makers, lengthy judicial procedures, and the accumulation of cases due to poor case management.

Additional challenges include the narrow definition of the plaintiff's interest and standing, the lack of or difficulty in enforcing environmental judgments (Pring and Pring, 2009), low environmental awareness, prioritization of private interests over public ones, and the general limitations of the legal system, which extend to the environmental field as well (Shafayat, 2025).

Accordingly, the following section discusses the adoption of innovative judicial systems to address the major procedural challenges facing environmental litigation, as outlined above. It will then examine the issue of class actions in environmental litigation, proposing solutions that enhance the effectiveness of judicial processes in achieving the objectives of environmental laws. Finally, it will explore environmental judicial rulings and their enforcement, aiming to avoid the problem of environmental laws becoming merely formal measures that fail to achieve their intended purpose.

Adopting Innovative Judicial Systems: An Urgent Need

Innovation in the field of litigation means adopting new working methodologies and fostering both intellectual and practical development to implement ideas and conduct processes more effectively than before (Anttiroiko et al., 2011). One of the most influential areas of innovation is service innovation, and the judiciary is a central component of the public services provided by states at both national and international levels (Sousa et al., 2015). Therefore, innovation naturally extends to the judicial system, with each field requiring its own tailored and creative methods of improvement (Bekkers et al., 2011).

The traditional threefold classification of legal cases, administrative, civil, and criminal, poses a challenge in environmental litigation. Environmental disputes frequently involve overlapping administrative, civil and criminal aspects, which can reduce the effectiveness of environmental courts. Hence, there is a need to adopt a modern and integrated judicial system that enhances the efficiency of specialized environmental courts by introducing a "3 in 1" trial model, integrating all three dimensions (Qi et al., 2023). This approach would enable a tri-governance framework capable of addressing environmental issues from criminal, civil and administrative perspectives simultaneously (Chen and He, 2025).

The high cost of litigation is another barrier to environmental justice. Financial constraints can prevent many individuals, especially those who are poor or marginalized, from seeking judicial protection. The situation becomes even more challenging in environmental disputes, as their litigation costs are often higher than those of other cases due to the complex nature of claims and evidence required. Moreover, there is no guarantee that such cases will achieve their intended outcome of environmental justice and adequate compensation. Consequently, third-party financing of environmental litigation may be considered as an alternative, where an external party covers litigation expenses in exchange for a portion of the compensation awarded, though the outcome of such arrangements remains difficult to predict (2025).

Although one of the most common beliefs is that litigation is prohibitively expensive and therefore discourages individuals from resorting to the courts, and that many disputes could be resolved outside the courts through alternative dispute-resolution mechanisms, some view litigation as a form of investment: an investment of time and money. Resorting to the courts can, in fact, save disputing parties time and money in pursuing their claims. Yet, like any investment, litigation is inherently uncertain; its outcomes cannot be predicted early on. It therefore carries a degree of risk, particularly in environmental disputes (Trubek et al., 1983).

Poor case management, which results in slow judicial procedures and the accumulation of cases in courts, is another major challenge in environmental litigation and beyond. This issue can be mitigated, or at least reduced, through the adoption of an electronic litigation system.

From all of the above, we can see that the issue of environmental litigation isn't just about the lack of substantive rules; it also lies in the unsuitability of procedural rules for this type of disputes. Traditional procedural rules don't fit the special nature of environmental disputes, considering the multiple parties involved, overlapping interests, and the long-term effects. Traditional rules assume actual, specific, and personal damage that can be proven using legally prescribed methods, which doesn't match the nature of environmental damage—it's collective, not personal, and future or potential rather than immediate, requiring complex and non-traditional methods of proof. Also, the long duration of litigation itself can worsen environmental damage, which in turn could make the resulting court judgments practically useless. That's why some comparative legal systems have adopted exceptional procedural rules in environmental disputes to match their nature and achieve their purpose.

The introduction of a smart litigation system, as a more advanced phase beyond electronic litigation, is even more effective in addressing these challenges. This system operates without human intervention, or with very minimal intervention, and enables rapid processing without being constrained by working hours or fixed timelines, while minimizing errors. Such a system can be used for case registration, case preparation, and even the hearing and adjudication of disputes. However, a smart litigation system requires its own legislation to provide the legal basis for its use and to establish all substantive and procedural rules governing its adoption (Alsandal, 2025). It is also essential to establish uniform and mandatory procedural standards for all similar claims to ensure fairness and consistency across cases (Higgins, 2019).

The integration of modern technology into litigation procedures, within the framework of smart courts, by shifting all judicial processes into the online environment and enabling litigants to participate virtually and through electronic procedures, especially in cross-border litigation where physical presence is difficult, helps overcome many challenges faced by litigants. Smart litigation reduces obstacles related to attendance and litigation costs, thereby improving judicial efficiency and ensuring access to justice for all (Tang, 2023).

Artificial intelligence (AI), as one of the most significant technological advancements worldwide and as a tool already widely used across countless sectors, naturally has a role to play in the judicial system as well. The judiciary is in great need of technology capable of processing legal data and information rapidly while preserving the core foundations and principles of justice such as the protection of human rights, judicial neutrality, and ensuring fairness for all. These objectives can indeed be supported through the use of AI tools when properly integrated into judicial work (Kovalchuk et al., 2024).

Class Actions in Environmental Litigation

As litigants around the world increasingly invoke the right to life and the right to a healthy environment in environmental cases, judges have begun to adopt these rights as foundational principles in adjudicating such disputes. In fact, the right to life has been regarded as a stronger legal basis than the right to a healthy environment in environmental litigation. Environmental cases have therefore broadened the scope of the right to life far more than the right to a healthy environment, to the point that governments' failure to address climate change and its harmful effects on the environment is now considered a serious threat to the right to life (Maboudi and D'Amico, 2025).

It is well established that the responsibility for protecting the environment and ensuring its sustainability is a collective one. It does not fall on individuals alone, but extends equally to both public and private entities. Governments, too, are responsible for achieving environmental sustainability as part of their domestic and international obligations. Thus, the pursuit of a safe and healthy environment is a shared societal responsibility toward present and future generations.

Nevertheless, it is unrealistic to expect all parties to voluntarily comply with environmental protection standards and abstain from causing harm to the environment. Legal rules are therefore necessary to deter violations of environmental law, and legal restrictions and sanctions must be imposed on offenders, whether individuals, corporations, or even states to achieve the ultimate goal of sustainable environmental justice.

Effective legal tools have a significant impact on achieving corporate social responsibility and corporate governance, including in the environmental sphere (Yang et al., 2024). It is evident that companies operating in sectors with potential negative environmental impacts tend to act with greater caution when they are certain that they may be exposed to environmental litigation. Such companies make considerable efforts to reduce the environmental harm arising from their activities, unlike other companies, particularly when severe penalties are imposed under environmental legislation (Hossain et al., 2025).

In this context, legal philosophy firmly recognizes that holders of private rights governed by private-law provisions are totally free to waive those rights and refrain from pursuing them. However, a classification supporting the concept of restorative justice in private law grants civil plaintiffs broad discretionary authority. Under this classification, private-law rights may be divided into three categories: rights that may be freely waived, rights that may not be waived, and rights that may be waived only to a limited extent (Psarras, 2023).

Requiring the conditions of legal interest and standing in the traditional litigation system conflicts with the unique nature of environmental disputes, due to the difference between environmental damages and other legal damages, as detailed earlier on this matter. So, we wonder: is it necessary to have the conditions of interest and standing in environmental lawsuits?

Requiring the traditional elements of legal interest and standing, as defined above, to be strictly applied in environmental cases would create a significant barrier to environmental litigation. Environmental harms cannot typically be confined to a specific individual or a narrowly defined group such that a private, individualized interest is formed to justify bringing a claim. Environmental impacts are widespread particularly in cases involving trans-boundary harm. Therefore, the concept of public interest must be adopted in environmental litigation. The benefits sought through environmental lawsuits serve a broad public interest: their purpose is to secure judicial protection for a large segment of society, not merely for the individual plaintiff or plaintiffs. As a result, such actions may be brought by a single person or by a group of individuals on behalf of others.

Therefore, public-interest environmental litigation is considered one of the innovative mechanisms for protecting vulnerable groups that are unable to seek judicial relief on their own and for addressing widespread environmental problems. This type of litigation also helps remove several other barriers to environmental justice, including the high cost of environmental lawsuits. Ultimately, it ensures environmental justice that is not limited to a specific individual or group, but rather extends to society as a whole (Li and Song, 2024).

Since there are no entities that represent or safeguard the public interest that could otherwise initiate litigation in matters aimed at protecting the general welfare at either the national or international level, China developed a judicial system specifically designed to serve the public environmental interest (Shen, 2026). This system has had significant financial repercussions for companies that are major polluters, as it increased their operational risks and placed continuous pressure on them to comply with environmental standards. As a result, it has contributed both to the development of environmental litigation and to strengthening its role in achieving sustainable environmental development (Liu et al., 2025). Under this system, public prosecutors, administrative bodies, and non-governmental organizations (NGOs) are allowed to file environmental lawsuits (Cao and Wang, 2011). Consequently, NGOs have the right to bring environmental claims even when their own private interests have not been affected or when they have suffered no direct harm from the environmentally damaging actions in question. The purpose of such litigation is to serve the public interest as a whole not the private interests of the organizations bringing the claims (Gao et al., 2023).

An example of that on 6, November, 2020, a lawsuit has been filed against the governments of Tanzania and Uganda before the East African Court of Justice, seeking an injunction to halt the East African Crude Oil Pipeline project. The plaintiffs claim that the two governments, without objection from the Secretary-General of the East African Community, responsible for overseeing the East African Community Treaty, signed agreements to construct the pipeline without conducting proper assessments of environmental, social, human rights, and climate impacts. The plaintiffs, who are the Centre for Food and Decent Livelihoods, the African Institute for Energy Governance, Friends of Nature, and the Strategic Litigation Centre in Tanzania, based their case on Ugandan national law, the East African Community Treaty, and its protocols. The East African Court of Justice dismissed the case on procedural grounds on 23, November, 2023. The NGOs filed an appeal, contending that their claims — including allegations of human rights violations, environmental harm, and lack of public participation — deserve judicial consideration. This case is still pending.(2020)

In Indonesia, citizens have a constitutional right to live in a healthy environment. Accordingly, they are entitled to file environmental lawsuits affecting that right, provided they have a sufficient legal interest to bring the claim as an interest derived from their status as citizens asserting their right to a healthy environment, without the need to demonstrate a private, individualized interest (Purwendah et al., 2022). Although class actions are legally recognized and constitute an effective tool for enforcing environmental law, their practical application still falls short of expectations. Environmental claims continue to be rejected by the courts due to failure to meet procedural legal requirements (Kurniawan et al., 2025).

In India, public-interest litigation has helped strengthen environmental justice (Bandyopadhyay, 2017). Legal innovations have further expanded the role of the judiciary by broadening the scope of public-interest litigation to include environmental cases. As a result, the Supreme Court of India has become one of the most active courts in developing and advancing this type of litigation (Mate, 2011).

In UAE, despite the existence of environmental legislation and the establishment of a specialized Ministry of Climate Change and Environment, the country's level of environmental litigation does not correspond to the strength of its legislative and executive frameworks. As noted earlier, there is only one dedicated environmental court in the UAE, located in Abu Dhabi, and no equivalent court exists in any of the other emirates. This means that environmental cases in the UAE are adjudicated either by specialized circles, where they exist, or by the ordinary judiciary. The concept of public-interest litigation is not recognized in the UAE Civil Procedure Law. Traditional criteria for admissibility, particularly legal interest and standing, remain dominant and are consistently applied by the UAE courts. Public-interest litigation appears only in a limited form within the Criminal Procedure Law, which authorizes the Public Prosecution to file certain criminal cases in the public interest. Judges may also consider the public interest when assessing the legality of administrative actions, but this remains far from a comprehensive public-interest litigation framework.

However, since the public interest encompasses both individual and collective interests, as well as present and future interests, and seeks to achieve security, stability and justice, private interest in this context becomes intertwined with the public interest. This makes the concept of environmental litigation based on the public interest neither unusual nor objectionable. Nevertheless, due to the formalistic nature of procedural laws, it is difficult to rely on the public-interest concept in environmental disputes. Accordingly, there must be legal foundations or at least judicial principles to support the recognition of public-interest environmental litigation.

Nevertheless, the Global Climate Litigation Report issued in 2023 confirms a continued increase in the number of climate-related lawsuits being brought before courts and judicial bodies at the regional and international levels (Burger and Tigre, 2023).

As for the requirement of standing, it is necessary to adopt the approach of allowing a single individual or a single entity to file environmental lawsuits on behalf of other individuals or groups affected by the environmental harm in question. Such an entity could be, for example, an environmental protection organization or a consumer protection association, provided it is legally authorized to act in this representative capacity. In this case, standing applies to the plaintiff or representative, whether an association, authority or other qualified body, and not to the affected individuals themselves. This approach not only satisfies one of the key admissibility requirements by ensuring that the claim reflects the shared public interests of the affected group, but also facilitates access to justice and enhances the ability to achieve it efficiently.

This procedural shift in environmental disputes can be seen as the necessary legal response to the unique nature of these types of disputes, and it is actually a fundamental condition for achieving environmental justice. It gives the judiciary the power to intervene effectively to prevent environmental damage from worsening, rather than just addressing it afterward, and it represents a natural development in the principle of the right to litigation guaranteed to everyone.

It appears from comparative judicial experiences that the expansion of the standing and interest requirements in environmental lawsuits resulted from judicial intervention rather than legislative action. In other words, this expansion came through judicial measures that addressed traditional procedural litigation mechanisms. For example, the Supreme Court of India established the idea of public interest litigation, allowing individuals and NGOs to file environmental lawsuits on behalf of affected groups. In contrast, in the Chinese experience, the public prosecutor and administrative bodies were given the authority to file environmental lawsuits, possibly to avoid the abuse of such lawsuits. From this, we can conclude that expanding the standing requirement reflects the judiciary's role in protecting the environment and creating a balance between allowing individuals and NGOs to pursue environmental litigation and ensuring that environmental lawsuits are not misused. Therefore, each legal system can choose the most suitable approach while giving the judiciary the space to perform its creative functional role in providing environmental protection.

Environmental Judicial Rulings and Their Enforcement

As noted earlier, the judiciary has emerged as one of the key mechanisms for environmental protection, particularly in compensating for the failure of states to fulfill their responsibilities in this area. For environmental rulings and judgments to achieve their intended purpose, they must be effectively enforced. Nevertheless, there remains an urgent need to establish a strong link between the judiciary on one side and the legislative and executive branches on the other to secure a healthy and safe environment for present and future generations (Martínez Navarro, 2025). Environmental protection is a shared responsibility (Fürst et al., 2024).

There is no doubt that judicial decisions play a crucial and significant role in reducing environmental harm through the environmental cases brought before the courts. Equally important is the role of the judiciary in reaching solutions and settlements to environmental disputes and in imposing legal penalties on all those who violate environmental laws.

One example of local judicial applications in this regard is the ruling of the Supreme Court of the Philippines in a case filed in 2010 by a group called Global Legal Action on Climate Change (GLACC) against various government departments including the Climate Change Commission, the Department of Public Works and Highways, the Department of Interior, and Local Government. The lawsuit aims to mitigate the impact of flooding, exacerbated by climate change, which affects the right of Filipinos to live in a healthy environment. The plaintiffs in the lawsuit requested adherence to two laws: Act 6716 "the Rainwater Collector and Springs

Development Law" which requires each Filipino locality to have and maintain a rainwater collector to ensure fresh drinking water is available in times of flooding; and Act 7160 or the Local Code of 1991, which establishes the responsibilities of local governments in relation to water management. The case was settled when the defendant government departments signed a Memorandum of Understanding and submitted a work plan committing to carry out the required works (2010).

In the UAE, the Abu Dhabi Court of Cassation, in Appeal No. 385 of 2017, upheld a judgment imposing a fine of AED 50,000 and ordering the confiscation of the fishing boat and the fishing tools used in a case in which four individuals were accused of committing acts that negatively affected the local environment. The violation consisted of using prohibited fishing equipment and fishing in a restricted area.

As part of the commitment of the UAE local judiciary to international law and instruments, the Ras Al Khaimah Court of Cassation, in Appeal No. 16 of 2007, issued on 21 October 2007, upheld a ruling affirming the jurisdiction of UAE courts over a case involving the crime of marine environmental pollution brought before them. In its decision, the court considered the principle that the territorial scope of criminal law is aligned with the territorial sovereignty of the State. The judgment also relied on the United Nations Convention on the Law of the Sea (UNCLOS), signed on 10 December 1982 in Montego Bay, Jamaica, which provides in Articles 2 and 3 that a coastal state's sovereignty extends beyond its land territory and internal waters to an adjacent belt of sea known as the territorial sea. Each State may determine the breadth of its territorial sea up to a limit not exceeding 12 nautical miles, measured from baselines established in accordance with the Convention (Article 3). The Convention further stipulates that a State's exercise of sovereignty over its territorial sea is subject to the provisions of UNCLOS and the rules of international law (Article 1/3). Since the United Arab Emirates is one of the signatories to the Convention at the time of its adoption, it has adhered to its provisions and set its territorial sea at 12 nautical miles, as provided for in Article 4 of Federal Law No. 19 of 1993 for Defining the Maritime Areas of the UAE. During the first session of the Third United Nations Conference on the Law of the Sea, held in New York from 2 to 24 December 1973, the Second Main Committee, upon the proposal of developing countries, introduced the concept of the Exclusive Economic Zone (EEZ). Despite being relatively new, the idea received broad support, recognizing the right of coastal States to an EEZ extending up to 200 nautical miles, within which the State exercises exclusive rights to exploit living and non-living resources. This led to the inclusion of the concept within the 1982 UNCLOS, set out in Part V ("Exclusive Economic Zone") (Articles 55–75). These provisions define the EEZ as an area extending up to 200 nautical miles from the baselines from which the breadth of the territorial sea is measured. They grant coastal States sovereign rights for the purposes of exploring, exploiting, conserving, and managing natural resources, both living and non-living, of the waters above the seabed, the seabed itself, and its subsoil. They also grant rights regarding other economic activities, such as producing energy from water, currents, and winds. Moreover, UNCLOS confers jurisdiction—within the limits set out in the Convention over activities related to the protection and preservation of the marine environment. It places particular emphasis on marine environmental protection, obliging States to safeguard and preserve it, and lists measures for preventing, reducing, and controlling marine pollution from any source. States must ensure that activities under their jurisdiction or control do not cause pollution damage to other States or spread beyond areas where they exercise sovereign rights. The Convention lists all sources of pollution, including pollution from ships caused by intentional or unintentional discharges of harmful substances. It also authorizes coastal States to enact laws and regulations to prevent marine pollution by vessels flying their flag or operating in their EEZ. Where clear objective evidence exists that a vessel navigating within the EEZ has violated applicable international rules or standards for preventing, reducing, and controlling pollution—or violated national laws consistent with those rules—the coastal State may, where warranted by the evidence, initiate legal proceedings, including the detention of the vessel. The Convention further permits imposing penalties on offenders for violations committed beyond the territorial sea, provided due process rights are respected. The UAE issued Federal Law No. 24 of 1999 Concerning the Protection and Development of the Environment, which defines the "Marine Environment" as: "The marine waters and their contents of natural resources, plants, fishes, other marine creatures, and the above atmosphere, as well as fixed and movable installations and projects established in the marine environment. The boundaries of the Marine Environment extend to the economic zone of the State." From this definition, it follows that the UAE's EEZ falls within the scope of the marine environment as defined by the Environmental Protection Law. Accordingly, the marine environment extends up to 200 nautical miles measured from the baseline, or in practical terms, 188 nautical miles beyond the territorial sea, given that the territorial sea itself covers 12 nautical miles, pursuant to Articles 4 and 12 of Federal Law No. 19 of 1993. Furthermore, Article 17 of the same law—found in Chapter I ("Scope of Environmental Protection") of Part II ("Protection of the Aquatic Environment") provides that the protection of the aquatic environment from pollution aims to achieve the following objectives: "1. … 2. Protecting the marine environment and its natural resources, both living and non-living, by preventing, reducing, and controlling pollution from any source; and 3.… " This clearly indicates that the scope of protection of the aquatic environment, including criminalization and penalties, extends to the outer limits of the UAE's EEZ with respect to preventing, reducing, and controlling marine pollution. This is fully aligned with the provisions of the 1982 United Nations Convention on the Law of the Sea.

Based on the foregoing, it can be inferred that local courts play a crucial role in achieving environmental sustainability and environmental justice in all countries through the application of both national environmental laws and international instruments. This applies whether a state adopts specialized environmental courts, environmental divisions, or judges specialized in environmental matters within ordinary courts. However, the role of the judiciary does not end with the issuance of judicial decisions. Such decisions alone do not necessarily ensure the achievement of the intended goals of environmental protection. It is therefore essential to activate and operationalize these rulings through the establishment of an effective mechanism for their implementation.

The case of Notre Affaire à Tous v. France is one of the leading judicial examples arising from government inaction in addressing climate change, amounting to a violation of both national and international laws. It clearly illustrates the importance of enforcing environmental judicial decisions. The lawsuit was brought in 2018 against the French government by four non-profit organizations: l'Homme (FNH), Greenpeace France, Notre Affaire à Tous and Oxfam France.

In 2021, the Paris Administrative Court ruled that the French State must pay financial compensation of €2,000 to each of the four claimant organizations. The Court also ordered the Prime Minister and the relevant ministers to take all necessary measures to remedy the climate-related harm identified in the lawsuit and to prevent its aggravation. The matter did not end there. On 21 May 2023, the four claimant organizations filed an appeal before the Conseil d'État due to the State's failure to comply with the judgment. In their appeal, the organizations argued that the French Administrative Court had not devoted the necessary effort, time or resources to ensure the enforcement of its ruling. The claimant organizations contended that this failure amounted to a violation of their right to a fair trial (Database, 2021).

From the above, we observe that the stage of enforcing judicial rulings is no less important than the stage of adjudicating environmental disputes with the issuance of the rulings themselves. The role of the judiciary does not end with the issuance of a judgment; its enforcement is a more sensitive stage, as a ruling loses its value without enforcement. Judicial protection must extend to the stage of enforcing environmental judgments, which have impacts on the public interest, and the need for the intervention of various executive authorities in the state to implement these judgments, since enforcement cannot be reduced to a single procedure. This complex nature of environmental judicial rulings, which assumes cooperation in their enforcement, may lead to a decline in the effectiveness of environmental judicial rulings in the absence of mechanisms to monitor their implementation. This requires cooperation between the judicial and executive authorities and the creation of advanced enforcement means suitable for implementing judgments in this type of disputes, such as the establishment of judicial monitoring committees and the imposition of periodic reporting by the entities subject to the rulings, showing the progress made in implementing those judgments.

Comparative judicial experiences highlight the importance of the judgment enforcement stage. In the Philippine experience, the Philippine judiciary developed a mechanism that allows the court to monitor the enforcement of environmental judgments through ongoing judicial orders. In contrast, the Indian judiciary granted the court the power of continuous judicial supervision over the enforcement of environmental judgments for long periods, requiring the relevant authorities to provide periodic reports reflecting their progress in implementing the environmental judgments. In other legal systems, the role of the judiciary was limited to merely issuing environmental judgments and leaving their enforcement to the executive authority without oversight, which affects their role in protecting the environment. Therefore, it is necessary to reconsider the traditional function of the judge of only issuing judgments and grant them the authority to follow up on their implementation. Even though the UAE judiciary has issued judgments that show determination in protecting the environment, it is still governed by traditional rules that separate the judgment issuance phase from the enforcement phase. The gap is not in the substantive rules or weakness of the judiciary but in the procedural enforcement system, which requires creating enforcement tools to ensure environmental protection.

CONCLUSION

This study has demonstrated that the responsibility for protecting the environment cannot rest solely on states; rather, it is a shared societal responsibility borne by governments, institutions, companies, and individuals alike. It is a responsibility with far-reaching effects—both present and future. The study also underscored the importance of having a specialized environmental judiciary to adjudicate environmental disputes and to enforce national and international environmental laws, thereby achieving environmental sustainability on the one hand and environmental justice on the other. Achieving these goals requires careful consideration of the following points:

- There must be a legal basis for establishing specialized judicial courts. As seen in the legal frameworks of the Philippines, Indonesia, and India, explicit legislative authority exists for the creation of specialized courts. By contrast, the United Arab Emirates lacks such a legal basis at the federal level, which presents a barrier to adopting a system of specialized environmental courts designed to advance environmental sustainability and justice.

- There is a need to adopt procedural legal frameworks that reflect the unique nature of environmental litigation and broaden access to this type of litigation for all. These frameworks must be innovative and capable of addressing environmental challenges more effectively. They should include: a three-tier environmental litigation system covering criminal, civil and administrative aspects; an expanded interpretation of the requirements of legal interest and standing as essential conditions for the admissibility of all claims, including environmental ones; recognition of public-interest litigation in environmental matters; recognition of collective standing for groups affected by environmental harm; the adoption of electronic litigation systems and smart litigation mechanisms to reduce the high costs associated with environmental disputes; and the implementation of class-action systems.
- It is essential to adopt judicial systems that actively enforce environmental laws. Many countries possess national environmental legislation in addition to the international environmental instruments they have ratified. The laws must be activated and applied broadly through the judiciary, by issuing judicial decisions that implement them and ensuring that those decisions are effectively enforced in practice. Moreover, reliance on only one form of environmental litigation, whether administrative or criminal, cannot achieve the intended objectives. Instead, administrative, criminal, and civil environmental litigation must all be activated together to ensure robust enforcement of national and international environmental laws and to guarantee environmental sustainability and justice for all.

REFERENCES

Abdel Shahid, D. S. F. (2001). Corporate governance is a global pursuit: What could be done in Egypt? *Cairo & Alexandria Stock Exchanges Working Paper*.

Ackermann, W., & Bastard, B. (1995). The diffusion of innovation in the judicial system. In *De Gruyter Studies In Organization* (pp. 73–88).

Albaheth, H. E. (2020). Establishing specialized economic courts in Saudi Arabia to boost investments. *Utopía y Praxis Latinoamericana*, *25*, 278–288.

Alobaidi, H. A., Jasim, A. K., Hadi, A. K. M., Al-Bayati, H. N. A., Hafedh, M. A., & Amini, A. (2025). Climate litigation and judicial activism in environmental protection. *Water and Soil Management and Modeling*, *5*, 88–100.

Alsandal, M. A. (2025). The development of civil procedural legislation to address emerging changes. *Journal of Legal Sciences. University of Sharjah*, *22*, 257–278.

Amirante, D. (2011a). Environmental courts in comparative perspective: Preliminary reflections on the National Green Tribunal of India. *Pace Environmental Law Review*, *29*(2), 441. DOI: 10.58948/0738-6206.1693

Amirante, D. (2011b). Environmental courts in comparative perspective: Preliminary reflections on the National Green Tribunal of India. *Pace Environmental Law Review*, •••, 29.

Anttiroiko, A.-V., Bailey, S. J., & Valkama, P. (2011). Innovations in public governance in the western world. In *Innovations in public governance*. IOS Press. DOI: 10.3233/978-1-60750-727-7-1

Bandyopadhyay, J. (2017). Environmental struggles and innovations in India: An historical perspective.

Bekkers, V., Edelenbos, J., & Steijn, B. (2011). An innovative public sector? Embarking on the innovation journey. In *Innovation in the public sector* (pp. 197–221). DOI: 10.1057/9780230307520_10

Burger, M., & Tigre, M. A. (2023). *Global climate litigation report: 2023 status review*.

Cao, M., & Wang, F. (2011). Environmental public interest litigation in China. *Asia Pacific Law Review*, *19*(2), 217–235. DOI: 10.1080/10192557.2011.11788249

Chen, L., & He, Y. (2025). From value recognition to effective implementation: The main problems and strengthening paths of law enforcement to combat IUU fishing. [Note: Replace with actual article number if available]. *Frontiers in Marine Science*, *12*, 1234567. DOI: 10.3389/fmars.2025.1596492

Cid Teles, M., & Casaleiro, P. (2025). Judicial perceptions on working time: Examining quality and intensity of work in Portuguese first instance courts. *Sortuz*, *15*(1), 269–293. DOI: 10.35295/sz.iisl/2011

Database, T. C. L. (2021). *Notre Affaire à Tous and Others v. France*. Administrative Court of Paris. The Climate Litigation Database.

Desai, B. H., & Sidhu, B. (2010). On the quest for green courts in India. *Journal of Court Innovation*, *3*, 79.

Doherty, J. M., Page, J. C., Hanle, P. A., Austin, J., & Thiam, S. N. (2025). Assessing science-based decision-making in US climate change lawsuits. *Journal of Environmental Law*, *37*(1), 117–140. DOI: 10.1093/jel/eqaf002

Frankel, M. S. (2001). The evolving role of scientific experts in the courts. *The Journal of Philosophy, Science & Law*, *1*(1), 50–56. DOI: 10.5840/jpsl2001113

Freiberg, A. (2001). Problem-oriented courts: Innovative solutions to intractable problems? *Journal of Judicial Administration*, *11*, 8–27.

Fürst, H., Madalena, S. R. T., & Cosenza, G. A. (2024). Environmental litigation within the Brazilian Supreme Court towards the defence of environmental public policies: The ADPF 743 case study. *Novos Estudos Juridicos*, *29*, 900–931.

Gao, Q., Quan, K., & Zhao, X. (2023). The standing to sue of NGOs in environmental public interest litigation in China: A doctrinal analysis of laws and cases.

Heyvaert, V. (2024). Dislocation of environmental litigation – New developments in corporate liability for environmental harm. *European Business Law Review*, *35*(3/4), 403–428. DOI: 10.54648/EULR2024025

Higgins, A. A. (2019). The rule of law case against inconsistency and in favour of mandatory civil legal process. *Oxford Journal of Legal Studies*, *39*, 725–750. DOI: 10.1093/ojls/gqz015

Hossain, A., Masum, A. A., & Rjiba, H. (2025). Fear of fairness in courts: The impact of liberal judge ideology on corporate carbon emissions. [Replace with actual article number if available]. *International Review of Financial Analysis*, *101*, 103456. DOI: 10.1016/j.irfa.2025.104022

Hussein, M. F. M. (2024). Towards the establishment of green courts for the settlement of environmental disputes: A study in the light of the report of the United Nations Environment Programme on environmental courts – Legal and economic dimensions of the litigation system in the twenty-first century.

Jiming, Y. (2020). The establishment of an overarching intellectual property judicial system. *Contemporary Social Science*, •••, 72–102.

Kang, D., & Hong, S. E. (2025). Legal education reform and medical litigation: Improved access but delayed justice in plastic surgery malpractice cases in South Korea. [Replace with actual issue/article if available]. *PLoS One*, *20*(4), e0312345. DOI: 10.1371/journal.pone.0323059 PMID: 40440320

Kaplan, L. A. (2006). Experts in the courthouse: Problems and opportunities.

Karim, M. R. (2025). Pursuing climate justice through public interest litigation: Theories, practices, and prospects. *International Journal of Innovative Research and Scientific Studies.*

Kovalchuk, O., Kolesnikov, A., Proskurniak, I., Halunko, V., & Kovach, Y. (2024). Implementation of innovative information technologies in judicial proceedings: Organizational and legal aspect. *Amazonia Investiga*, *13*(76), 53–62. DOI: 10.34069/AI/2024.76.04.4

Kurniawan, I. D., Septiningsih, I., Handayani, F., & Ikrimah, A. (2025). Formal requirements for class action lawsuits in environmental cases in Indonesia: Problems and solutions. *Journal of Law. Environmental Justice*, *3*(1), 79–103. DOI: 10.62264/jlej.v3i1.114

Li, X., & Song, Z. (2024). A critical examination of environmental public interest litigation in China - Reflection on China's environmental authoritarianism. *Humanities & Social Sciences Communications*, *11*(1), 123. DOI: 10.1057/s41599-024-03047-9

Liu, D., Chen, C., Yao, S., & Yu, N. (2025). Environmental litigation risk premium in corporate equity financing costs. *Energy Economics*, *143*, 108123. DOI: 10.1016/j.eneco.2025.108229

Maboudi, T., & D'Amico, E. (2025). Life, liberty, and litigation: Do constitutional rights impact climate litigation outcome? *Political Research Quarterly.*

Maboudi, T., & D'Amico, E. (2026). Life, liberty, and litigation: Do constitutional rights impact climate litigation outcome? *Political Research Quarterly*, *79*(1), 126–139. DOI: 10.1177/10659129251380572

Martínez Navarro, J. A. M. (2025). The right to environmental protection through the judicialization of climate change: Environmental principles and judicial activism. *Estudios de Deusto, 73*, 269–298.

Mate, M. (2011). Public interest litigation and the transformation of the Supreme Court of India.

Mayo-Anda, G. (2025). Environmental courts and tribunals in the Philippines: Opportunities and challenges in environmental litigation. In *Brill's Asian Law Series*. Brill.

Mendes, L. F., Diniz, N. J. A., & Gonzales Filho, A. G. (2025). Human rights and climate justice: The role of the Inter-American Court in recognising the right to a healthy environment. *Actualidad Juridica Ambiental, 2025*(July), 138–181.

Minchun, Z., & Bao, Z. (2012). Specialised environmental courts in China: Status quo, challenges and responses. *Journal of Energy & Natural Resources Law, 30*(4), 361–390. DOI: 10.1080/02646811.2012.11435303

Negandhi, P., Acharya, S., Khan, F., & Kumar, S. (2026). Climate change and human health: A medical perspective on emerging challenges and mitigation: A comprehensive review. *Multidisciplinary Reviews, 9.*

Popielarski, M. (2016). Climate change litigation: Regulatory pathways to cleaner energy. *Law Library Journal, 108*, 301.

Preston, B. J. (2012). Benefits of judicial specialization in environmental law: The Land and Environment Court of New South Wales as a case study. *Pace Environmental Law Review, 29*(2), 396. DOI: 10.58948/0738-6206.1692

Preston, B. J. (2014). Characteristics of successful environmental courts and tribunals. *Journal of Environmental Law, 26*(3), 365–393. DOI: 10.1093/jel/equ019

Pring, G. R., & Pring, C. K. (2009). *Greening justice: Creating and improving environmental courts and tribunals*. The Access Initiative & World Resources Institute.

Psarras, H. (2023). Private law rights and powers of waiver.

Purwendah, E. K., Awaludin, A., & Triana, I. D. S. (2022). Public participation in environmental protection: Citizen law suits in the Indonesian civil justice system. *IOP Conference Series. Earth and Environmental Science, 1030*(1), 012022. DOI: 10.1088/1755-1315/1030/1/012022

Qi, X., Wu, Z., Xu, J., & Shan, B. (2023). Environmental justice and green innovation: A quasi-natural experiment based on the establishment of environmental courts in China. *Ecological Economics, 205*, 107700. DOI: 10.1016/j.ecolecon.2022.107700

Rengarajan, S., Palaniyappan, D., Ramachandran, P., & Ramachandran, R. (2018). National Green Tribunal of India—An observation from environmental judgements. *Environmental Science and Pollution Research International*, *25*(12), 11313–11318. DOI: 10.1007/s11356-018-1763-2 PMID: 29572740

Rydberg, A. V. (2024). Climate change litigation: General perspectives and emerging trends. *International Community Law Review*, *26*(4), 347–366. DOI: 10.1163/18719732-12341504

Sayem, S. N. (2025). Ensuring a 'tomorrow': The International Environmental Court. *North South Journal of Peace and Global Studies, 2*, 10.47126.

Shafayat, S. M. (2025). Environmental tort regime under the Environment Court Act 2010 in Bangladesh: A critical review. *Australian Journal of Asian Law*, *25*, 85–101.

Sharma, Y., Bhuvaneswari, D., Shukla, D., Tripathi, S., Roy, S., & Sharma, J. (2025). The role of environmental law in addressing climate change: A global perspective. *Journal of Environmental Protection and Ecology*, *26*, 700–712.

Shelton, D., & Kiss, A. C. (2005). *Judicial handbook on environmental law*. UNEP/Earthprint.

Shen, H. (2026). Guardians of the green? A critical examination of procuratorial public interest litigation in China's environmental governance. *Journal of Environmental Policy and Planning*, 1–18. DOI: 10.1080/1523908X.2026.2625845

Smith, D. C. (2018). *Environmental courts and tribunals: Changing environmental and natural resources law around the globe*. Taylor & Francis.

Sousa, M. D. M., Ferreira, V. D. R. S., Najberg, E., & Medeiros, J. J. (2015). Portraying innovation in the public service of Brazil: Frameworks, systematization and characterization. *Revista de Administração (São Paulo)*, *50*(4), 460–476. DOI: 10.5700/rausp1213

Stern, R. E. (2014). The political logic of China's new environmental courts. *China Journal (Canberra, A.C.T.)*, *72*, 53–74. DOI: 10.1086/677051

Sulyok, K. (2024). Transforming the rule of law in environmental and climate litigation: Prohibiting the arbitrary treatment of future generations. *Transnational Environmental Law*, *13*(3), 475–501. DOI: 10.1017/S2047102524000116

Tang, Z. (2023). Smart courts in cross-border litigation. *Rabels Zeitschrift für ausländisches und internationales Privatrecht / The Rabel Journal of Comparative and International Private Law*, 118–143.

Tao, Q., Jun, A., & Ma, S. (2025). Environmental engagement: An effective way in mitigation of firms' litigation risks in China. *Accounting and Finance, 65*(3), 2589–2608. DOI: 10.1111/acfi.70007

Tigre, M. A., Murcott, M. J., & Samuel, S. A. (2025). *Climate litigation and vulnerabilities: Global South perspectives.*

Trubek, D. M., Sarat, A., Felstiner, W. L., & Kritzer, H. M. (1983). The costs of ordinary litigation. *UCLA Law Review. University of California, Los Angeles. School of Law, 31*, 72.

Vellore Citizens Welfare Forum v. Union of India, (1996). Supreme Court of India.

Wahid, A. Y., & Magassing, A. M. (2025). Establishing an environmental court in Indonesia: Addressing legal challenges for environmental justice in line with SDG 13 and SDG 16. *Journal of Lifestyle and SDGs Review, 5*, Article e03043.

Wang, L., Pan, M., Qian, X., & Lv, K. (2025). Do specialized courts matter? Environmental judiciary and corporate emissions in China. *Energy Policy, 199*, 114532. DOI: 10.1016/j.enpol.2025.114532

Yang, Y., Wang, B., & Su, X. (2024). Judicial quality and corporate ESG performance: Evidence from the establishment of circuit courts. [Replace with actual article number if available]. *Finance Research Letters, 65*, 105678. DOI: 10.1016/j.frl.2024.105610

Yanti Sulistiawati, L., Bouquelle, F., Lin, J., Lavrysen, L., Ortega, M., Pereira, R., & Tseng, S. (2022). *Environmental courts and tribunals: A guide for policymakers*. United Nations Environment Programme.

Zhang, J. (2019). A thesis on the establishment of specialized environmental and resources courts in China [Conference session]. *IOP Conference Series: Earth and Environmental Science*. IOP Publishing.

Chapter 9
Human Rights and Climate Change:
A Judicial Response From India and Maldives

Naresh Prajapati
http://orcid.org/0000-0002-0752-3651
School of Law, Bennett University, Greater Noida, India

Deepak Kumar Chauhan
http://orcid.org/0000-0003-2344-1215
School of Legal Studies, Central University of Punjab, Bathinda, India

Bindu Jindal
http://orcid.org/0009-0004-8907-0773
Maharishi Markandeshwar (Deemed to be) University, India

Preeti Jain
http://orcid.org/0009-0003-7881-8857
Kurukshetra University, India

ABSTRACT

This chapter analyses a comparative examination of legal measures in India and the Maldives, addressing the link between human rights and climate change. In the face of climate catastrophes like global warming, cloudbursts, and dry winters, it explores judicial pronouncements and legal systems that recognise climate rights as fundamental human rights in these vulnerable nations. It studies how both judiciaries interpret and enforce constitutional and international laws to shield communities from climate impacts, highlighting India's incorporation of human rights into climate law and urging the Maldives to enhance its legal framework through

DOI: 10.4018/979-8-3373-6635-7.ch009

specialised climate courts and global cooperation. The chapter also contributes to academic discourse and offers policy suggestions to strengthen judicial capacity in addressing complex climate-related challenges.

INTRODUCTION

Climate change is one of the challenges of the 21st century, not only as an environmental crisis but also a threat to human rights. It interferes with the livelihoods, endangers food and water availability, heightens displacement and is disproportionate to vulnerable groups (Prajapati & Chauhan, 2025). In the case of such nations as India and the Maldives, climate change and human rights have not only crossed paths but become daily facts. Being a low-lying island state, Maldives is at an existential risk due to the increasing sea level and shoreline erosions, which threaten the country with a loss of its territorial integrity. Although geographically and economically diverse, India is a country that is experiencing several climate effects, such as heatwaves, floods, crop failures, and water shortages. In either case, climate change contributes to inequality, the dilution of development benefits, and an inability of the legal framework of nations to serve as a source of human rights. It is on this background that there has been an increasing role of the judiciary in the two countries. Making courts interpret environmental laws is not confined anymore; they now also directly interact with constitutional rights, international duties, and climate justice considerations. In India, judicial activism has resulted in a broader understanding of the right to life of Article 21 to include the right to a clean and safe environment. Even though judicial activism in the Maldives has been relatively low, over the recent years, judicial rulings indicate a corresponding increase in regard to the necessity of legal responsibility in environmental management (Gill, 2015).

Although the global rate of climate-related litigation is on the rise, there is still a significant gap in the perception of how small island developing states (SIDS), like Maldives, react to the issues with their legal and judicial system, especially in comparison with other large and populous democracies, like India. The chapter bridges that gap by providing a comparative study of how two structurally and politically different countries tackle the combined challenge of climate change and human rights through their courts (Global Trends in Climate Change Litigation: 2024 Snapshot - Grantham Research Institute on Climate Change and the Environment, 2024).

The chapter discussed in the context of the general change in legal thinking on the international level from 2022 to 2025, during which the issue of climate change has been solidly established as a human rights threat. The historic milestones include continuing to operationalise the Loss and Damage Fund at COP27 and the UN Human Rights Council acknowledging the right to a clean, healthy, and sustainable environment in 2022. In 2025, the International Court of Justice passed a landmark advisory opinion that stated that states have binding obligations to avert and reduce harm to climate particularly when it poses a threat to vulnerable groups and future generations. The Inter-American Court of Human Rights, in 2025, made the climate crisis an explicit threat to human rights safeguarded through the American Convention, and urged states to make climate action rights-based and grounded in science. In the meantime, a coalition of African lawyers asked the African Court to make its own decision, oriented towards corporate responsibility, just transition, and climate justice. As the world keeps ceding populations to the encroachment of fossil fuel projects, the global legal community, with the support of the International Tribunal on the Law of the Sea (2024), now acknowledges that climate protection has become a fundamental human rights requirement (Phelan et al., 2025).

The chapter examines the judicial reaction of India and Maldives to climate change based on human rights. It examines the essential legal instruments, landmark cases, national policy, as well as international commitments that inform the jurisprudence of environmental law in both nations. It evaluates the role of the judiciary as the means of legal redress and moral authority so that the climate action is founded on science, policy, and justice. The chapter adds to the overall discussion of climate law and human rights by pointing out the achievements, the failures, and the legal innovations. By so doing, it highlights the essential responsibility of the existence of responsive law institutions, coherent policies and regional cooperation so that the battle against climate change becomes a battle also in favour of human dignity, equity and survival.

BRIEF OVERVIEW OF THE JUDICIAL SYSTEMS IN INDIA AND MALDIVES

The Indian judicial system, as a result of its colonial past, adapts the hierarchical system modelled to coordinate the legal issues at the grassroot level up to the national level. At the top of hierarchy is the Supreme Court, which was inaugurated on January 28, 1950 and is the last court of appeal as well as the protector of the Constitution. Under it, there are High Courts which work in 25 states on civil and criminal appeals. On the local level, district courts operate under state control in order to process ordinary litigation (Singh and Kumar, 2019).

India also responded to the increasing environmental pressures by the formation of the National Green Tribunal (NGT) in 2010. This is a specialised organ that deals specifically with environmental disputes that include the issues of forest protection, pollution and biodiversity. The NGT is central to the enforcement of environmental rights and principles like sustainable development and precautionary approach. It can substitute the conventional judicial system by providing a faster hearing process and specialised adjudication and reinforced access to environmental justice (Gill, 2015).

Maldives, on the contrary, has a three-level judicial system, and this includes lower courts, the High Court, and a special appellate system that includes the President. Courts under the Ministry of Justice are the general courts in inhabited islands and the four specialised courts, namely Criminal, Civil, Family, and Juvenile, located in the capital of Malé (Jeffery, 2024). The Chief Justice headed the High Court, which appeals and also hears particular first-instance cases. The lower courts may automatically appeal to strengthen the usage of procedures to access justice (Shafie and Aziz, 2019).

Nevertheless, Maldives faces singificant weaknesses in terms of climate change and the application of human rights in judicial affairs. Interestingly, it is essential to noting that the lack of specialised environmental court or tribunal restricts the effectiveness of the system to handle complex and science-based disputes relating to climate. And the last appeal system sits on written submissions and consults the President, not by an open hearing, but by administrative review, by a Judicial Committee in the office of the President. This arrangement has elicited incessant questions on institutional independence, especially since, until recently, the Attorney General, who is also the chief prosecutor in the state, was a member of this committee (Nanwani & Ayus, 2012). Such confusion of executive control and judicial adjudication may degrade the separation of powers, posing danger in terms of rightful and objective management of rights-based claims over the environment. With the escalating climate crisis, these structural vices highlight why there is a dire need to reform the judiciary to boost its autonomy and to improve institutional capacity to deliver climate justice within Maldives.

LITERATURE REVIEW

This section discusses the implications of climate change that are complex and the changing role of judicial intervention in protecting human rights, especially for the vulnerable island states and the developing countries. Luetz (2017) explains how climate change can cause mass migration in the small island countries, including Maldives, and requests that proactive policy frames be implemented to guarantee fair results. Jolly & Trivedi (2025) discuss climate adaptation as a global public good, but admit the restriction of international cooperation in providing effective responses. Ambiguity Derman (2020) examines the connection between climate change, human rights and legal mobilisation using the examples of advocacy of the Inuit Circumpolar Council and Maldives. In the same manner, Mitra and Sanghi

(2023) emphasise the ethical and legal concerns of artificial islands, including Hulhumale, in the context of human mobility and climate risk.

Within the Indian context, Sengupta (2015) points out that the judiciary actively contributes to dealing with climate problems, calling for more active international cooperation and technology development. Following the development of climate litigation all over the globe, Preston (2018) emphasises how environmental rights are gaining prominence in the new legal traditions. Setzer and Benjamin (2019) place climate litigation in the Global South as a means of getting beyond constraints of the larger climate justice movement. The article by Pattajoshi (2019) offers a narrow case study about Indian climate litigation and the changing role of the National Green Tribunal. Chaturvedi (2021) emphasises the role of the judiciary in ensuring international environmental standards. Ram-Mohan et al. (2022) also put Indian litigation practices into context through a comparison of global litigation practices to establish their overall influence.

Fernando (2018) analyses the justiciability of climate issues in Canada and the United States and reports some significant differences in legal strategies used in the two countries. Pozzo (2021) provides a comparative legal perspective on climate change litigation, with the focus on its global nature. Busch and Judick (2021) also report on changing narratives within climate-sceptic think tanks both in the United States and Germany, which is an indication of a larger shift in the discourse. McDave and Dagadu (2023) assess the response given by international human rights institutions towards climate misfortunes of vulnerable populations. Nesje et al. (2023) examine the ethical values of climate policy concerning intergenerational fairness. Limon (2010) places emphasis on the human rights approach to climate governance, especially by the Malé Declaration, which is a critical document by small and vulnerable island states. Olsen et al. (2013) suggest that physical and social sciences need to collaborate more on the fight against climate change. The structural adaptation strategies suggested by Brown et al. (2023) include the elevation of land and the right to territory advantages of the island nations.

Recent developments enrich this. Jolly and Trivedi (2025) discuss the Indian Supreme Court's discretion to acknowledge the right to part with climate harm, whereas Rajwanshi and Sanni (2024) discuss Maldives advocating legal obligations of worldwide collaboration and fair energy changes in the ICJ Climate Advisory proceedings. Reports by UNEP (2025) indicate that there are more than 3,000 cases of climate lawsuits in 55 jurisdictions, and the increasing trend of rights-based climate litigation by marginalised populations. These works combined emphasise the rising role played by courts and other instruments of the law in regulating climate responsibility and advancing justice internationally.

This evaluation gives handy information on the complicated nature of the issue of climate change. It also points to the urgency of active policy development, law enforcement, global integration, legislation, and the variety of research methods in the context of managing the impact of climate change, particularly on the populations of vulnerable islands.

RESEARCH METHODOLOGY

The chapter utilises the comparative analysis approach to study the way in which human rights protection by the judiciary of India and Maldives is tackled in respect to climate change. The sources of data collection are the major court cases and relevant texts of legislation. The text of analysis has included remarkable work by Indian legislation, passing specific legislation and policies to protect the environment, taking into consideration climate change issues, and making judicial contributions. The work in Maldives is being analysed in the same manner. Thus, the issue of historical views has been an issue in this case as well.

LAW AND POLICY FRAMEWORK IN INDIA AND MALDIVES

The legislation in India provides a holistic and stratified system of environmental protection. A cornerstone is the Environmental Protection Act, 1986, which provides central and state governments with an opportunity to promote the quality of the environment and stop pollution (Kaur et al., 2022). In addition to this, the Air (Prevention and Control of Pollution) Act, 1981, and the Water Act, 1977, provide a technical set of standards and authorise pollution control boards to carry out inspection and take action against polluting agents (Deshpande et al., 2023). The National Green Tribunal Act, 2010, in the meantime, established a special tribunal to hasten the environmental cases, with binding jurisdiction and a very high record of coming up with landmark rulings on environmental control and justice (Gupta et al., 2025).

The Energy Conservation Act, 2001 also contributes to the move towards efficiency in India by controlling energy consumption in industries. Legislative measures to preserve biodiversity and control the utilisation of natural resources are the Forest Conservation Act, 1980 and the Wildlife Protection Act, 1972. Remarkably, the Compensatory Afforestation Fund Act (2016) compels the afforestation in order to offset any loss of forest due to development projects. These are supplemented by Biological Diversity Act, 2002, which ensures fair mechanisms of benefit sharing. Another, more recent policy proposal is the Green Hydrogen/Green Ammonia Policy (2022) that seeks to decarbonise the industry and transport using clean energy manufacturing.

The mitigation of climate is also informed in wider policies like the National Action Plan on Climate Change (2008) which is an action plan that formulates climate governance by creating thematic missions in the field of energy, water, urban development and agriculture. The coordinated responses to climate-linked disasters are provided with the help of legal requirements of the Disaster Management Act, 2005. Article 21 (right to life) of the Constitution, and the Protection of Human Rights Act (1993) work together to create a strong legal foundation where the judiciary can perceive environmental protection as a fundamental right.

Climate change has been known to pose an existential threat in Maldives, and its legal system is both urgent and vulnerable. One of the key legal frameworks is the Climate Emergency Act, 2021. It proposes mandatory sustainability reporting, institutionalises the Climate Change Directorate and establishes the Special Envoy for Climate Change (*The President's Office*, 2021). Notably, it made a bold national target of attaining net-zero carbon emissions by the year 2030. Although this is aspirational, the bindability of this target is of concern because there is no specific climate court or more legally binding carbon budgeting frameworks. By comparison, other Pacific Island countries, including Fiji, have more formal accountability mechanisms by having a Climate Change Act, 2021, which encompasses independent accountability bodies and binding adaptation planning. The existence of this gap stresses the fact that Maldives needs to establish better institutional checks to achieve the goals of its climate law.

On an institutional level, enforcement is mainly funnelled to the Ministry of Environment, Climate Change and Technology (MoECCT), the Environmental Protection Agency (EPA) and the National Disaster Management Authority (NDMA). Although these are the bodies to implement the policies, enforcement is not even. Despite the fact that EPA mandates it to conduct environmental impact assessment and control compliance, it experiences high capacity and resource-related limitations. On the same note, NDMA is given the responsibility of disaster preparedness and response. Nonetheless, it does not have a rapid legal process of responding to disasters caused by climatic conditions, and this restricts its capacity to respond to an emergency.

The Environmental Protection and Preservation Act, 2008, gives extensive authority to halt any harmful activity, which has been rather reactive. The Environmental Levy (Budget Amendment) Act, 2017 and the associated Environment and Climate Adaptation Fund are a substantial move in the area of domestic climate funding. Nevertheless, fund distribution and impact are not as transparent (Malatesta and Schmidt di Friedberg, 2017).

Maldives climate change policy framework, 2015 provides six strategic pillars, which are intergenerational equity, resilience, and international cooperation (Maldives, 2024). Although logically correct, the structure largely depends on foreign aid and fails to have cumbersome liabilities and legal redress to breach of the agreement. This framework is strengthened by the 2008 Constitution, which entrenched environmental protection and human rights in the national law. Article 22 requires the state to take action to conserve the environment, and Article 23(d) considers a healthy and balanced environment as the fundamental right of the citizens. Human Rights Commission Act, 2006 supports these constitutional guarantees because it allows the Commission to eliminate regulatory barriers, offer legal advice, and ensure compliance with international standards (Falaah, 2016).

Even though there is no specialised environmental court, Maldives has demonstrated its commitment through climate diplomacy, especially in the International Court of Justice. Nevertheless, lacking a judicial system to challenge the enforceability of its net-zero commitments or promptly adjudicate environmental complaints, a lot of its regulation is still theoretical. The intensification of the legal recourse, the institutional capacity development, and the introduction of the enforcement schemes of the peer island countries would help to ensure a substantial increase in the climate governance results.

NATIONAL CLIMATE ACTION PLAN

The National Action Plan on Climate Change (NAPCC), 2008, in India, provides plans for combating climate change while guaranteeing long-term development. Eight national missions handle different climate challenges. The National Solar Mission aims to increase solar energy utilisation and make India a leader in solar power. The National Mission for Enhanced Energy Efficiency enhances energy efficiency through market-based mechanisms in various economic sectors (Pandve, 2009). The objectives of the National Mission for Sustainable Habitat include energy-efficient urban design, the use of sustainable building materials, and the development of public transportation. The National Water Mission aims to preserve, manage, and enhance water resource utilisation. The National Mission for Sustaining Himalayan Ecology researches and implements sustainable approaches to enhance the fragile ecosystem of the Himalayas. The National Mission for a Green India aims to increase and restore forest cover while promoting sustainable forest management. The National Mission for Sustainable Agriculture promotes sustainable agriculture and climate change resiliency. The National Mission on Strategic Knowledge for Climate Change fosters an understanding of climate change, its impacts, and potential solutions. The NAPCC demonstrates India's commitment to reducing climate change, promoting sustainable development, and protecting the environment (Sebopela & Odeku, 2014). In order to make the framework more human rights relevant, every mission has a varying relationship with the idea of rights: an example is the National Water Mission, which directly contributes to the right to access water. Meanwhile, participation in urban planning is encouraged by the Sustainable Habitat Mission. Nevertheless, other rights like equality and accessibility of the marginalised communities may be incorporated further in all missions.

The Strategic National Action Plan on Disaster Risk Reduction and Climate Change Adaptation (SNAP) 2010-2020 of Maldives is designed to deal with the issue of climate change and build disaster resilience. Maldives' low elevation puts it at risk of sea-level rise, shoreline erosion, and extreme weather events. This plan aims to mitigate catastrophe risk and support the country's adaptation to climate change. It supports risk and vulnerability assessments for coastal areas, water resources, infrastructure, tourism, and fisheries (*Disaster et al. in the Republic of Maldives*, 2019). Knowing these threats enables Maldives to safeguard its most vulnerable areas. This includes water sustainability, climate-resilient agriculture, and protection of critical infrastructure. Coral reefs are essential for tourism and fishing; thus, they must be conserved.

SNAP is based on governance and institutional transformations. It necessitates strengthening institutions and governance to reduce catastrophe risk and adapt to climate change. Strengthening government collaboration and incorporating risk mitigation into national and local development plans (Wen et al., 2023). Good governance ensures that such projects are resilient and prosperous. SNAP is built on the foundation of increasing capacity and awareness. A plan teaches government officials, community leaders, and the public about climate risks and adaptation. The strategy encourages climate resilience activities. The SNAP seeks global finance. This will solve significant financial and technical challenges. This strategy adheres to global standards such as the Hyogo Framework for Action. Maldives seeks help from foreign countries and climate funding. This promotes successful worldwide strategic cooperation. The SNAP will allow Maldivians to build positive and holistic resilience. It will help narrow the divide between disaster risk in the short term and climate change impacts in the long term, thus saving Maldivian livelihoods and ensuring sustainable development (Maldives: Strategic National Action Plan (SNAP) to disaster risk reduction and climate change adaptation 2010-2020, 2018).

Nevertheless, due to the expiration of SNAP in 2020, where there appears to be no follow-up strategy, there has been an institutional continuity consideration issue. The adoption of the Maldives is very much dependent on the international climate finance, especially on the Green Climate Fund and bilateral financing, and this creates the question of long-term financial independence and ownership of the adaptation. Nevertheless, Maldives is still in need of assistance from international means to maintain its national adaptation agenda.

ROLE OF THE JUDICIARY TOWARDS CLIMATE CHANGE AND HUMAN RIGHTS PROTECTION

The April 5, 2024, in *M.K. Ranjitsinh v. Union of India* (2024 INSC 280), Supreme Court of India, the recognition of the right to the unfavourable effects of climate change, is the turning point in Indian jurisprudence. Issued by a three-judge bench presided over by Chief Justice DY Chandrachud when a case was brought up on the conservation of the Great Indian Bustard and the Lesser Florican, this ruling was made under the umbrella notion of the right to life and equality as ensured under the Indian Constitution (Tripathi, 2024). A 2021 ruling, which presented a blanket prohibition of overhead power lines over 99,000 square kilometres across Gujarat and Rajasthan, was in the background of this case so as to protect these endangered birds. Nevertheless, the court noted that such a limitation of the power lines that would only require underground cables would provide a substantial obstacle to the potential of the region in exploiting wind and solar energy, thus undermining the efforts of India to achieve its goals of clean energy development and taking action in mitigating global climate change. This was the finishing line of the long-term environmental jurisprudence history of the court, as outlined in Article 21. However, it provided a more precise focus by specifically addressing climate change as a human rights issue on its own. This prevented the ambiguity of doctrines and gave lower courts and enforcers of the law a more grounded legal norm.

Traditionally, the Indian judicial system has been progressively acknowledging the innate bond between the environment and the right to life (Khanna and Law, 2024). Such landmark cases include *M.C. Mehta v. Kamal Nath & Ors* (AIRONLINE 1996 SC 711), and *Virendra Gaur and Ors v. State of Haryana and Ors* (1994 SUPPL. (6) SCR 78), the state of Haryana made it a constitutional obligation for the state to ensure environmental equilibrium and a clean environment as a part of the right to life. Karnataka Industrial Areas Development Board v. C. Kenchappa and Ors (AIR 2006 Supreme Court 2038) also mentioned the serious dangers of the increase in sea level and global temperature, and the first signs of the growing number of judicial decisions which recognise the problem of climate change. Even though, starting in 2018, there is a significant number of rulings that uphold the right to a clean environment, the right to the absence of the negative impact of climate change was first clearly expressed in 2024 by the Supreme Court (Preston, 2018). This understanding recognises the interdependence of a sustainable and clean environment with human rights, which are the most basic ones, such as health and diseases that are influenced in a negative way by factors like air pollution, climate-related diseases, and lack of food and water. These problems are concentrated among disadvantaged communities, which underscores the overlap of the social equity with environmental justice: the fact that communities in the Lakshadweep Islands are particularly susceptible to climate change, as compared to those in the inland areas, such as in Madhya Pradesh, points out the general implications of climate change on the right to equality. Introduced minority and tribal groups tend to be affected severely when their economic and cultural foundations are closely linked to the environmental conditions (Mitkidis and Valkanou, 2020). Nonetheless, instead of repeating the scope of Article 21, the rulings of the Supreme Court, including the *State of Telangana* v. *Mohd. Abdul Qasim (Died) Per Lrs.* (2024 INSC 310), highlight the judiciary's adherence to environmental justice as it acknowledges the constitutional imperative to safeguard against climate change under Articles 14 and 21. In these cases, the idea of striking a balance between environmental conservation and sustainable development is highlighted, usually by referring to India's commitments under international agreements, including the Paris Agreement, to support domestic environmental governance. The effects of such judicial rulings are far-reaching and ensure that environmental rights are not just things on paper but the norms that must be followed in the developmental path the nation takes. *Ridhima Pandey v. Union of India* (Civil Appeal No(s). 388/2021), climate change remains a severe challenge; judicial activism is an essential element of governance, which protects the environment and supports fair and sustainable development (Ghosh, 2020). This changing legal environment underscores the active participation of the judiciary in environmental regulation, whereby a legal framework should constantly evolve in response to the dynamic nature of the climate change challenges (Dash, 2021). It highlights the importance of the judiciary in striking a balance between development and the ecological sustainability of India so that development should not be at the cost of the environmental responsibilities and health and welfare of the citizens. Conclusively, incorporating climate change concerns in the constitutional rights of the Supreme Court of India adds more weight to the state with its environmental justice policy and is in line with the global agenda (Chaturvedi, 2021). Some of these decisions have had direct policy and legislative implications. As a result of the judicial nudges, the EIA Notification amendments were made, and the renewable purchase obligations (RPOs) by the Ministry of Power indicated that the judiciary was influencing the courtroom.

The current legal wrangles in Maldives can serve as an example of how hard it can be to find a balance between economic growth and environmental preservation as a small island nation. As an example of this, in 2015, the Maldives Environmental Protection Association (MEPA) was able to appeal against the action of the Ministry of Fisheries and Agriculture to allow the use of some pesticides in the fishing sector, claiming that the chemicals were highly hazardous to marine organisms. The Maldives High Court decided in favour of MEPA, and the usage of these pesticides was halted, which established a decisive precedent for the environmental assessment procedures in the country. Another instance, in 2014, the Environment Protection

Authority (EPA) of the Maldives sued the State Trading Organisation (STO) over a major development project on a populated island claiming that the EPA had not fully consulted the community and that the development would have a catastrophic effect on the environment. This case resulted in the Maldives Supreme Court ruling to suspend the permit and supports the importance of social consultation and strict environmental control in development projects (Techera & Cannell-Lunn, 2019). Another recent Supreme Court ruling is a legal challenge brought by environmental activist Humaida A. Ghafoor against the government to halt the Gulhifalhu Port Development project, which is a major land reclamation project pushed by former President Abdulla Yameen. Although an Environmental Impact Assessment (EIA) had raised several environmental concerns, the project was hurriedly implemented, thus attracting a lot of criticism on the grounds of non-observation of the proper procedures and governance and accountability challenges about the environmental concerns. The High Court gave its initial verdict in favor of the environmental issues raised by Ghafoor, when the state went to court, however, on May 15 2024, the Supreme Court reversed the High Court injunction directing the suspension of the K. Gulhifalhu reclamation project (Mohamed, 2024). Such instances highlight the bigger dilemma of economic ambitions and environmental sustainability that Maldives, as most of the island countries, must face. The country is highly susceptible to environmental degradation since the country depends on tourism and fishing, and the economic stability of the nation and the health of the population may be jeopardised in the long term (Gossman, 2023). The Gulhifalhu project is an example of the traps of focusing more on the quick economic development rather than environmental protection, and it should be used as a reminder of the environmental and social consequences that such practices may have. These court cases provided critical precedents in further development projects in Maldives and the like with the consideration of a balanced approach in development that provides rigorous environmental assessment, a true involvement of the people and strict environmental standards. These actions determine legal and regulatory environment of the Maldives and they are also a part of worldwide debates on sustainable development policies, especially in eco-sensitive areas. The Maldives experience shows the great importance of judicial systems in enforcing environmental laws and the significance of the common people in the environmental governance and provides a very important lesson to the world in terms of implementing the environmental factor in economic planning and development (Supreme Court Overrules Injunction Ordering Suspension of Gulhifalhu Reclamation, 2024). Other key decisions include: *EPA v. Ministry of Housing and Infrastructure* (2018), in which the court stopped the construction of the housing project without the impact studies of the coral reefs. *Rashfa v. City Council of Malé* (2016), judicial decision that maintained the entitlement of the citizens to take part in the policy-making regarding waste management, City Council of Malé (2016) has made the municipal authority reconsider its consultation model. Such cases have given concrete results. After judicial interventions, EIA regulation at EPA was revised in 2020, and Disaster Management Authority introduced new transparency requirements on emergency protocols. Capacity-building workshops after 2018 were also introduced by MoECCT to bring the ministries to the standards of environmental compliance.

The juxtaposition of the environmental law of jurisprudence and legal systems of both India and the Maldives can also provide some thought-provoking knowledge on how law and the judiciary have developed over time and contributed to the ecological issues being addressed in the two nations. The history of environmental law in India and in Maldives, since 1950 and 1965 respectively, represents the importance of legal systems in developing policies of sustainable development (Divan and Rosencranz, 2002).

The environmental legal framework in India has grown considerably throughout the decades as the country has shifted its focus towards having a few specific environmental regulations to a full-fledged statutory and policy regime (Leelakrishnan, 2008). The Indian judicial system, especially the Supreme Court, has been a key participant in this development. The M.C. Mehta cases have led to landmark cases that broaden the meaning of Article 21 of the Indian Constitution in declaring that the right to a healthy and clean environment is a part of the right to life (Singh, 2007). This judicial activism has brought the formation of the National Green Tribunal in 2010, which is a specialised court, which has been directly appointed to hear only environmental cases, making the process easier and environmental legislation more enforced. The courts have been instrumental in adopting the idea of applying the Public Trust Doctrine and the Polluter Pays Principle into the Indian environmental law, and by so doing, environmental consideration has been woven into the legal ruling (Puthucherril, 2010). The proactive steps taken by the judiciary have paved the way to govern the environment and it has become mandatory that the government and industries should prioritise environmental health in their operations and projects (Gadgil, 2014).

The subject of environmental law has been given an enhanced significance in Maldives owing to a particular vulnerability of the nation to climate change, especially the rise in the sea level and maintenance of the marine ecosystems in the country (Haseena & Saleem, 2017). In the cases, e.g. *Maldives Environmental Protection Association v. The Ministry of Fisheries and Agriculture* (2015), emphasise the role of the judiciary in the protection of these delicate balances within the environment (Mustafa, 2016). The case created a prohibition on the use of dangerous pesticides that posed a threat to marine organisms in recognition of the important relationship between the health of the environment and the welfare of the country on the bench. On the same note, in *Environment Protection Authority v. State Trading Organisa-*

tion (2014), he role of the judiciary as a guarantor of environmental integrity was highlighted by the intervention by the in the Maldives Supreme Court to prevent a development project because of poor community consultation and environmental assessment. This is indicative of a wider trend to grow more environmentally responsible and sustainable planning (Naseer, 2018).

The judiciary in India and Maldives has been found to play transformative roles in environmental protection. The magnitude and the variety of the environmental issues in India have resulted in a vibrant development of environmental jurisprudence. By contrast, the smaller size but high susceptibility of Maldives to environmental risks has made its legal activities more limited to issues, including marine protection and climate change resiliency (Jairaj & Shivakumar, 2019). The activist approach of the two courts underlines the universal necessity to have a strong, vigilant, and ecologically-sensitive legal system capable of responding to the challenges of the local and global ecological environment. Although India offers an example of a proactive environmental jurisprudence that covers numerous problems, such as air pollution, to conservation, Maldives exemplifies the need for specialised focus areas, especially in the law of marine and climate change-related laws, which are critical to small island countries (Ali and Hameed, 2016). To sum up, both countries are taking positive steps through judicial activism and legal reforms in the context of the imperativeness of the law and the judiciary in managing and anticipating environmental issues. Such a comparative approach emphasises the importance of legal framework in ensuring the environmental regulation and fostering a balance between development and environmental protection (Jayakumar, 2017).

LEGAL ENFORCEMENT

Various laws and regulations in India impose environmental regulations and pollution control. The central tools are the Central Pollution Control Board (CPCB) and State Pollution Control Boards (SPCBs) that act as an advisory organ to the Central Government and organise the work of SPCBs. CPCB assists the central government, and SPCBs implement laws covering pollution control, industrial permits, the monitoring of pollution and decisive measures against the violators (CPCB Central Pollution Control Board, 1974). The National Green Tribunal (NGT), which was formed in 2010 with the purpose of solving environmental conflicts, ensures that the conflicts are disposed of swiftly and rapidly, and forests and other natural resources are safeguarded. The NGT is a strong enforcement instrument since it can put specific direct remedial measures on environmental laws, which it interprets. Public Interest Litigation (PIL) is a very important legal process through which a person, a group of individuals or a non-governmental organisation (NGO) can appeal to a court to seek relief or protect a public interest, especially in environmental cases. It has also led to several court interventions, which have culminated into the effective execution of the environmental laws (Vasudevan, 2021). The Ministry of Environment, Forest and Climate Change (MoEFCC) formulate policies and programmes and monitors the environmental programmes. Environmental litigation or claims have been created through the formation of specialised courts and tribunals. Additionally, the civil society and non-governmental organisations proactively aid in environmental monitoring and PILs, as well as awareness production, creating a participatory enforcement model that is specific to India. This ecosystem will encourage checks and balances, and the public will be involved in environmental governance (Climate Change, Official Website of Ministry of Environment, Forest and Climate Change, Government of India, 2020).

Environmental protection in Maldives is more of a policy-based activity that has minimal judicial and civil society involvement. The major organisation engaged in the development and implementation of the environmental policy of the country is the Ministry of Environment, Climate Change and Technology (MoECCT). The Environmental Protection Agency (EPA) is tasked with enforcing the laws, preparing EIAs, monitoring and enforcing the regulations, and taking harsh measures against the violators (Bamboo for Climate Mitigation: A Giant Leap Towards Carbon Neutrality in Maldives 2024). The National Environment Action Plan sets the strategic guideline on how the environmental issues and concerns of the country are to be handled, such as policy formulation, EIAs, enforcement of the policy, and punitive measures according to the disputes. The major environmental guardians are the government, policies, and laws (Environmental Protection Agency, 2023).

The EPA in the Maldives experiences a lot of difficulties in enforcement because of the limited number of staff, technical capacity, and dependence on executive action which leads to poor regulation. In contrast to India, Maldives do not have a PIL system, and civil society involvement in the enforcement of the law is still low, which lowers the level of accountability. The development of an autonomous environmental court or expedited court can help improve dispute resolution. Environmental defender and whistleblower laws are also important as they require a person to take risks during large-scale development projects. Civic participation can be promoted by enhancing the protection of the law, access to counsel and anti-intimidation protections. These include institutional reforms, increasing funding, technical capacity-building, especially as environmental conflicts related to tourism and infrastructure increase. Though India has a strong legal system, a strong judiciary, and an active civil society, Maldives must work on institutional and participatory gaps to implement its environmental policies and promote the meaningful protection of the environment. The implementation of stricter enforcement and engagement of the population contributes to the attainment of environmental justice (Transparency Maldives, 2025).

INNOVATIVE LEGAL APPROACHES

Several important judgments in India have highlighted the right to a healthy environment. In a novel ruling, the Supreme Court in *MK Ranjitsinh and Ors. v. Union of India and Ors.*, recognised the fundamental right, right to life, must be free from adverse effects of climate change (Zaidi, 2024). Previously, in *Subhash Kumar v. State of Bihar* (A.I.R. 1991 S.C.), the Supreme Court ruled that the right to life under Article 21 includes the right to pollution-free water and air for the full enjoyment of life. *M.C. Mehta v. Union of India* (1987 AIR) is a landmark decision that addressed the oleum gas leak and established the principle of absolute responsibility for companies engaged in hazardous operations. A clean environment was recognised as a fundamental right. In *Indian Council for Enviro-Legal Action v. Union of India* (1996 AIR 1446), the Supreme Court ruled that companies must compensate for environmental harm caused by them, thereby reinforcing the "polluter pays" principle (Gill, 2015).

Maldives' Constitution guarantees everyone the right to a healthy environment. Article 21 of the Maldivian Constitution expressly recognises the right to life, which has led to court interpretations of environmental rights. This is consistent with worldwide trends in which courts define the right to life as including environmental rights (Maldives' Constitution, 2008).

TECHNOLOGICAL INNOVATIONS

India and Maldives have launched numerous schemes and technological innovations to combat climate change and protect their citizens from its impacts, including renewable energy and smart city developments, which contribute to sustainability and resilience initiatives. Bifacial solar panels and perovskite solar cells have enhanced photovoltaic (PV) cell efficiency and reduced costs, making solar energy more affordable and effective (NREL, 2022). Thermal storage advancements have enabled solar thermal energy to offer continuous power (IRENA, 2021). Larger, more efficient offshore wind farms collect stronger winds (Global Wind Energy Council, 2022) while floating wind turbines expand wind energy generating areas (IEA, 2022). Lithium-ion and solid-state batteries have increased energy density, cost, and safety (DOE, 2021). Flow batteries can store a large quantity of energy for an extended period (Sandia National Laboratories, 2022), whilst hydrogen storage can produce clean fuel from excess renewable energy (Hydrogen Council, 2021). Smart grid technologies use digital technology to monitor, regulate, and optimise electricity production and delivery. Demand response systems minimise peak power plant use by regulating energy demand (EPRI, 2021). India is using smart cities to modernise its urban infrastructure and governance. Energy-efficient building designs, electric buses, and metro rail projects all help to enhance public transit (BEE, 2022; MoHUA, 2021). IoT sensors enhance digital connectivity, enabling more efficient resource management and streamlined online government services. Microgrids and solar rooftop projects (MNRE, 2021) facilitate the integration of renewable energy. Technological solutions track water availability and quality (Jal Shakti Ministry, 2021) and convert waste into energy. Promoting electric vehicles and improved traffic management systems (MoRTH, 2021) enhances urban mobility. Community planning and smart city apps increase public engagement. Innovations enhance Maldives' infrastructure against sea-level rise. Rising buildings and floating architecture are being created (UNDP, 2021). Coastal preservation measures include coastal barriers and mangrove restoration (ADB, 2021; WWF, 2021). Artificial islands and land reclamation operations provide higher-elevation housing (World Bank, 2021), while geoengineering provides resilient land masses (MIT Technology Review). Sustainable urban design incorporates green spaces and smart technologies (ICLEI, 2021). Renewable energy and efficiency programmes include energy-efficient building designs and investments in solar and wind energy. WHO (2021) and UNDP (2021) suggest desalination and rainwater gathering. Community involvement and education are enhanced with the help of awareness campaigns and participatory planning (UNICEF, 2021; ADB, 2021). The use of advanced forecasting and early warning helps to speed up evacuations and pre-preparations (NOAA, 2021; UNDRR, 2019).

In spite of these developments, in legal and institutional formations, technology is usually ahead of the curves. Regulatory loopholes concerning large scale renewables in the national grid remain in Maldives. The lack of clear procedures of permitting, the existence of obsolescent land use rules and regulations concerning renewable infrastructure and poor coordination among agencies thwart broader adoption. The Energy Act is also obsolete in responding to more decentralised generation, e.g. solar rooftops or microgrids. Despite being more advanced, India experiences bottlenecks in environmental clearances, land acquisition and obsolete building codes that slow down adoption of clean technologies. Reforms to eliminate these legal barriers (fast-tracked) would allow solutions to be more effectively deployed and scaled up (Akpobome, 2024).

Technological solutions are not only environmental, but they are also directly at the crossroads of human rights. The fundamental rights to economic and social rights are fulfilled by access to clean energy, safe water, and climate-resilient housing. Desalination, rainwater harvesting, and solar electrification in remote or underserved communities can have a serious positive impact on public health, gender equity, and education in both India and Maldives. Similarly, early warning systems and climate-resilient infrastructures safeguard the rights to life, homes, and livelihoods, especially of the coastal and vulnerable communities. With the right-based frameworks, these innovations can assure a situation of inclusive and equitable climate adaptation (Al-Emran and Griffy-Brown, 2023).

CONCLUSION

India and Maldives have done a commendable job in dealing with climate change using the legal, judicial and policy approaches. The judiciary system and, in more detail, the Supreme Court in India has been of central concern as it identified the right to be spared the negative impacts of climate change in Article 21 of the Constitution. This has broadened the concept of environmental protection as an inherent part of right to life and social justice, especially to the vulnerable groups. Simultaneously, Maldives has shown a legislative initiative with the Climate Emergency Act of 2021 that pledges the nation to become a net-zero carbon emitter by 2030. Courts intervening to stop the operations that are damaging to the environment have only increased this legislative intention and the significance of strict environmental analysis.

Although the two nations appreciate the relationship between climate change and human rights, a comparative study indicates that they have significant gaps in their comprehension. The institutional position of India, such as special environmental tribunals, public interest litigation and thriving civil society, supports this, but judicial knowledge of climate science is patchy. Maldives, on the other hand, does not have an environmental court; enforcement is limited by the institutional capacity and the participation of the community in the legal proceedings. It is essential to strengthen such areas.

India needs to follow up by aligning its climate-related missions with the Sustainable Development Goals and investing in the development of judicial capacity on climate matters to move forward. Maldives ought to create an environmental tribunal, reform its abandoned SNAP system, and create official channels of community law access. These reforms combined will be able to favour an inclusive climate governance that upholds both ecosystem and human dignity.

REFERENCES

Akpobome, O. (2024). The Impact of Emerging Technologies on Legal Frameworks: A Model for Adaptive Regulation. *International Journal of Research Publication and Reviews*, *5*(10), 5046–5060. DOI: 10.55248/gengpi.5.1024.3012

Al-Emran, M., & Griffy-Brown, C. (2023). The role of technology adoption in sustainable development: Overview, opportunities, challenges, and future research agendas. *Technology in Society*, *73*, 102240. DOI: 10.1016/j.techsoc.2023.102240

Ali, S. A., & Hameed, S. (2016). Marine conservation and environmental law in the Maldives. *Ocean and Coastal Management*, *119*, 64–72.

Asian Development Bank (ADB). (2021). *Coastal protection and climate resilience*.

Bamboo for Climate Mitigation: A Giant Leap Towards Carbon Neutrality in Maldives. (2024). The Maldives Journal. https://themaldivesjournal.com/50903

Bureau of Energy Efficiency (BEE). (2022). *Green building initiatives*.

Chaturvedi, E. (2021, December). Climate Change Litigation: Indian Perspective. *German Law Journal*, *22*(8), 1459–1470. DOI: 10.1017/glj.2021.85

CPCB | Central Pollution Control Board. (1974). CPCB. https://cpcb.nic.in/spcbs-pccs/

Dash, P. C.. (2021, April 3). Role of Judiciary for the Protection of Environment in India. *EPRA International Journal of Multidisciplinary Research (IJMR)*, 9–16. DOI: 10.36713/epra6627

Deshpande, T., Mukherji, R., & Sastry, M. L. (2023). Policy styles and India's national action plan on climate change (NAPCC). *Policy Studies*, •••, 1–18. DOI: 10.1080/01442872.2023.2261388

Disaster risk reduction in Republic of Maldives. (2019, October 17). UNDRR. https://www.undrr.org/publication/disaster-risk-reduction-republic-maldives

Divan, S., & Rosencranz, A. (2002). *Environmental law and policy in India: Cases, materials, and statutes*. Oxford University Press.

Electric Power Research Institute (EPRI). (2021). *Demand response systems*.

Environmental Levy (Budget Amendment) Act 2017 (No. 36 of 2017) - Climate Change Laws of the World. (2017). https://climate-laws.org/document/environmental-levy-budget-amendment-act-2017-no-36-of-2017_17b0

Environmental Protection Agency. (2023, October 11). EPA. https://en.epa.gov.mv/

Falaah, S. (2016, January 1). Islamic Shari'ah and the Rights of Women: The Maldives' Thirst for International Human Rights Norms of Equality and Non-Discrimination. *Muslim World Journal of Human Rights, 13*(1), 101–130. Advance online publication. DOI: 10.1515/mwjhr-2016-0009

Gadgil, M. (2014). *Ecology is for the people: A methodology manual for people's biodiversity register*. National Biodiversity Authority.

Ghosh, S. (2020, January 1). *Litigating Climate Claims in India*. AJIL Unbound. DOI: 10.1017/aju.2020.5

Gill, G. N. (2015, July 31). *Human rights and environment protection in India: the judicial journey from public interest litigation to the National Green Tribunal*. Edward Elgar Publishing eBooks. DOI: 10.4337/9781784711337.00013

Global trends in climate change litigation: 2024 snapshot - Grantham Research Institute on climate change and the environment. (2024, September 2). Grantham Research Institute on Climate Change and the Environment.

Global Wind Energy Council. (2022). *Offshore wind energy*.

Gossman, P. (2023, October 27). *"We Still Haven't Recovered."* Human Rights Watch. https://www.hrw.org/report/2023/10/18/we-still-havent-recovered/local-communities-harmed-reclamation-projects-maldives

Gupta, S., Kumar, R., Shrivastava, R., Prajapati, N., Balasubramani, R., & Chauhan, D. K. (2025). Citation dynamics and their relationship with social attention score: A case of climate change law. *Global Knowledge, Memory and Communication*. Advance online publication. DOI: 10.1108/GKMC-09-2024-0614

Haseena, M., & Saleem, A. (2017). Environmental challenges in Maldives: A small island developing state's perspective. *Asian Journal of Environment and Ecology, 4*(2), 1–12.

HRCM. (2008). Human Rights Commission. https://hrcm.org.mv/en

Hydrogen Council. (2021). *Hydrogen storage and clean fuel*.

ICLEI - Local Governments for Sustainability. (2021). *Climate-resilient urban design*.

International Energy Agency (IEA). (2022). *Floating wind turbines*.

International Renewable Energy Agency (IRENA). (2021). *Solar thermal energy*.

Jairaj, B., & Shivakumar, K. M. (2019). *Environmental jurisprudence in India.* Regal Publications.

Jal Shakti Ministry. (2021). *Smart water management initiatives.*

Jayakumar, S. (2017). *Sustainable development and environmental law: A critical review of Indian and international perspectives.* Cambridge Scholars Publishing.

Jeffery, R. (2024, April 29). Human Rights and Transitional Justice in the Maldives: Closing the Door, Once and For All? *Human Rights Review (Piscataway, N.J.), 25*(2), 233–256. Advance online publication. DOI: 10.1007/s12142-024-00716-9

Jolly, S., & Trivedi, A. (2025). *M K Ranjitsinh v Union of India: Climate litigation as a game-changer to place human concerns at the core of climate crisis in India. Chinese Journal of Environmental Law*, Advance online publication.

Kaur, S., Kaur, D., Tiwana, A. S., & Gupta, S. (2022, January 1). *Climate Change-Related Governance and Policies in Indian Himalayas.* Springer Climate. DOI: 10.1007/978-3-030-92782-0_16

Khanna, G., & Law, L. (2024, April 8). *Supreme Court Recognises Right to Be Free from Climate Change Adverse Effects.* Live Law. https://www.livelaw.in/supreme-court/for-first-time-supreme-court-recognizes-right-to-be-free-from-adverse-effects-of-climate-change-254638

Leelakrishnan, P. (2008). *Environmental law in India.* LexisNexis.

Malatesta, S., & Schmidt di Friedberg, M. (2017, May). Environmental policy and climate change vulnerability in the Maldives: From the 'lexicon of risk' to social response to change. *Island Studies Journal, 12*(1), 53–70. DOI: 10.24043/isj.5

Transparency Maldives. (2025). *Findings on the Existing Mechanisms For Environmental Impact Assessment.*

Maldives. (2023, January 20). Human Rights Watch. https://www.hrw.org/world-report/2023/country-chapters/maldives

Maldives. (2024, January 12). Human Rights Watch. https://www.hrw.org/world-report/2024/country-chapters/maldives

Maldives: Strategic National Action Plan (SNAP) for Disaster Risk Reduction and Climate Change Adaptation 2010-2020. (2018, August 21). PreventionWeb. https://www.preventionweb.net/publication/maldives-strategic-national-action-plan-snap-disaster-risk-reduction-and-climate-change

McDave, K. E., & Dagadu, P. P. (2023). Reconsidering the status and rights of climate refugees under international law. *International Journal of Law and Society, 6*(2), 168–172.

Ministry of Housing and Urban Affairs (MoHUA). (2021). *Smart Cities Mission.*

Ministry of New and Renewable Energy (MNRE). (2021). *Solar rooftop programs.*

Ministry of Road Transport and Highways (MoRTH). (2021). *Intelligent traffic management.*

Mitkidis, K., & Valkanou, T. N. (2020, March 1). *Climate Change Litigation: Trends, Policy Implications and the Way Forward.* Transnational Environmental Law. DOI: 10.1017/S2047102519000414

Mitra, R., & Sanghi, S. (2023). The small island states in the Indo-Pacific: Sovereignty lost? *Asia Pacific Law Review, 31*(2), 428–450. DOI: 10.1080/10192557.2023.2181806

Mohamed. (2024, May 15). Supreme Court overturns injunction ordering suspension of Gulhifalhu reclamation. *Https://En.Sun.Mv/89461*. Retrieved May 19, 2024, from https://en.sun.mv/89461

Mustafa, A. (2016). Judicial activism and environmental law in the Maldives. *Maldives Law Journal, 5*(1), 45–67.

Nanwani, S., & Ayus, A. M. (2012, July 15). *Strengthening Legal Education And Judicial Training in the Maldives: A Case Study.* IIUM Law Journal. DOI: 10.31436/iiumlj.v15i2.67

Naseer, A. (2018). *Environmental protection in the Maldives: Legal perspectives and policies.* Maldives Law Publishers.

National Oceanic and Atmospheric Administration (NOAA). (2021). *Weather and sea-level monitoring.*

National Renewable Energy Laboratory (NREL). (2022). *Photovoltaic cell efficiency.*

Olsen, N., Friis-Christensen, E., Floberghagen, R., Alken, P., Beggan, C. D., Chulliat, A., Doornbos, E., da Encarnação, J. T., Hamilton, B., Hulot, G., van den IJssel, J., Kuvshinov, A., Lesur, V., Lühr, H., Macmillan, S., Maus, S., Noja, M., Olsen, P. E. H., Park, J., & Visser, P. N. (2013). The SWARM satellite constellation application and research facility (SCARF) and SWARM data products. *Earth, Planets, and Space, 65*(11), 1189–1200. DOI: 10.5047/eps.2013.07.001

Pandve, H. (2009, January 1). *India's National Action Plan on Climate Change*. The Indian Journal of Occupational & Environmental Medicine/the œIndian. *Journal of Occupational and Environmental Medicine, 13*(1), 17. Advance online publication. DOI: 10.4103/0019-5278.50718 PMID: 20165607

Phelan, A. L., Meier, B. M., Patterson, D. W., Hesselman, M., Tahzib, F., & Gostin, L. O. (2025). The ICJ Advisory Opinion: A legal mandate for planetary health. *Lancet, 406*(10508), 1068–1070. DOI: 10.1016/S0140-6736(25)01725-8 PMID: 40886723

Prajapati, N., & Chauhan, D. K. (2025). *Promoting Women's leadership in climate resilience: a strategy for South Asia*. Virtual Commons - Bridgewater State University. https://vc.bridgew.edu/jiws/vol27/iss2/5

Preston, B. J. (2018, November 19). The Evolving Role of Environmental Rights in Climate Change Litigation. *Chinese Journal of Environmental Law, 2*(2), 131–164. DOI: 10.1163/24686042-12340030

Puthucherril, T. G. (2010). *Towards sustainable coastal development: Institutionalising integrated coastal zone management and strategic environmental assessment in South Asia*. Transnational Publishers.

Rajwanshi, H., & Sanni, T. (2024, December 29). *Maldives' position and contribution to the ICJ advisory opinion case on climate change*. Indian Blog of International Law. https://airrlaw.com/maldives-position-and-contribution-to-the-icj-advisory-opinion-case-on-climate-change-indian-blog-of-international-law/

Ram-Mohan, N., Kim, D., Zudock, E. J., Hashemi, M. M., Tjandra, K. C., Rogers, A. J., & Yang, S. (2022). SARS-CoV-2 RNAemia predicts clinical deterioration and extrapulmonary complications from COVID-19. *Clinical Infectious Diseases, 74*(2), 218–226. DOI: 10.1093/cid/ciab394 PMID: 33949665

Sebopela, M. C., & Odeku, K. O. (2014, September 29). *A Comparative Analysis of Regulatory Frameworks on Climate Change*. DOI: 10.5901/mjss.2014.v5n20p3078

Shafie, A. A., & Aziz, S. A. (2019, December 19). Supreme Court's Judicial Review Authority Vs. Constitutional Supremacy in the Maldives: Does the Constitution Still Reign Supreme? *IIUM Law Journal, 27*(2), 469–499. DOI: 10.31436/iiumlj.v27i2.417

Singh, G. (2007). *Environmental law in India*. Macmillan India Ltd.

Singh, M. P., & Kumar, N. (2019, February 14). *Tracing the History of the Legal System in India*. Oxford University Press eBooks. DOI: 10.1093/oso/9780199489879.003.0001

Techera, E., & Cannell-Lunn, M. (2019, November 1). A review of environmental law in Maldives with respect to conservation, biodiversity, fisheries and tourism. *Asia Pacific Journal of Environmental Law*, *22*(2), 228–256. Advance online publication. DOI: 10.4337/apjel.2019.02.03

The President's Office. (2021). *President ratifies Climate Emergency Act*. The President's Office. https://presidency.gov.mv/Press/Article/24678

Tripathi, A. (2024, April 10). *Climate change recognised as a fundamental right in India: Implications and solutions*. Times of India Blog. https://timesofindia.indiatimes.com/blogs/agyeya/climate-change-recognized-as-a-fundamental-right-in-india-implications-and-solutions/

Vasudevan, P. (2021, September 13). *Public Interest Litigation and Protection of the Environment | Law column*. Law Column. https://lawcolumn.in/public-interest-litigation-and-protection-of-the-environment/#google_vignette

Wen, J., Wan, C., Ye, Q., Yan, J., & Li, W. (2023, February 1). *Disaster Risk Reduction, Climate Change Adaptation and Their Linkages with Sustainable Development over the Past 30 Years: A Review*. International Journal of Disaster Risk Science. *International Journal of Disaster Risk Science*, *14*(1), 1–13. Advance online publication. DOI: 10.1007/s13753-023-00472-3

Zaidi, F. (2024), *Analysis: What does the new Supreme Court judgment mean for climate action in India?* https://www.downtoearth.org.in/news/governance/analysis-what-does-the-new-supreme-court-judgment-mean-for-climate-action-in-india--95462

Chapter 10
Indigenous Rights and Climate Justice:
A Comparative Study Under International Human Rights and Environmental Law

Driss Ed. Daran
http://orcid.org/0000-0001-9954-7249
College of Law, United Arab Emirates University, UAE & Faculté Polydisciplinaire de Khouribga, Sultan Moulay Slimane University, Beni Mellal, Morocco

Riad Al Ajlani
College of Law, United Arab Emirates University, UAE

Zia-d-Din Malik
http://orcid.org/0000-0001-6015-0914
Islamia University of Bahawalpur, Pakistan

Amine Assaid
Université Sidi Mohamed Ben Abdellah, Fes, Morocco

Magdi Shouaib
College of Law, United Arab Emirates University, UAE

ABSTRACT

Climate change poses a severe legal and ethical challenge, disproportionately impacting Indigenous peoples who contribute least to global emissions. This chapter offers a comparative analysis of Indigenous rights and climate justice under inter-

DOI: 10.4018/979-8-3373-6635-7.ch010

national human rights and environmental law. It scrutinizes key instruments like UNDRIP, ILO Convention 169, the UNFCCC, and the Paris Agreement, alongside pivotal cases such as Saramaka v. Suriname and Teitiota v. New Zealand. The findings highlight a critical gap: human rights law protects Indigenous land and culture but lacks climate-specific mandates, while environmental law promotes climate action without recognizing Indigenous peoples as binding right-holders. To reconcile these regimes, the chapter proposes integrating Indigenous rights and judicial accountability into climate governance and formally institutionalizing Indigenous ecological knowledge for a just and effective global response.

1. INTRODUCTION

Climate change has emerged as the defining challenge of the twenty-first century, shaping not only environmental and economic realities but also the deeper questions of justice, equity, and human rights. The accelerating pace of sea-level rise, extreme weather events, biodiversity loss, and desertification poses grave threats to human survival and global stability (Jha & Dev, 2024). Yet these harms are not distributed evenly: certain communities bear a disproportionate share of the impacts despite contributing least to global greenhouse gas emissions (Abate & Kronk, 2013). This asymmetry places climate change squarely within the discourse of *climate justice*, which seeks equitable sustainability and protection for historically marginalized groups (Schlosberg & Collins, 2014).

The native people are placed in an exceptional place in this world story. They are the stewards of Custodians of some of the most biodiversic and resilient ecosystems on the planet and their cultural and spiritual identities cannot be separated to the natural environment. Therefore, ecological degradation poses a direct threat on their existence and self-determination. Sea level height increases are threatening the Pacific Island communities, Amazonian peoples are being threatened by deforestation, and Arctic ice melting is jeopardizing Inuit lives (Gerhardt, 2023). The effects of these climatic factors are exacerbated by the existence of the legacy of the colonialism, economic marginalization, and political exclusion. Consequently, Indigenous rights protection is not an ethical requirement, but a condition of vibrant and successful climate management (Giacomini, 2022).

Even though these challenges are rather urgent, the international law has been reacting disjointedly. The issue is regulated by two major legal regimes international human right law and international environmental law, which tend to work in isolation. The human rights law, including the United Nations Declaration on the Rights of Indigenous Peoples (UNDRIP, 2007), ILO Convention No. 169 (1989) and the core laws such as the International Covenant on Civil and Political Rights (ICCPR,

1966) recognizes Indigenous rights to land, culture, and participation (Barnabas, 2017). Jurisprudence has strengthened these protections, as has been provided by the Inter-American Court of Human Rights and the African Court on Human and Peoples' Rights, which have combined environmental harm and human rights violations (Perez-Leon-Acevedo, 2024). Nevertheless, these frameworks do not have clear obligation to climate and Indigenous people become exposed to transboundary and systemic threats.

On the other hand, the international environmental law with its basis in the "United Nations Framework Convention on Climate Change (UNFCCC, 1992) and the "Paris Agreement (2015) is oriented on collective action, which is mitigation, adaptation, and finance (Leal-Arcas, 2025; Khan, 2024). Even though these tools mention equity and Indigenous knowledge, these are symbolic tools, which are preambulatory in a text, but not obligatory in the text (Gunn, 2021). The outcome is a fragmentation in which the human rights law offers enforceable though reactive solutions, whereas the environmental law encourages the proactive measure but does not offer binding Indigenous protections (Okedele et al., 2024).

The chapter critically looks at the intersection of international human rights and environmental regimes in the context of the Indigenous rights in climate governance. It appraises their own advantages and limitations, on a doctrinal and comparative legal basis, with the aid of a treaty analysis, jurisprudence, and commentary, of academic commentators. The cases of *Saramaka People v. Suriname (2007), Ogiek v. Kenya (2017), Urgenda v. Netherlands (2019), and Teitiota v. New Zealand (2020)* shows that courts are struggling with the issue of rights and climate obligations interaction. Collectively with other recent changes e.g. the recent recognition of the right to a healthy environment by the UN General Assembly (2022) and forthcoming advisory opinion by ICJ on climate change, these demonstrate the existence of a rapidly changing legal environment.

Finally, the chapter concludes that to have meaningful climate justice, there is a need to bridge the gap between human rights as well as the environmental law. It goes to the point that Indigenous people should not be treated as peripheral stakeholders but as right holders and bearers of ecological knowledge at the center of global sustainability. The global community can only attain environmental integrity as well as social justice by incorporating their rights, knowledge systems, and governance practices into the international law.

2. INDIGENOUS RIGHTS IN INTERNATIONAL HUMAN RIGHTS LAW

2.1 Foundational Instruments

The international human rights law has gone a long way to recognize the use of Indigenous rights over the last four decades, yet it has some significant gaps that need to be addressed. One of the key milestones is the United Nations Declaration on the Rights of Indigenous Peoples (UNDRIP, 2007) passed by the UN General Assembly that recognizes the right to self-determination, cultural integrity, land, and decision-making. Article 26 acknowledges the right to lands and resources owned or used traditionally by the Indigenous people and Article 32 demands that the states should ensure Free, Prior and Informed Consent (FPIC) is received prior to any project consent under Article 32. Even though UNDRIP is not a treaty, but a declaration, it is becoming a more valid reflector of customary international law and courts and human rights entities often refer to it (Craft et al., 2018).

The other important tool is ILO Convention No. 169 (1989) which is the only binding multilateral treaty that covers specifically Indigenous and tribal peoples. It covers land tenure, consultation, cultural preservation and governance of natural resources (Articles 1415), but is limited in scope since, as of April 2024, it is not followed by many states (just 111 states) and four of the biggest settler nations have not (Canada, the United States, and Australia) (Verde, 2024). Additional protections are also provided by wider human rights treaties (ICCPR, 1966; ICESCR, 1966) or even regional regimes (ECHR, 1950; African Charter, 1981; American Declaration, 2016), although are not explicitly invoked where Indigenous peoples are not specifically mentioned.

2.2 Jurisprudence and Case Studies

Inter-American System

The Inter-American Court of Human Rights (IACtHR) has been the most progressive in linking Indigenous rights with environmental protection.

- *Saramaka People v. Suriname* (2007): The Court held that granting concessions for logging and mining on tribal lands without consent violated collective property rights. It required states to ensure consultation and, in cases of major impact, consent. This judgment is widely cited for establishing FPIC as a legal requirement.

- *Kichwa Indigenous People of Sarayaku v. Ecuador* (2012): The Court reaffirmed FPIC, holding Ecuador responsible for permitting oil exploration on Sarayaku territory without consultation.
- *Lhaka Honhat v. Argentina (2020):* It was the first time the Court considered that the right to a healthy environment was on its own and enforceable, along with the rights to food, water, and cultural identity.

These instances demonstrate how the Inter-American system addresses the issues of environmental degradation as the direct infringement of Indigenous rights.

African System

- The African court on human and peoples right has also made historic decisions.
- *Ogiek v. Kenya (2017)*: The Court decided that the Ogiek were not entitled to be evicted in the Mau Forest as this interfered with their rights to property, culture, religion, and development, as provided in the African Charter. It stressed the inseparable nature of land, cultural survival and environmental stewardship.
- *Endorois v. Kenya (African Commission, 2010)*: The Commission stated that the relocation of the Endorois to facilitate tourism development interfered with several rights, and made the key precedents on consultation and benefit-sharing.

Both cases highlight how African institutions view Indigenous peoples as environmental custodians whose rights cannot be subordinated to development projects.

UN Treaty Bodies

UN treaty bodies have also advanced Indigenous environmental rights.

- *Teitiota v. New Zealand* (Human Rights Committee, 2020): While rejecting the applicant's asylum claim, the Committee recognised that climate-induced displacement could, in principle, engage the right to life under Article 6 of the ICCPR. This was a breakthrough in linking climate change with human rights obligations.
- CERD Urgent Action Procedures: The Committee on the Elimination of Racial Discrimination has repeatedly urged states to suspend extractive projects threatening Indigenous communities, treating environmental harm as racial discrimination (Kleiva, 2017).

European Context

Although the European Court of Human Rights (ECtHR) has not developed Indigenous-specific jurisprudence, its recent climate rulings expand the scope of environmental protection. In *Verein KlimaSeniorinnen Schweiz v. Switzerland* (2024), the Court found that insufficient climate action violated the right to family life (Article 8 ECHR). Although it is not an Indigenous case, it is an indication of a broader judicial readiness to construe human rights broadly in the context of climate change, which would be helpful in the context of Indigenous claims in Europe and elsewhere.

2.3. Recent Developments (2020–2024)

In recent years a further recognition of environmental rights relating to Indigenous peoples has taken place:

- UN General Assembly Resolution (2022): Stated that the right to a clean, healthy, and sustainable environment is a universal human right. It is not binding but empowers the Indigenous peoples in the international fora (Kastlunger, 2024).
- Human Rights Council Resolutions (2021–2023): Appointed a Special Rapporteur on climate change and human rights, who has repeatedly highlighted Indigenous peoples as disproportionately affected groups (Luporini, 2025).
- Regional Climate Litigation: Courts in Colombia, Canada, and the Philippines have increasingly referenced Indigenous rights in environmental rulings, reinforcing global trends (Tigre, 2022).

2.4 Strengths of Human Rights Law

The human rights law possesses a number of strengths in terms of promoting protection of the Indigenous people and their habitats. The main contribution is its normative clearness: documents like the United Nations Declaration on the Rights of Indigenous Peoples (UNDRIP) and the ILO Convention No. 169 define specific requirements on how to protect the land, culture, and meaningful involvement in the decision-making process (Tomaselli, 2020). It provides ways of enforceability as well because regional human rights courts can make binding decisions that connect Indigenous rights to environmental protection, which will provide communities with legal means to protect their lands (Petersmann, 2021). The other significant strength lies in the awareness of Free, Prior, and Informed Consent (FPIC), which is currently being addressed as a significant preventive measure to the adverse impacts of extractive and development projects. Lastly, environmental rights are becoming appreciated in the human rights jurisprudence as seen in *Lhaka Honhat v. Argentina*, which stated that a healthy environment was a human right that needed protection (Mardikian, 2023). Collectively, these characteristics reflect that the human rights law has the potential of promoting Indigenous and environmental justice.

2.5 Limitations of Human Rights Law

Although the international human rights law has some strengths, it has a number of weaknesses that restrain its effectiveness in dealing with climate challenges affecting the Indigenous people. One of the biggest gaps is a lack of climate-related commitments because the current treaties do not directly address climate concerns but rather only indirectly address them by requiring the general norms of rights to be used by Indigenous peoples (Mardikian, 2023). The process of implementation is also not comprehensive, as the interpretation of treaties is inconsistent, and enforcement is generally absent, as even judicial decisions are often ignored by states (Yoshida, 2024). The limited adoption of documents such as ILO Convention No. 169 also slims down the world protection, and the financial and procedural access discourages Indigenous rights to justice (Simporios, 2024). Collectively, these shortcomings reveal the need to integrate human rights law with stronger, enforceable commitments under international environmental law to safeguard Indigenous rights amid climate change.

3. CLIMATE JUSTICE IN INTERNATIONAL ENVIRONMENTAL LAW

3.1 Foundational Treaties and Principles

Modern international environmental law is based on two seminal treaties, namely, the United Nations Framework Convention on Climate Change (UNFCCC, 1992) and the Paris Agreement (2015). The two of them form the legal foundation of global climate cooperation on the principle of equity and collective responsibility between the states. UNFCCC brought the principle of Common but Differentiated Responsibilities (CBDR) when all the countries are supposed to mitigate climate change considering that the developed countries have higher responsibilities because of their historical emissions and the superior financial and technological ability (Fajardo et al., 2024). Gradually, the idea has extended in climate justice discussions to demand more protections and support vulnerable communities, such as Indigenous people

(Habib et al., 2022). The Paris Agreement supported climate governance with Nationally Determined Contributions (NDCs) and the recognition of Indigenous rights and traditional knowledge, which are not binding (Tran and Salamanca, 2023). The conservation of Indigenous knowledge and fair benefit-sharing is also supported by such complementary instruments as the Convention on Biological Diversity (CBD, 1992) or its Nagoya Protocol (2010) (Colella et al., 2023).

3.2 COP Outcomes and Indigenous Inclusion (2021–2024)

Recent Conferences of the Parties (COPs) of the UNFCCC can be followed in the gradual but discontinuous recognition of Indigenous rights in global climate negotiations. Indigenous knowledge was recognized in the Glasgow Climate Pact (COP26, 2021) during adaptation strategies and promised support to forest protection and biodiversity efforts (Bulkeley et al., 2023). With such promises, Indigenous observers condemned the fact that their voices were not represented in major discussions. During COP27 (2022), the creation of the Loss and Damage Fund was an improvement of this, as it compensates the vulnerable states due to climate damage (Obergassel et al., 2023). Nevertheless, Indigenous representatives cautioned that when there is no outright inclusion mechanisms, the fund would fail to consider those communities who are most at risk due to climate change.

Global Stocktake, which was initiated at COP28 (2023) in Dubai, evaluated the progress on the Paris Agreement. Participation in the process was highlighted by indigenous participants as lacking binding guarantees to be part thereof, they indicated that the Local Communities and Indigenous Peoples Platform (LCIPP) had made the dialogue better, but power relations remained asymmetrical (Frizzarin Bassa and Schneider, 2025). Finance and equity Geneva In COP29 (2024) in Baku, Indigenous activists insisted on direct access to climate funding to enhance local control (Rodríguez, 2024). Together, these COPs demonstrate the increased awareness about the Indigenous issues, but their presence is more of a symbolic gesture, rather than a concrete measure (Redvers et al., 2023).

3.3 Principles of Equity and Justice

International environmental law is based on the principle of equity which is manifested in a number of interrelated doctrines at its core. The idea of Common but Differentiated Responsibilities (CBDR) which was initially used to differentiate between developed and developing countries has transformed into a pillar of climate justice by compensating historic injustice and current exposure (GROMEA, 2025). Indigenous peoples claim that CBDR can be applied in the states as well to make sure that the least responsible emitters are supported and turned to. The principle of the polluter pays expands on this argument and means those who benefit by owning carbon-intensive industries to pay their environmental damages (Fullerton & Muehlegger, 2019). Precautionary principle requires preventive measures when scientific knowledge is unclear and this is critical to the Indigenous people depending on delicate ecosystems. Equally, intergenerational equity focuses on the responsibilities to the future, reiterating the Indigenous worldviews based on stewardship (Park et al., 2023). Although they are strong, these principles are not fully applied and the realization of the Indigenous rights is still more of an oratory than a practical one.

3.4 Weaknesses in Legal Protection

Although the international environmental law has improved through normative progressions, it has a number of weaknesses that compromise the protection of Indigenous rights:

i. Soft-Law Language: The mention of Indigenous peoples in the Paris Agreement is merely non-binding, in the form of preamble text and voluntary mechanisms. States, which do not have operational provisions, cannot be brought to justice (Abidin, 2023).
ii. State-Centrism: The environmental law is constructed on the state sovereignty, and the Indigenous people rely on the state representation. This institutional marginalization marginalizes the Indigenous voices during the international negotiations and their right to directly access climate finance is prohibited.
iii. Fragmentation: Environmental treaties tend to overlap, and not to be coherent. An example is that biodiversity agreements recognize Indigenous knowledge whereas climate agreements do not, which results in unequal protection.
iv. Implementation Gaps: There is a poor enforcement of the Indigenous rights even in those areas where they have been identified. Even the formation of the Loss and Damage Fund, which is pioneering, is an example of the disconnection between theory and practice, which is that the money will be taken by the state governments without reaching Indigenous communities.
v. Displacement and Energy Transition Conflicts: Ironically, certain climate policies made to hasten the process of leaving fossil fuels behind have done harm to Indigenous people. Megaprojects in renewable energy (like wind farms and hydroelectric dams) have been established on Indigenous land without their approval that recreates the model of dispossession under the auspices of a green label (O'Neill, et al., 2019).

3.5 Emerging Trends and Case Law

There are some emerging trends of the increasing use of rights-based approaches to international environmental law. One of them is the acknowledgment of the rights of nature. One example is the inclusion of legal rights of nature within the constitutions of Ecuador and Bolivia based on the Indigenous cosmologies that consider ecosystems as living beings but not resources (Tanasescu, 2022). The second trend is the evolution of climate litigation in which strategic lawsuits will present the failure of states to act effectively in regards to climate change as a breach of human rights. The country has been at the forefront, with the Torres Strait Islanders having petitioned the UN Human Rights Committee to the same effect in which in 2022, the committee concluded that Australia breached the rights of people to climate harm by not implementing measures to protect the people (Schuijers, 2023).

Advisory opinions also reflect this concurrence. The 2017 advisory opinion by the Inter-American Court of Human Rights stated that states have the duty to eliminate environmental damage that would compromise the human rights. In perspective, an advisory opinion sought by the International Court of Justice in 2023 (treated in 2025) can be used to reinforce the argument at the international level. The point is that generally, the overlap is increasing, yet there is still a need to implement it stronger.

4. COMPARATIVE ANALYSIS OF THE TWO REGIMES

4.1 Areas of Convergence

Despite the different origins and purposes of the international human rights law and the international environmental law, they are becoming more and more interrelated based on their similar principles that are especially applicable to the Indigenous people. This is converged in three main areas.

Equity and Justice

The two frameworks acknowledge justice as the key to solving global issues. The vulnerable groups are safeguarded by human rights law as it outlaws discrimination and guarantees participation and environmental law is responsive by recognizing the differentiated responsibilities through the principle of Common but Differentiated Responsibilities (CBDR) and intergenerational equity. Collectively, the principles facilitate fairness in the allocation of environmental burdens and benefits, which is consistent with the ambitions of Indigenous people to address the historical injustices and preserve the cultural survival (Nicholas, 2022).

Participation and Consultation

Participation is one of the pillars of the two regimes. The Free, Prior and Informed Consent (FPIC) principle is a principle of UNDRIP and ILO Convention No. 169, which focuses on the Indigenous agency in decision-making. In the same way, there has been an increased incorporation of participatory tools and measures in environmental law as with the so-called Local Communities and Indigenous Peoples Platform (LCIPP) under the UNFCCC. Nonetheless, its mandate is more consultative and prevents the Inuit from having an impact on it beyond the discussion to an equal power (Belfer et al., 2019).

Environmental Harm as a Human Rights Violation

An expanding body of jurisprudence affirms that environmental degradation infringes fundamental rights. In *Lhaka Honhat v. Argentina* (IACtHR, 2020), the Court recognized the autonomous right to a healthy environment, while *Teitiota v. New Zealand* (2020) linked climate-induced displacement to the right to life. These cases illustrate how human rights and environmental protection increasingly overlap, particularly for Indigenous communities.

4.2 Areas of Divergence

However, even with all these points of convergence, there are still areas of major divergences that still restrict the functioning of Indigenous people within the framework of international law.

Strict vs. Soft Protections

The human rights law also provides a binding legal protection in terms of treaties like the ICCPR and ICESCR and enforceable regional judicial decisions. Environmental law, on the contrary, is to a great extent based on soft-law tools and voluntary commitments (Jalil, 2025). Even such historic documents as the Paris Agreement (2015) can include only oblique allusions to Indigenous rights in preambular and knowledge-sharing statements. Therefore, Indigenous communities lack options of holding accountable through environmental law as compared to the modalities through human rights law (Giacomini, 2022).

Rights-Holders vs. Stakeholders

According to the human rights law, Indigenous peoples are the rights-holders and are entitled to enforcing their claims. Nonetheless, in most instances, they are approached as stakeholders in the context of the environmental law, i.e., participants who are not legally binding and whose views are considered. Such framing poses a threat to weaken Indigenous agency and reduce communities to a consultant but not a decision-making position (Jull et al., 2023).

State Sovereignty vs. Community Rights

Environmental law places more focus on the sovereignty of the state, with inter-state obligations being highly prioritized. In comparison, human rights law provides protection to individuals or groups of people even in cases of the state (Donnelly and Whelan, 2020). This gap allows states to implement climate or development policies that can jeopardize the rights of Indigenous people. These forms of structural inequalities based on the colonial past strengthen the marginalization of the Indigenous peoples in global environmental governance.

4.3 Lessons from Case Law

Jurisprudence is a better way to explain the comparative analysis where the two regimes overlap.

Urgenda v. Netherlands (2019)

The Dutch Supreme Court in this landmark case found that insufficient climate action infringed Articles 2 and 8 of the European Convention on Human Rights (ECHR), which guarantees the right to life and family life (Olsson, 2024). While not an Indigenous case, *Urgenda* is significant because it used human rights law to enforce climate obligations that environmental treaties could not. The Paris Agreement provided context but no enforceable duty; the Court grounded its ruling in human rights. This illustrates how rights law can fill accountability gaps in environmental law.

Torres Strait Islanders v. Australia (UN Human Rights Committee, 2022)

This complaint marked the first successful climate-related case before the Committee. The Islanders claimed that Australia had infringed their rights by being unable to safeguard them against rising oceans within the confines of the ICCPR. The Committee was agreeable, and saw violation of rights to culture, family life, and home. More importantly, the case concerned an Aboriginal community and established that the failure to act in relation to climate could be a breach of Indigenous rights (Nicholas, 2022). In this case, human rights law presented direct solutions that could rectify deficient provisions of environmental treaties which only promote adaptation solutions without any enforceable limitations.

Teitiota v. New Zealand (HRC, 2020)

This case had a precedent though it failed on fact that taking people back to places that were rendered inhabitable by climate change would be against the right to life. To the Indigenous communities that are threatened with extinction due to climate change, this argument creates possibilities of future protection, although it has not yet taken legal status in environmental treaties (Teaiwa, 2019).

Lhaka Honhat v. Argentina (IACtHR, 2020)

The Court realised the autonomous right to healthy environment at the first instance. This decision demonstrates the expansion of the substantive protection by human rights courts in the region in a pattern not obvious in the case of the environmental treaties. The Indigenous claims are boosted by the establishment of the environmental rights in the human rights systems.

Verein KlimaSeniorinnen Schweiz v. Switzerland (ECtHR, 2024)

The European Court discovered that poor climate policies contravened Article 8 of the ECHR. Though the case was brought by elderly women, the ruling shows that the human rights law can offer positive responsibility to states in the climate context (Olsson, 2024). Such jurisprudence can be used by the indigenous communities to seek the same protections.

ICJ Advisory Opinion Request (2023–present)

The advisory opinion request on the duty of states in relation to climate change by UN General Assembly on behalf of Vanuatu and backed by more 130 states, might have far-reaching consequences ((International Court of Justice, 2025). Provided the ICJ confirms binding commitments to avoid transboundary climate harm, Indigenous people would indirectly benefit. Although they do not have direct standing to the ICJ, the states acting on behalf of Indigenous peoples may invoke such commitments to reinforce the global normative order.

4.4 Comparative Synthesis

The comparative discussion shows that there is a paradox in the current legal frameworks. The human rights law offers binding guarantees and the potential to be legally enforceable, but this was not initially established to meet the demands of the climate change issue. On the other hand, environmental law is much more comprehensive in scope and provides more institutional frameworks, as it lacks proper enforcement mechanisms and fails to consider Indigenous rights (Saharkhiz, 2024). Such disconnect offers a fragmented legal practice according to which human rights law enables Indigenous peoples to engage in a challenge to climate-related harms in a reactive and often case-specific manner and environmental law facilitates collective state action at the expense of silenced Indigenous voices and limited opportunities of holding anyone to account. This separation is structural inequality that keeps the Indigenous people as rhetorical categories that are still at the systemic level, albeit poorly embedded in practice (Balint, et al., 2014). It is, therefore, not only topical but also important to seal the gap in between those two frameworks in order to achieve equitable and effective climate governance.

4.5 Scholarly Perspectives

Analysts feel that there is need to have additional integration of the two regimes. Knox mentions that the issue of environmental responsibility should be perceived through the prism of human rights in order to ensure that the vulnerable population could be adequately protected (2020). According to Peel and Osofsky (2018), one of the areas where these two regimes intersect is in the emergence of climate litigation. At the same time, Schlosberg and Collins (2014) note that climate justice should not only achieve distributive fairness but also recognition and participation, and in this aspect Indigenous peoples are excluded. These views highlight the importance of legal reform. Unless this is harmonized, the Indigenous peoples will remain vulnerable to fragmented protection as they are compelled to traverse parallel systems that do not fully satisfy their vulnerabilities.

5. CHALLENGES AND GAPS

5.1 Barriers to Access to Justice for Indigenous Communities

In spite of procedural avenues to redress by the Indigenous peoples offered by international human rights law, the access to justice is still limited in practice. International and regional institutions, such as Inter-American Court of Human Rights (IACtHR) and the UN Human Rights Committee, are commonly slow, expensive and technological hurdles that many communities cannot easily get over (Pegram & Rodríguez, 2018). Cases can take years or decades and by the time they are ruled, it may be too late and damage done to the environment.

Even when rulings are issued, enforcement is weak. For example, despite landmark judgments in *Saramaka People v. Suriname (2007)* and *Ogiek v. Kenya (2017)*, implementation has been inconsistent due to state resistance, competing development priorities, or lack of resources. Lack of binding compliance measures in the international law implies that Indigenous people frequently make empty victories which do not translate on the ground.

Even legal systems at the domestic level might not offer sufficient protection at that level. Courts are not always independent, states can criminalize Indigenous activism, and corporations can have an inappropriate influence on policy (Eichler, & Barnier-Khawam, 2021). This compels the indigenous communities to resort to transnational advocacy, but the dependence on the outside allies highlights the marginality of the indigenous communities in the formal legal systems.

5.2 Conflicts Between Development Projects and Indigenous Rights

One of the most pressing challenges in today's development and climate agenda is the tension between state-led development projects and the protection of Indigenous rights. Governments often justify extractive industries and large-scale infrastructure projects in the name of national interest, energy security, or climate finance. Yet these initiatives frequently come at the direct expense of Indigenous peoples' lands, cultures, and livelihoods. Oil exploration and logging in the Amazon in Ecuador and Peru, have threatened the Waorani and Achuar peoples, apparently without seeking Free, Prior, and Informed Consent (FPIC) as mandated by UNDRIP (2007) and national legislation.

Whereas individual court cases have made certain projects come to a halting, deeply rooted political and economic interests are still encouraging it (Knox, 2020). In Canada, the Wet'suwet'en Nation has opposed pipelines in their ancestral territory despite the fact that the Indigenous land rights are accepted in the courts of law (Peel & Osofsky, 2018). Hydroelectric projects in Brazil and India are also the source of similar conflicts that are promoted as green energy but repeat the colonial processes of dispossession (Schlosberg and Collins, 2014). Arctic communities including the Inuits are exposed to climate effects as well as increased resource extraction. These instances reveal a fundamental paradox in which climate and development projects not paying attention to Indigenous consent may replicate the history of environmental colonialism as understood in Saramaka v. Suriname (2007) and the Ogiek case (2017).

5.3 Fragmentation of International Law

The international law continues to be structurally fragmented particularly between the human rights and the environmental regimes. Human rights legislation is quite strong in the recognition of the Indigenous rights but does not have climate-specific obligations. Environmental law formulates climate processes but largely considers Indigenous people as stakeholders instead of when they are right holders. Such a gap gives rise to disproportional, even contradictory protection (Koskenniemi, 2006).

Major instruments depict the gap. The Paris Agreement recognizes the knowledge of Indigenous people primarily in the preamble and it does not establish any binding requirements of Free, prior, and Informed Consent (FPIC) (UNFCCC, 2015). In comparison, the FPIC is explicitly defined within the context of the law as presented by UNDRIP and ILO Convention No. 169 (ILO, 1989; United Nations, 2007). The disparity permits states to follow through on environmental commitments at the expense of Indigenous rights and compel communities to go through various forums with scarce resources (Knox, 2020).

Fragmentation promotes selective compliance and forum shopping as well, undermining accountability and strengthening power structures (Peel and Osofsky, 2018; Schlosberg and Collins, 2014).

5.4 Political and Structural Constraints

In addition to a lack of legislation, political facts limit the efficacy of protections. The international law is still based on state sovereignty, and Indigenous people cannot directly engage in the negotiations of the treaties, but instead have to allow states to speak on their behalf, which proves to be ineffective due to the economic interests of states and the history of marginalization. During climate negotiations, the role of Indigenous representatives is an observer, and the actual decision-making authority is assigned to states. Even such mechanisms as the Local Communities and Indigenous Peoples Platform (LCIPP), despite its usefulness as a platform of knowledge-exchange and capacity building, is consultative and limited to non-binding structures, which restricts its effectiveness (In the absence of binding power, the contribution of the indigenous people will be tokenistic.

Additionally, climate governance is determined by geopolitical inequalities. The negotiations are dominated by powerful states who usually give priority to the energy security and economic competitiveness, hence Indigenous voices, especially of Global South are put on the backburner. This replicates power inequalities globally in climate governance, which places Indigenous people in a dual state of marginalization, not only to their states but also to the international system. These inequalities are enhanced by colonial legacies. Most of Indigenous people never agreed to the establishment of states that represent them at the international levels. Structural exclusion of Indigenous sovereignty is still largely undiscussed in international law which continues the trends of dispossession.

5.5 Climate-Specific Gaps

Despite growing recognition of environmental harm in human rights law, there is still no universal, binding right to a healthy environment across all international treaties. The UN General Assembly declared this right universal in 2022, but its resolution remains non-binding (UNGA, 2022). Some regional courts have gone further—*Lhaka Honhat v. Argentina* (IACtHR, 2020) recognized the right to a healthy environment as directly enforceable—but global implementation remains fragmented (Boyd, 2021).

Climate-related obligations are similarly underdeveloped. Major human rights treaties were never designed to address transboundary climate issues (Knox, 2020). Indigenous groups and their advocates have used strategic litigation, framing state inaction as violations of existing rights such as life, health, and culture. This has been evident in *Urgenda v. Netherlands* (2019), *KlimaSeniorinnen v. Switzerland* (ECHR, 2024), and the Torres Strait Islanders' case before the UNHRC (2022), where courts affirmed states' human rights duties in climate action (Peel & Osofsky, 2018).

Access to climate finance remains another gap. Mechanisms like the Green Climate Fund and the Loss and Damage Fund are state-managed (Roberts & Weikmans, 2017), leaving Indigenous communities without direct access, even though Indigenous-led land stewardship is among the most effective responses to biodiversity loss and climate change (IPBES, 2019; Garnett et al., 2018).

5.6 Criminalization and Violence Against Indigenous Defenders

Another obstacle, which is frequently not discussed in the context of legal terminology, is the incrimination of the Indigenous environmental activists. In Latin America, Africa, and Asia, leaders of the Indigenous populations who oppose extractive projects are harassed, prosecuted, or even killed. According to reports by Global Witness (2023), Indigenous people represent an overrepresented percentage of the environmental defenders killed all over the world. Protection under the international law is not strong: as human rights agencies denounce such violence, the enforcement is poor, and they usually receive impunity. Such fear environment compromises participation of the Indigenous people in the climate governance and propagates structural injustice.

5.7 Emerging but Uneven Recognition

All these notwithstanding, there are indications of improvement. The courts are becoming more open to attachment of climate commitments to the human rights. A case against Australia filed by the Torres Strait Islanders (HRC, 2022) was the first of its kind in making a state responsible in the failure to safeguard an Indigenous community due to the effects of climate change. The KlimaSeniorinnen case of ECtHR decision (2024) opens the opportunities of rights-based climate claims even further. Such developments are however skewed within regions and rely on judicial activism. In the absence of universal recognition and enforcing measures, Indigenous protections are easily weakened. Profits in one jurisdiction do not necessarily have the same effect on the rest of the world, and many communities remain vulnerable.

6. TOWARDS A FUTURE FRAMEWORK FOR CLIMATE JUSTICE

The comparison between the international human rights law and international environmental law shows the evident development in the recognition of Indigenous people as the significant agents in climate governance. However, neither of the two regimes has been found to be adequate: human rights law offers powerful but disjointed protections, whereas environmental law enshrines equity principles, but with no enforcement measures (Okedele, et al., 2024). In order to reach significant climate justice, the reforms should be directed at alignment of these structures and integration of Indigenous rights in binding international duties.

6.1 Embedding Indigenous Rights in Climate Treaties

A crucial reform in international climate governance is moving beyond symbolic acknowledgment toward binding obligations to protect Indigenous rights. The *Paris Agreement (2015)* refers to Indigenous peoples only in its preamble and non-binding provisions on knowledge-sharing, allowing states to overlook these rights in climate planning and Nationally Determined Contributions (NDCs). Future UNFCCC negotiations or a new protocol should embed *Free, Prior and Informed Consent (FPIC)* as a binding requirement, ensuring Indigenous consent for projects such as renewable energy or conservation (Ewell, 2022). Aligning climate law with human rights frameworks like *UNDRIP* and *ILO Convention No. 169* would strengthen coherence. Moreover, Indigenous representation under the UNFCCC must advance from consultative to co-governance roles. Recognizing Indigenous peoples as rights-holders, not merely stakeholders, would ensure enforceable protections and promote climate policy grounded in both ecological and cultural integrity (Gunn, 2021).

6.2 Strengthening International Judicial Pathways

Responsibility in the law. Although the human rights courts are the pioneers of climate jurisprudence, they have limited jurisdiction. Increasing the size of international courts and tribal institutions may offer greater protection to the Indigenous people.

i. ***International Court of Justice (ICJ):*** The advisory opinion on climate obligations that the UN General Assembly has requested in 2023 is an unprecedented opportunity. In case the ICJ confirms that states are obliged to reduce climate damage, this would enable Indigenous communities to make states answerable by the domestic and regional courts (Rajamani, 2025). Some of the options to take in the future may involve allowing Indigenous groups to indirectly be involved in the proceedings of the ICJ through state sponsorship.
ii. ***International Tribunal of the Law of the Sea (ITLOS)***: ITLOS is in the process of deliberating on advisory opinion of climate change and law of the sea on behalf of the small island states. Awareness of the need to safeguard marine environments would directly increase the benefits of the coastal Indigenous populations, including the Torres Strait Islanders (Smyth and Isherwood, 2016). The decision to enable ITLOS to have these opportunities to expand its jurisdiction to clearly deal with Indigenous claims can bring new leeways in justice.
iii. ***Regional Courts***: It is important to strengthen the ability of Inter-American and African Courts to implement climate obligations. These courts have already recognized Indigenous environmental rights (*Ogiek, Lhaka Honhat*), but implementation gaps persist (Bagni, 2024). Creating compliance-monitoring bodies could improve enforcement.
iv. ***Domestic Incorporation:*** The states are expected to entrench international commitments within the domestic law whereby the indigenous peoples may seek redress by not necessarily resorting to international litigation. Indicatively, the implementation of matauranga Māori in the climate policy by New Zealand presents an example of domestic integration (Taiapa, et al., 2025).

6.3 Institutionalizing Indigenous Knowledge in Governance

Climate adaptation and biodiversity conservation rely on the Indigenous ecological knowledge that has developed over a long period of sustainable management of land and other resources. It is however more complied with than practiced in international law. Reforms are required to overcome this gap. States ought to clearly incorporate Indigenous knowledge in their Nationally Determined Contributions (NDCs) so as to guarantee that it has a place in mitigation and adaptation plans. Green Climate Fund and the Loss and Damage Fund schemes are initiatives that should focus on Indigenous-led projects because studies have revealed that Indigenous-managed lands are performing better than state-controlled lands in protecting biodiversity and in storing carbon (Zhang et al., 2023). Permanent institutions such as the LCIPP should be transformed into real policymaking institutions that have co-decision powers and the Indigenous governance systems should be legalized as legitimate environmental institutions. By institutionalizing these systems, Indigenous knowledge would cease to be symbolically recognized but rather actually shared in the global environmental governance in a power-sharing manner.

6.4 Harmonizing Human Rights and Environmental Law

The prevailing gap between human rights and environmental law remains to be a setback to the protection of the Indigenous people and thus harmonization is paramount. First, the right to a healthy environment, despite being proclaimed by the UN general assembly in 2022, should be made binding and enshrined in treaties like the ICCPR, ICESCR or a new environmental convention to enhance the Indigenous legal claims. Second, there should be cross-referencing of environmental and human rights treaties to ensure that one framework supports the other to ensure coherence. Third, collaboration between the Human Rights Council and the UNFCCC in terms of joint monitoring would help to enhance the supervision of adherence in states to their Indigenous and climate-related commitments. Lastly, a hybrid between the two regimes in climate litigation would establish a more reliable jurisprudence and avoid piecemeal interpretation. This harmonization would develop a more integrated, enforceable system, which would guarantee more robust, rights-based protection of Indigenous peoples in global environmental governance.

6.5 Ensuring Climate Finance Access and Benefit-Sharing

A major unresolved gap in global climate governance is that the Indigenous communities lack direct access to finance. When states allocate billions of climate funds, a small portion of it goes directly to Indigenous peoples. In order to stem this unfairness, a number of reforms are required. The Green Climate Fund and the Loss and Damage Fund should have direct access mechanisms to facilitate Indigenous-led efforts. The obligation of sharing benefits should compel states and corporations to fairly share the benefits of the climate projects which are carried out on the Indigenous lands. Also, it is important to have capacity-building to assist Indigenous communities in designing, implementing, and monitoring their own climate projects. Financial justice would enable the Indigenous peoples to cease being mere spectators of the state programs and become empowered participants in the global climate change and sustainable development.

6.6 Expanding Recognition of Rights of Nature

One of the potential reforms based on Indigenous cosmologies is the right to nature, which perceives ecosystems as living beings with their rights. Ecuador and Bolivia under the influence of such ideas have already provided such rights in their constitutions. By acknowledging nature as a rights-holding being, the legal protection of lands and waters that are essential to the existence of Indigenous people would be enhanced, Indigenous people would have more tools to control the presence of destructive projects, and new legal mechanisms would be established to oppose such endeavors. Making the right to nature a part of the international law, either via treaties or in resolutions of UN thus, would bring the global norms to the worldviews of Indigenous peoples and broaden the environmental law (Daran and Din, 2025).

6.7 Future Research and Policy Directions

The research and policy in the future should be centered on the implementation of these reforms excellently. Some of these questions are how it is possible to monitor Free, Prior and Informed Consent (FPIC) in the context of international climate governance, how international treaties could include models of Indigenous co-management, and how international financial institutions can become more equitable and accessible. Also, new jurisprudence should be considered which can bring together fragmented legal regimes including an upcoming climate obligation advisory opinion of the ICJ. These are complex matters that need an interdisciplinary approach that includes legal scholarship and political science, Indigenous studies, and climate science to establish more just and inclusive systems of governance.

7. CONCLUSION

The changing system of the international environmental law demonstrates the progress and ongoing constraints of the international climate governance in the context of the realization of Indigenous rights. The United Nations Declaration on the Rights of Indigenous Peoples (UNDRIP) and the ILO Convention No. 169 are increasingly acknowledging Indigenous rights, and so-called participation spaces are widened through the Local Communities and Indigenous Peoples Platform (LCIPP). However, implementation is still not even. The structural obstacles continue to restrict access by Indigenous peoples to decision making, financial resources and fair play in policy development in their lands, cultures, and future. A remedy to these divides would involve legal innovation and a real shift in the ways in which the global institutions regard the Indigenous worldviews and knowledge systems.

Reform should be practical and should be the subject of a future research and policy. Some of the most significant questions are the need to effectively monitor Free, Prior and Informed Consent (FPIC) in the climate governance, the ability to include Indigenous co-management models in global treaties, and the ability to make sure financial institutions provide direct and fair access to climate funding. An imminent International Court of Justice advisory opinion on climate commitments could define the overlapping jurisdiction of human-rights and environmental obligations, and will be an opportunity to balance the world jurisprudence.

These issues require an interdisciplinary effort of legal, political science, Indigenous studies, and climate science in answering. Indigenization of knowledge, financial justice and right to nature are not marginal reforms but fundamental in fair climate governance. By acknowledging Indigenous systems of governance as legitimate environmental authorities, it will be guaranteed that their principles of stewardship are used in the formulation of international policy.

Finally, it is ethically and practically imperative to reconcile human-rights and environmental regimes. It would enhance accountability through integration, lower the legal fragmentation, and produce more equitable results to vulnerable communities. By making Indigenous people equal partners and actors in climate change, the process makes climate action more inclusive and sustainable, shifting their rights and knowledge to the center of global sustainability.

REFERENCES

Abate, R., & Kronk, E. A. (Eds.). (2013). *Climate change and indigenous peoples: The search for legal remedies*. Edward Elgar Publishing. DOI: 10.4337/9781781001806

Abidin, H. (2023). The Rights of Indigenous Peoples in Nationally Determined Contributions. *International Human Rights Law Review*, *12*(1), 90–111. DOI: 10.1163/22131035-12010006

Bagni, S. (2024). Indigenous people and the global environmental movement: Can the language of "rights" really capture their claims? In *Minority Rights and Social Change* (pp. 109–130). Routledge. DOI: 10.4324/9781003394280-8

Balint, J., Evans, J., & McMillan, N. (2014). Rethinking transitional justice, redressing indigenous harm: A new conceptual approach. *The International Journal of Transitional Justice*, *8*(2), 194–216. DOI: 10.1093/ijtj/iju004

Barnabas, S. G. (2017). The legal status of the United Nations Declaration on the Rights of Indigenous Peoples (2007) in contemporary international human rights law. *International Human Rights Law Review*, *6*(2), 242–261. DOI: 10.1163/22131035-00602006

Belfer, E., Ford, J. D., Maillet, M., Araos, M., & Flynn, M. (2019). Pursuing an indigenous platform: Exploring opportunities and constraints for indigenous participation in the UNFCCC. *Global Environmental Politics*, *19*(1), 12–33. DOI: 10.1162/glep_a_00489

Boyd, D. R. (2021). *The environmental rights revolution: A global study of constitutions, human rights, and the environment*. University of British Columbia Press.

Bulkeley, H., Chan, S., Fransen, A., Landry, J., Seddon, N., Deprez, A., & Kok, M. (2023). Building Synergies Between Climate & Biodiversity Governance: A Primer For COP28.

Colella, J. P., Silvestri, L., Súzan, G., Weksler, M., Cook, J. A., & Lessa, E. P. (2023). Engaging with the Nagoya Protocol on Access and Benefit-Sharing: Recommendations for noncommercial biodiversity researchers. *Journal of Mammalogy*, *104*(3), 430–443. DOI: 10.1093/jmammal/gyac122

Craft, A., Gunn, B. L., Knockwood, C., Christie, G., Askew, H., Borrows, J., ... & Lightfoot, S. (2018). UNDRIP implementation: more reflections on the braiding of international, domestic and indigenous laws.

Daran, D. E., & Din, M. Z. U. (2025). Environmental Human Rights: An Innovative Approach to Achieving Sustainable Development. In *Legal Frameworks and Educational Strategies for Sustainable Development* (pp. 125-142). IGI Global.

Donnelly, J., & Whelan, D. J. (2020). *International human rights*. Routledge. DOI: 10.4324/9780429266072

Eichler, J., & Barnier-Khawam, P. (2021). Criminalization, securitization and other forms of illegalizing indigenous contestations in Chile: Responses from constitutional law and inter-American jurisprudence on Mapuche people's rights. *Journal of Human Rights Practice*, *13*(2), 357–385. DOI: 10.1093/jhuman/huab041

Eichler, J., & Navarro, F. V. M. (2023). Proceduralising indigenous peoples' demands: Indigenous environmental rights and legal pluralism in contemporary jurisprudence. *Legal Pluralism and Critical Social Analysis*, *55*(1), 5–34. DOI: 10.1080/27706869.2023.2194846

European Court of Human Rights. (2024). *Verein KlimaSeniorinnen Schweiz and Others v. Switzerland*. Application No. 53600/20, Judgment of April 9, 2024 (Grand Chamber).

Ewell, C. (2022). Reimagining the renewable energy transition: The potential for mandatory corporate due diligence to ensure respect for the right to free, prior, and informed consent. *Yale J. Int'l L.*, *47*, 309.

Fajardo, V., Huffman, A., & Villa, L. Z. (2024). *Reassessing Common But Differentiated Responsibilities and Respective Capabilities: Climate Finance and the Paris Agreement at COP29*. Völkerrechtsblog.

Frizzarin Bassa, G., & Maren Schneider, J. (2025). Indigenous Peoples' Collaborative Leadership in Global Environmental Governance at the UN Conference of Climate Change: Decolonial-Collaborative Leadership Principles.

Fullerton, D., & Muehlegger, E. (2019). Who bears the economic burdens of environmental regulations? *Review of Environmental Economics and Policy*, *13*(1), 62–82. DOI: 10.1093/reep/rey023

Garnett, S. T., Burgess, N. D., Fa, J. E., Fernández-Llamazares, Á., Molnár, Z., Robinson, C. J., & Watson, J. E. M. (2018). A spatial overview of the global importance of Indigenous lands for conservation. *Nature Sustainability*, *1*(7), 369–374. DOI: 10.1038/s41893-018-0100-6

Gerhardt, C. (2023). *Sea Change: an atlas of islands in a rising ocean*. Univ of California Press.

Giacomini, G. (2022). *Indigenous peoples and climate justice: a critical analysis of international human rights law and governance*. Springer Nature. DOI: 10.1007/978-3-031-09508-5

GROMEA. J. V. (2025). *The Cost of Climate Justice: Applying the Principle of Common but Differentiated Responsibilities and Respective Capabilities* (Doctoral dissertation, UNIVERSITY OF ST. LA SALLE).

Gunn, B. L. (2021). Protecting Indigenous Peoples' Rights though Indigenous Peoples' Participation in Decision-Making: A Climate Change Example. *McGill J. Sust. Dev. L.*, *17*, 1.

Habib, R. I., Altaf, H. S. S., & Razzaq, M. U. (2022). The Function of the European Union and its Member Nations in Upholding Maritime Human Rights. [PJSS]. *Pakistan Journal of Social Sciences*, *42*(2).

Human Rights Committee. (2020). *Ioane Teitiota v. New Zealand*. Communication No. 2728/2016, U.N. Doc. CCPR/C/127/D/2728/2016 (Views adopted on January 7, 2020).

Inter-American Court of Human Rights. (2007, November 28). *Case of the Saramaka People v. Suriname (Preliminary objections, merits, reparations, and costs)* (IACHR Series C No. 172).

Inter-American Court of Human Rights. (2012). *Kichwa Indigenous People of Sarayaku v. Ecuador*. Series C, No. 245.

Inter-American Court of Human Rights. (2020, February 6). *Lhaka Honhat (Our Land) v. Argentina*. Series C, No. 400.

International Court of Justice. (2025, July 23). *Obligations of States in respect of climate change: The Court gives its Advisory Opinion and responds to the questions posed by the General Assembly (Press release 2025/36)*. International Court of Justice. https://www.icj-cij.org/node/115844 (Accessed on 10-09-2025).

International Labour Organization. (1989). *Indigenous and Tribal Peoples Convention, C169*. https://www.ilo.org/dyn/normlex/en/f?p=NORMLEXPUB:12100:0:NO:P12100_ILO_CODE:C169

IPBES. (2019). *Global assessment report on biodiversity and ecosystem services*. Intergovernmental Science-Policy Platform on Biodiversity and Ecosystem Services. https://doi.org/DOI: 10.5281/zenodo.3831673

Jalil, S. (2025). A REVIEW OF THE ROLE OF SOFT INSTRUMENTS IN INTERNATIONAL ENVIRONMENTAL LAW. *Hacettepe Hukuk Fakültesi Dergisi*, *15*(2), 824–868. DOI: 10.32957/hacettepehdf.1672865

Jha, M. K., & Dev, M. (2024). Impacts of climate change. In *Smart internet of things for environment and healthcare* (pp. 139–159). Springer Nature Switzerland. DOI: 10.1007/978-3-031-70102-3_10

Jull, J., Fairman, K., Oliver, S., Hesmer, B., & Pullattayil, A. K. (2023). Interventions for Indigenous Peoples making health decisions: A systematic review. *Archives of Public Health = Archives Belges de Santé Publique*, *81*(1), 174. DOI: 10.1186/s13690-023-01177-1 PMID: 37759336

Kastlunger, J. (2024). *UNGA Resolution 76/300-the human right to a clean, healthy and sustainable environment, the possible impact on the establishment of a human right to a healthy environment* (Doctoral dissertation).

Khan, B. Z. (2024). Common but Differentiated Responsibilities: A Shield for Equity or Barrier to Global Climate Action? *Journal of Asian Development Studies*, *13*(4), 1159–1166. DOI: 10.62345/jads.2024.13.4.94

Kleiva, K. (2017). *EXTRACTIVE INDUSTRIES, ENVIRONMENTAL RACISM AND INDIGENOUS RIGHTS IN THE AMERICAS* (Doctoral dissertation, Central European University).

Knox, J. H. (2020). Human rights principles and climate change. In Jodoin, S., & Seck, S. C. (Eds.), *Human rights and the environment: Legality, indivisibility, dignity and geography* (pp. 75–92). Cambridge University Press., DOI: 10.1017/9781108765015.006

Koskenniemi, M. (2006). *Fragmentation of international law: Difficulties arising from the diversification and expansion of international law*. Report of the Study Group of the International Law Commission. United Nations. https://legal.un.org/ilc/documentation/english/a_cn4_l682.pdf

Leal-Arcas, R. (2025). Effectiveness of the UN Framework Convention on Climate Change and the Paris Agreement. *Connecticut Journal of International Law*, *41*(1). Advance online publication. DOI: 10.2139/ssrn.5396993

Luporini, R. (2025). Climate Change Adaptation and Disaster Risk Reduction in International Human Rights Law. In *Climate Change Adaptation, Disaster Risk Reduction and Human Rights in International Law* (pp. 67–103). Springer. DOI: 10.1007/978-3-031-94596-0_4

Mardikian, L. (2023). The right to a healthy environment before the inter-american court of human rights. *The International and Comparative Law Quarterly*, *72*(4), 945–975. DOI: 10.1017/S0020589323000416

Nicholas, G. (2022). Protecting Indigenous heritage objects, places, and values: Challenges, responses, and responsibilities. *International Journal of Heritage Studies*, *28*(3), 400–422. DOI: 10.1080/13527258.2021.2009539

O'Neill, L., Thorburn, K., & Hunt, J. (2019). Ensuring Indigenous benefit from large-scale renewable energy projects: Drawing on experience from extractive industry agreement making.

Obergassel, W., Arens, C., Beuermann, C., Elsner, C., Hermwille, L., Kreibich, N., ... & Spitzner, M. (2023). Last call for 1.5 degrees: COP27 in Sharm El Sheikh schedules fund for loss and damage for departure while mitigation stays grounded.

Okedele, P. O., Aziza, O. R., Oduro, P., & Ishola, A. O. (2024). Human rights, climate justice, and environmental law: Bridging international legal standards for social equity. *Human Rights (Chicago, Ill.)*, *20*(12), 232–241.

Olsson, E. (2024). Greening the European Convention on Human Rights A Study on the Intersection of Climate Change and Human Rights in the European Court of Human Rights.

Park, J., Wilcox, K., & Ineese-Nash, N. (2023). A World Fit for the Next Seven Generations: Upholding Indigenous Rights for the Foundation of a Sustainable Future. *Canadian Journal of Children's Rights/Revue canadienne des droits des enfants, 10*(2), 5-29.

Peel, J., & Osofsky, H. M. (2018). *Climate change litigation: Regulatory pathways to cleaner energy*. Cambridge University Press., DOI: 10.1017/9781108284646

Pegram, T., & Rodríguez, N. H. (2018). Bridging the Gap: National Human Rights Institutions and the Inter-American Human Rights System. In *The Inter-American Human Rights System: Impact Beyond Compliance* (pp. 167–198). Springer International Publishing.

Perez-Leon-Acevedo, J. P. (2024). Reparations in environmental cases: Should the International Criminal Court consider the Inter-American Court of Human Rights' jurisprudence? *Journal of International Dispute Settlement*, *15*(3), 377–403. DOI: 10.1093/jnlids/idae002

Petersmann, M. C. (2021). Contested indigeneity and traditionality in environmental litigation: The politics of expertise in regional human rights courts. *Human Rights Law Review*, *21*(1), 132–156. DOI: 10.1093/hrlr/ngaa054

Rajamani, L. (2025). Empowering International Law to Address Claims for Climate Reparations. *The American Journal of International Law*, *119*(3), 484–506. DOI: 10.1017/ajil.2025.10084

Redvers, N., Aubrey, P., Celidwen, Y., & Hill, K. (2023). Indigenous Peoples: Traditional knowledges, climate change, and health. *PLOS Global Public Health*, *3*(10), e0002474. DOI: 10.1371/journal.pgph.0002474 PMID: 37831713

Roberts, J. T., & Weikmans, R. (2017). Postface: Fragmentation, failing trust and enduring tensions over climate finance. *International Environmental Agreement: Politics, Law and Economics*, *17*(1), 129–137. DOI: 10.1007/s10784-016-9347-4

Rodríguez, A. M. (2024). *Environmental and Social Risks and Opportunities for Indigenous Peoples in Climate and Nature Finance* (Doctoral dissertation, Harvard University).

Saharkhiz, N. (2024). Legal Pathways for Environmental Justice in Indigenous Territories. *Journal of Human Rights. Law & Policy*, *2*(2), 25–32.

Schlosberg, D., & Collins, L. B. (2014). From environmental to climate justice: Climate change and the discourse of environmental justice. *Wiley Interdisciplinary Reviews: Climate Change*, *5*(3), 359–374. DOI: 10.1002/wcc.275

Schuijers, L. (2023). Australia's Inaction on Climate Change Is a Violation of Torres Strait Islanders' Human Rights: Billy v Australia. *Melb. J. Int'l L.*, *24*, 140.

Simporios, L. L. (2024). Silenced narratives: Assessing the human rights situations of indigenous peoples in the Philippines. *Journal of Interdisciplinary Perspectives*, *2*(5), 16–36. DOI: 10.69569/jip.2024.0062

Smyth, D., & Isherwood, M. (2016). Protecting sea country: Indigenous people and marine protected areas in Australia. *Big, Bold and Blue: Lessons from Australia's marine protected areas*, 307-325.

Taiapa, K., Moewaka Barnes, H., & Wright, S. (2025). Climate change and mātauranga Māori: Making sense of a western environmental construct. *Kotuitui*, *20*(4), 405–416. DOI: 10.1080/1177083X.2024.2350195

Tanasescu, M. (2022). *Understanding the rights of nature: A critical introduction* (p. 168). transcript Verlag.

Teaiwa, K. (2019). No distant future: Climate change as an existential threat. *Australian Foreign Affairs*, (6), 51–70.

Tigre, M. A. (2022). Climate Change and Indigenous Groups: The rise of Indigenous voices in climate litigation. *e-Publica*, *9*(3), 214.

Tomaselli, A. (2020). Political participation, the international labour organization, and Indigenous peoples: Convention 169 'participatory'rights. *International Journal of Human Rights*, *24*(2-3), 127–143. DOI: 10.1080/13642987.2019.1677612

Tran, M., & Salamanca, A. (2023). *Advancing climate policy: harnessing indigenous knowledge at the science-policy interface*. SEI Working Paper. Stockholm Environment Institute. . 029.DOI: <ALIGNMENT.qj></ALIGNMENT>10.51414/sei2023

UN Human Rights Committee. (2022, September 22). *Daniel Billy and others v. Australia* (Torres Strait Islanders case), Communication No. 3624/2019.

UN Human Rights Committee. (2020, January 7). *Teitiota v. New Zealand* (Communication No. 2728/2016)

United Nations. (1966). *International covenant on civil and political rights*. United Nations.

United Nations. (1992). *Convention on Biological Diversity*. 1760 U.N.T.S. 79.

United Nations. (2007). *United Nations declaration on the rights of Indigenous peoples*. United Nations. https://www.un.org/development/desa/indigenouspeoples/declaration-on-the-rights-of-indigenous-peoples.html

United Nations Framework Convention on Climate Change. (2015). *Paris Agreement*. https://unfccc.int/sites/default/files/english_paris_agreement.pdf

United Nations General Assembly. (2022, July 28). *The human right to a clean, healthy and sustainable environment* (Resolution A/RES/76/300). United Nations. https://digitallibrary.un.org/record/3982508

Urgenda Foundation v. State of the Netherlands, ECLI:NL:HR:2019:2007 (Supreme Court of the Netherlands, Dec. 20, 2019).

Verde, R. (2024). Improving the implementation of Article 3 of the UNDRIP: addressing the right to self-determination for Indigenous Peoples, to uphold their basic human rights'.

Verein KlimaSeniorinnen Schweiz and Others v. Switzerland, Applications nos. 53600/20 and 68209/20 (European Court of Human Rights, Apr. 9, 2024).

Yoshida, A. (2024). Beyond non-bindingness: States' implementation of UN human rights treaties bodies' concluding observations. *Nordic Journal of Human Rights*, *42*(2), 158–176. DOI: 10.1080/18918131.2023.2289763

Zhang, Y., West, P., Thakholi, L., Suryawanshi, K., Supuma, M., Straub, D., Sithole, S. S., Sharma, R., Schleicher, J., Ruli, B., Rodríguez-Rodríguez, D., Rasmussen, M. B., Ramenzoni, V. C., Qin, S., Pugley, D. D., Palfrey, R., Oldekop, J., Nuesiri, E. O., Nguyen, V. H. T., & Agyei, F. K. (2023). Governance and conservation effectiveness in protected areas and indigenous and locally managed areas. *Annual Review of Environment and Resources*, *48*(1), 559–588. DOI: 10.1146/annurev-environ-112321-081348

Chapter 11

Intergenerational Justice in Climate Litigation: Challenges of Standing, Causation, and Justiciability

Saeed Al Ali
http://orcid.org/0000-0003-3451-3130
United Arab Emirates University, UAE

ABSTRACT

The concept of intergenerational justice, and the right of future generations to a clean, healthy, and sustainable environment, involves the duty of the current generation to pass on the environment to future generations, based on the principles of comparable options, quality, and access. If this right is legally recognized, it must be protected through various legal means, especially litigation. This chapter confirms that intergenerational justice is a legal matter, as evidenced by its study of a number of judicial precedents in several jurisdictions, constitutional texts, international treaties, and declarations, in order to understand the challenges facing litigation in this field, such as challenges related to legal standing, justiciability, and causation. This chapter argues for expanding the concept of legal standing, adopting a precautionary and cumulative approach to causation, and for the explicit legal recognition of intergenerational justice in national laws.

I. INTRODUCTION

Intergenerational justice and the "future generations … right to a clean, healthy, ecologically balanced, and sustainable environment…" are considered among the most important legal solutions to fight climate change. (Weston, 2007). If this right

DOI: 10.4018/979-8-3373-6635-7.ch011

is recognized legally, it must be protected from violation through all available legal means, especially through litigation.

However, as the next sections will discuss, the use of litigation to protect this right may face several challenges. Before addressing these challenges, this section provides the necessary conceptual and historical background to explain the difficulties litigants encounter in accessing justice and pursuing litigation. It will also clarify the methodology of this chapter and outline its roadmap.

A. Conceptual & Historical Background

Intergenerational justice assumes a community that extends across time, in which "human species, hold the natural environment of our planet in common with all members of our species: past generations, the present generation, and future generations" (Weiss, 1990). All of these generations should work together to protect the environment and to act as both "trustee" and "beneficiary," with a duty to pass it on "in at least as good condition as we received," it under the principles of comparable options, comparable quality, and comparable access (Weiss, 2008). Within this definition, recognized intergenerational harms include irreparable environmental degradation by present generations that may now or in the future degrade the quality of the environment, reduce access to the Earth and its resources, or foreclose options for future generations, thereby undermining equity between generations (Weston, 2007).

These ideas appeared in several international instruments. For example, the 1972 Stockholm Declaration mentioned this idea explicitly when it stated that "Humanity bears a solemn responsibility to protect and improve the environment for present and future generations" (Handl, 2012). The Stockholm Declaration was not the only instrument to do so; the World Charter for Nature (WCN) of 1982 also stated that: "The present generations have the responsibility of ensuring that the needs and interests of present and future generations are fully safeguarded" (UN. General Assembly (37th sess.: 1982-1983), 1982).

In 1992, both the Rio Declaration on Environment and Development and the United Nations Framework Convention on Climate Change (UNFCCC) explicitly referred to intergenerational justice and emphasized the protection of "the climate system for the benefit of present and future generations of humankind, on the basis of equity" (United Nations Framework Convention on Climate Change, 2024). Both of them also consider the environmental degradation as a violation of intergenerational equity rather than merely a present injury (Weston, 2007). The UNFCCC further defines the "adverse effects of climate change" as "changes in the physical environment or biota" that have "significant deleterious effects on the composition, resilience or productivity of natural and managed ecosystems or on human health and welfare" (United Nations Framework Convention on Climate Change, 2024). All of the above-mentioned documents frame intergenerational justice as a duty imposed on states or present generations, rather than as a right directly claimable by future generations.

However, in 1981, the African Charter on Human and Peoples' Rights recognized this concept as a right by requiring states to guarantee the rights "of every person of present and future generations to live in an environment adequate to his or her health and well-being" (UNECE, 1998). The Aarhus Convention likewise addressed intergenerational justice as a rights-based concern, stating in its preamble that "Access to information, public participation and access to justice" are secured "for the benefit of present and future generations" (UNECE, 1998).

The recognition of intergenerational justice as a right implies the availability of a remedy; as the traditional legal maxim holds, "Where there's a right, there must be a remedy," (UNECE, 1998) and without remedy, rights are symbolic. It should be emphasized that this discussion is not about recognizing the general right to live in a healthy environment, as exists in some countries' constitutions such as the Philippines, but rather about the distinct right to intergenerational justice in the environmental domain, including future generations' right to live in a healthy environment (*1987 Philippine Constitution - The LawPhil Project.*, 1978).

As noted earlier, litigation is one of the most important means of protecting this right. However, litigation in the context of intergenerational justice raises several distinct challenges, including: First, standing, which can be defined as the "right to make a legal claim or seek judicial enforcement of a duty or right" (Garner, 2019). Second, causation, that can be defined as "the act or omission that produces an injury and without which the injury would not have occurred" (Garner, 2019). Finally, justiciability, "The quality or state of being appropriate or suitable for adjudication by a court" (Garner, 2019).

B. Methodology and Roadmap

In order to reach effective solutions to the challenges outlined above, this chapter will analyze and compare the relevant judicial precedents such as *Oposa, Urgenda, Milieudefensie, Neubauer, Future Generations,* and *Leghari*. The chapter will also relay in its analysis on constitutional texts from various countries such as the Constitution of Papua New Guinea, and the Philippines. The analysis will also cover international treaties and declarations such as the Rio Declaration, the Aarhus Convention, and the Escazú Agreement (Preparatory, 1992). The goal of this comparative analysis is to develop a comprehensive understanding of how different legal systems address the intergenerational justice right, the litigation challenges it raises, and the most effective responses to those challenges.

This chapter will proceed by addressing each challenge individually, standing, causation, and justiciability, and will demonstrate how existing legal and judicial efforts respond to these challenges in order to achieve the chapter's objectives and help fill the remaining gaps in current practice.

II. STANDING: WHO CAN SPEAK FOR THE UNBORN?

Standing can be defined as the right to "make a legal claim or seek judicial enforcement of a duty or right." It secures the first step to access to justice for both the current and the future generations (Garner, 2019). Traditionally, this right is granted if the plaintiff shows an interest that the plaintiff seeks to protect through the claim. This interest should be legal, directly and personally affecting the plaintiff (Drigo, 2024). This interest may involve a future harm the plaintiff seeks to prevent, a present harm the plaintiff seeks to stop, or a harm that has already occurred for which compensation is sought. However, the harm "must be concrete and particularized, and actual or imminent, not conjectural or hypothetical." not a "generally available grievance about government," (Lujan v. Defenders of Wildlife, 1992).

Therefore, limiting or expanding the concept of standing based on the plaintiff's interest correspondingly limits or expands access to justice for both current and future generations (Angstadt, 2015). Upon examining the issue of interest and the details mentioned above, it becomes apparent that proving standing in claims related to the right to intergenerational justice is, on its face, difficult. This is because such cases involve the rights of persons who have not yet been born, future rights, and interests or harms that are inherently challenging to establish. The remaining question, therefore, is who has the right to represent future generations in the present.

A. International and Regional Instruments

The Rio Declaration is a non-binding declaration that explicitly refers to future generations in its third principle. This principle states that "the right to development must be fulfilled so as to equitably meet developmental and environmental needs of present and future generations" (Preparatory, 1992). However, the Declaration does not establish a legal right or define standing. Instead, it articulates an expectation of access in Principle 10, which states that "effective access to judicial and administrative proceedings, including redress and remedy, shall be provided" (Preparatory, 1992).

By contrast, the Aarhus Convention and the Escazú Agreement are both binding regional treaties that refer explicitly to future generations and link environmental protection to access to justice, while granting access rights to the present generation (UNECE, 1998). The Aarhus Convention affirms in its preamble that environmental protection is directed toward "the benefit of present and future generations," (UNECE, 1998) and provides in Article 9 that "members of the public" who have "a sufficient interest" or claim a violation of a right "shall have access to a review procedure before a court of law or another independent and impartial body established by law," (UNECE, 1998).

The Escazú Agreement similarly recognizes in its preamble the importance of guaranteeing environmental rights "for present and future generations," (United Nations Treaties, 2018) and mandates in Article 8 that "Each Party shall guarantee access to judicial and administrative mechanisms to challenge and appeal decisions… in environmental matters" (United Nations Treaties, 2018). Neither treaty defines standing criteria directly; instead, both impose binding obligations on States to guarantee access to justice, while leaving the determination of standing and the assessment of what constitutes a sufficient interest to national law. Intergenerational justice is therefore addressed indirectly under both treaties as future generations are treated as beneficiaries of environmental protection, while access to justice rights granted to the present generation enable them to bring claims aimed at protecting environmental interests whose effects extend to the benefits of future generations (United Nations Treaties, 2018).

The African Charter on Human and Peoples' Rights goes further by considering environmental protection as a right. Article 24 recognizes that "all peoples shall have the right to a general satisfactory environment favourable to their development," (Hansungule, 2012) a provision that has been understood to expand the protection of environmental into the next generations (African Commission on Human and Peoples' Rights, 2024). Unlike the Rio Declaration, this Charter establishes a binding legal right, rather than a policy principle (Hansungule, 2012). Although the Charter does not define standing in its text, the Rules of Procedure of the African Commission on Human and Peoples' Rights allow communications to be submitted by "any person or group of persons, or by any nongovernmental organization," thereby enabling broad access to justice without requiring individualized or immediate harm (African Commission on Human and Peoples' Rights, 2024).

B. Constitutional and Statutory Texts

Rights may be established either in constitutions or in statutes, and the same is true for standing. At the constitutional level, section 24 of the South African Constitution provides a clear example. This constitution confirms that "everyone has the right to an environment that is not harmful to their health or well-being," and the environment must be protected "for the benefit of present and future generations."

[1] Consequently, the current generation have the ability to pursue judicial protection of environmental rights for the benefit of future generations, without the need to prove immediate or individualized harm.

A similar intergenerational idea appears in the National Goals and Directive Principles of the Constitution of Papua New Guinea, which describe the present generation as "trustee and custodian of the environment for future generations" (UNHCR, 1975). Although these principles are not directly justiciable, they provide a constitutional foundation that supports standing for the present generation seeking to protect environmental interests on behalf of future generations.

In the Philippines, the Constitution does not explicitly mention future generations, but it recognizes the "right of the people to a balanced and healthful ecology."[2] As will be shown in the following sections, Philippine courts have interpreted this provision expansively, extending its protection to future generations through judicial doctrine.

At the statutory level, India's National Green Tribunal Act (2010), dealt with environmental matters differently. This Act relaxed the traditional requirements of standing facilitated access to justice in these matters (India Code, 2015). This Act grants standing to "any person" with interest, "including any representative body or organization," even where the harm is public, long-term, or futuristic (India Code, 2015).

In New Zealand, section 274 of Resource Management Act structured the standing very broadly. It permits participation in proceedings by persons who have "an interest in the proceedings that is greater than the interest that the general public has" (Angstadt, 2015). Given that the Act is explicitly concerned with sustainable management of natural resources, including long-term environmental effects, this broad standing allows present actors to litigate environmental harms whose principal impacts may fall on future generations (Angstadt, 2015).

The U.S. National Environmental Policy Act (NEPA),adopts a procedural approach to environmental protection. It requires the concerned agencies to assess the environmental impacts of proposed actions before decisions are taken, a process that can indirectly protect future generations. [3] However, NEPA does not create substantive environmental rights (Robertson v. Methow Valley Citizens Council, 1989) Standing under NEPA is therefore procedural standing, that allows citizens to challenge deficiencies in the environmental impact assessment process, rather than the substantive merits of the project itself.

NEPA is applicable only to projects undertaken, approved, funded, or permitted by a federal agency, (U.S. Department of Energy, 2023) and it does not extend to purely private conduct. As a result, NEPA cannot be used to address alleged intergenerational environmental harms arising solely from private activity (Department of Transportation v. Public Citizen, 2004).

C. National Cases

There are several judicial precedents that address standing in different countries around the world. In this section, we will review the most significant cases to analysis how courts dealt with the challenge of standing arising in representing future generations.

1. Oposa v. Factoran (1993), Supreme Court of the Philippines

This case was between the Secretary of the Department of Environment and Natural Resources and minors that were joined by their parents against. These minors brought this case "for themselves, for others of their generation and for the succeeding generations," to challenge the large-scale logging that was authorized by the government through Timber License Agreements (TLAs). They argue that continued logging would "cause irreparable injury to the plaintiffs and to generations yet unborn" and would "adversely affect the ecological balance" necessary for human life (*Oposa v. Factoran, Jr.*, 1993).

The petitioners claimed that the government's acts violated their constitutional right to a balanced and healthful ecology and sought judicial intervention to prevent further environmental degradation "for the welfare of present and future generations." They demanded that the court require the government to cancel all existing timber license agreements and to prohibit the issuance of new licenses (*Oposa v. Factoran, Jr.*, 1993).

The Supreme Court in this case extended the protection of the constitutional "right to a balanced and healthful ecology" to protect future generations. The Court allowed minors to "represent their generation as well as generations yet unborn," with standing "for themselves, for others of their generation and for the succeeding generations" (*Oposa v. Factoran, Jr.*, 1993). In addition, the Court granted standing to persons with an expected benefit, thereby broadening standing rules in environmental litigation, and the recognition of intergenerational responsibility as a legal basis for access to courts (*Oposa v. Factoran, Jr.*, 1993).

2. Stichting Urgenda v. State of the Netherlands (2019)

This case was adjudicated in the Netherlands. It was between the Dutch state as a defendant and "the Urgenda Foundation, together with 886 individual citizens," (*Urgenda Foundation v. State of the Netherlands, 2024)* as plaintiffs. They alleged that the existing climate policies were insufficient to prevent dangerous climate change and that the State's greenhouse gas reduction goals failed to reflect scientific consensus, as evidenced by the projections in the reports of the Intergovernmental

Panel on Climate Change (IPCC) concerning the impacts of climate change on the Dutch population (*Urgenda Foundation v. State of the Netherlands, 2015*).

At first instance, the District Court (2015) emphasized that standing in matters related to future generations is an important and recurring concern. The court used the Dutch Civil Code provisions on public-interest suits to conclude that *Urgenda* also has standing on behalf of future generations, because the foundation's statutory purpose has an intergenerational dimension. The District Court held that the State's reduction goals were insufficient, as they were limited, which is considered a breach of the State's "duty of care to the Dutch people to preserve the climate" (*Urgenda Foundation v. State of the Netherlands, 2015*). On appeal, standing on behalf of future generations was not expressly granted; however, the court rejected the State's objection to *Urgenda's* representative capacity, thereby keeping the door open for future generations. The Supreme Court in 2019 upheld the findings of both the court of first instance and the court of appeal (*Urgenda Foundation v. State of the Netherlands, 2015*).

Accordingly, these cases are examples of judicial decisions that granted standing to plaintiffs and allowed the defense of future generations by permitting their representation by present generations and by expanding the concept of standing to include future interests or the interests of generations that have not yet been born.

3. Juliana v. United States (2020)

By contrast, the situation in the United States is different. In *Juliana*, the U.S. Court of Appeals for the Ninth Circuit acknowledged that the youth plaintiffs "alleged harms that befall or will befall them personally and to a greater extent than older segments of society" (*Juliana v. United States, 2020*). Yet held that the relief they sought, including a plan to quickly minimizing CO_2 emissions and to require the United States to "cease permitting, authorizing, and subsidizing fossil fuels" was "beyond our constitutional power," because such decisions rest with the political branches rather than the courts (*Juliana v. United States, 2020*).

The plaintiffs framed their harms as constitutional violations asserting that government conduct represent a "violations [of their] substantive rights … including the rights to life, liberty, and property" (*Juliana v. United States, 2020*) and that these harms were as claimed were by some of the plaintiffs are "psychological harm, others impairment to recreational interests, others exacerbated medical conditions, and others damage to property." thereby reflecting a present and future injury affects mainly younger persons and future generations (*Juliana v. United States, 2020*). The Juliana opinion itself does not expressly mention future generations. The Supreme Court later declined to review the case, so the Ninth Circuit's reasoning stands (*Juliana v. United States, 2020*).

However, the Ninth Circuit's refusal to grant relief for those to plaintiffs asserting personal injury suggests that representative claims on behalf of future generations would stand an even lower chance of judicial acceptance, given the court's conclusion that the requested remedy lay beyond the scope of its judicial authority.

D. Discussion and Recommendations

The above analysis of international instruments, constitutions, statutes, and case law shows that standing is the procedural gateway through which intergenerational justice may become legally actionable. Traditional standing doctrines, as mentioned earlier, require individualized, immediate, and concrete harm (*Lujan v. Defenders of Wildlife, 1992*). These requirements usually do not suit claims involving future generations, because their interests are future-oriented, cumulative, and arise gradually, and because the affected generations have not yet been born (Weston, 2007). When courts rigidly follow such requirements, access to justice for intergenerational claims is effectively barred. Standing in intergenerational justice claims can be achieved by adjusting existing standing requirements, without the need to recognize a new right or new right-holders; (Angstadt, 2015) the following paragraphs discuss these adjustments.

First, to enable access to justice in cases affecting future generations, standing must be articulated in broad terms and interpreted expansively by courts through public-interest standing in environmental matters. For example, the Aarhus Convention provides that "members of the public" who have "a sufficient interest" or allege impairment of a right "shall have access to a review procedure before a court of law" (UNECE, 1998). Similarly, the Escazú Agreement requires States to "guarantee access to judicial and administrative mechanisms to challenge and appeal decisions… in environmental matters" (United Nations Treaties, 2018). Both agreements use open-textured language on access to justice, allowing States to adopt broad standing rules and courts to interpret those rules flexibly. Which grants the present generation the necessary standing to protect environmental interests and benefits that extends to future generations (UNECE, 1998).

The Rules of Procedure of the African Commission on Human and Peoples' Rights provide another example of broadly articulated standing terms. These Rules clarify that claims or "communications" may be submitted by "any person or group of persons, or by any nongovernmental organization," without requiring proof of individualized, immediate, or personal harm (African Commission on Human and Peoples' Rights, 2024).

Second, because future generations are not yet born, protecting their environmental interests necessarily requires representative or trustee-based standing exercised by the present generation. This approach reflects the principle that "all generations are partners caring for and using the Earth" and must pass it "in at least as good condition as we received" it subject to the principles of comparable options, comparable quality, and comparable access (Weiss, 2008). Judicial precedent confirms this approach. In *Oposa v. Factoran,* the Supreme Court of the Philippines held that minors could "represent their generation as well as generations yet unborn," grounding standing in "the concept of intergenerational responsibility" and in the constitutional "right to a balanced and healthful ecology" (*Oposa v. Factoran, Jr.*, 1993). Rather than granting standing directly on future generations, this trustee-based approach secures judicial protection of their interests through present representatives.

Third, standing requirements are most easily fulfilled where they are based on a recognized substantive environmental right. Rights-based standing transforms access to justice from a procedural inquiry into whether a claimant can demonstrate a sufficient interest, into a claim alleging the violation of an existing legal right, which itself constitutes a legally protected interest. Article 24 of the African Charter on Human and Peoples' Rights is an example of this approach. This article states that "all peoples shall have the right to a general satisfactory environment favourable to their development," (Hansungule, 2012) which has been understood to extend beyond the present generation. This shows how recognizing environmental protection as a right lowers the burden of proof for standing and enables present actors to bring claims aimed at protecting the environmental interests of future generations.

Having examined standing, its definition, and the challenges it presents in intergenerational justice claims, as well as the legal solutions to those challenges, the next section will turn to causation. It follows the same approach by identifying the challenges and examining how existing legal texts, together with relevant case law, have addressed these challenges and what solutions are available.

III. CAUSATION: LINKING PRESENT CONDUCT TO FUTURE HARM

Standing is not the only procedural barrier that intergenerational justice cases face. Even where access to courts is secured, litigants encounter a further challenge in establishing causation, particularly because of difficulties related to proving responsibility under the "expansion of the temporal boundaries." between conduct and harm (Drigo, 2024). Causation can be defined as the link or the connection between the conduct of the defendant and the alleged harm (Garner, 2019).

Under the traditional approach, causation requires proof that the harm is the "fairly traceable" of the defendant's conduct (*Lujan v. Defenders of Wildlife*, 1992). This approach assumes an immediate harm resulting from a specific act and affecting identifiable persons (*City of Los Angeles v. Lyons*, 1983. Courts therefore typically require proof of actual damage, (*Lujan v. Defenders of Wildlife*, 1992) that has "Actual or imminent," is neither remote nor speculative, (*Allen v. Wright,* 1984) and demonstrates "a causal connection between the injury and the conduct complained of the injury has to be fairly traceable to the challenged action of the defendant," (*Lujan v. Defenders of Wildlife*, 1992).

Applying these requirements makes it "extremely difficult for claimants to demonstrate the necessary causal links (Drigo, 2024) in intergenerational justice claims, where the alleged harm is future-oriented and linked to generations that are not yet born (Weston, 2007). The uncertainty surrounding future harm, together with the cumulative nature of environmental damage and its distribution across populations, is often regarded as "too speculative" (*Allen v. Wright,* 1984) or "too attenuated" (*Juliana v. United States,* 2020) to satisfy traditional causation standards. This difficulty is further amplified in climate-related cases by the involvement of multiple actors, operating across different jurisdictions, and contributing through numerous actions to harm that occurs over long periods of time (Weiss, 2008).

A. International and Regional instruments

In response to these challenges, international environmental law has adopted a preventive, precautionary, and risk-based approach to causation, which is more suitable for intergenerational environmental protection. The Rio Declaration on Environment and Development (1992) is a clear example of this approach. Principle 15 states:

> *In order to protect the environment, the precautionary approach shall be widely applied by States according to their capabilities. Where there are threats of serious or irreversible damage, lack of full scientific certainty shall not be used as a reason for postponing cost-effective measures to prevent environmental degradation (Preparatory, 1992).*

This approach reconfigures causation by replacing proof of actual damage with foreseeable risk, allowing legal responsibility to arise before the actual harm. Article 3(1) of the United Nations Framework Convention on Climate Change adopts a similar approach when it stated that:

> *The Parties should take precautionary measures to anticipate, prevent or minimize the causes of climate change and mitigate its adverse effects. Where there are threats of serious or irreversible damage, lack of full scientific certainty should not be used as a reason for postponing such measures (United Nations Framework Convention on Climate Change, 2024).*

B. National Case Applications: Causation Accepted

In *Urgenda*, the court recognized the harm as the serious risk of irreversible consequences of climate change to the environment and "to the lives and welfare of Dutch residents," (*Urgenda Foundation v. State of the Netherlands, 2154).* The conduct challenged was the Dutch government's failure to fulfill its "duty of care to protect the residents of the Netherlands from the real threat of dangerous climate change," (*Urgenda Foundation v. State of the Netherlands, 2154)* specifically "by failing to reduce greenhouse gas emissions to the level necessary to prevent" such threats, as determined on the basis of the scientific findings contained in the IPCC reports. With respect to causation, the court held that every greenhouse gas emission contributes to increased atmospheric concentrations and thus to climate change, rejecting arguments based on the limited size of national emissions. Accordingly, the court stated that "There is a sufficiently close link between the Dutch emissions, global climate change and the serious risk of harm to the lives and welfare of residents of the Netherlands," (*Urgenda Foundation v. State of the Netherlands, 2154).*

In *Milieudefensie v. Shell*, the harm, the challenged conduct, and the causation analysis were largely similar to those in *Urgenda*; the principal difference was that the defendant was a private company rather than a government. The harm in this case was also the irreversible consequences of climate change, as the court recognized that "some future changes are unavoidable and/or irreversible but can be limited by deep, rapid, and sustained global greenhouse gas emissions reductions." (*Milieudefensie et al. v. Royal Dutch Shell plc,* 2019).

The defendant in this case was Royal Dutch Shell, and the challenged conduct concerned the company's corporate climate policy, which failed to adequately address its contribution to climate change, as the court held that "Shell's current policy is not sufficiently concrete, is subject to many reservations and does not cover the full scope of the Shell Group's emissions," (*Milieudefensie et al. v. Royal Dutch Shell plc,* 2019).

The court further stated that "the aggregate annual volume of CO_2 emissions into the atmosphere ... associated with the business activities and sold energy-carrying products of the Shell Group constitutes an unlawful act," (*Milieudefensie et al. v. Royal Dutch Shell plc,* 2019). The Court explicitly linked Shell's conduct to the harm by holding that "there is a direct, linear link between human-caused greenhouse gas emissions ... and global warming," (*Milieudefensie et al. v. Royal Dutch Shell plc,* 2019). RDS therefore contributes to climate change through its CO_2 emissions.

In *Neubauer v. Germany*, the primary difference between this case and *Urgenda* and *Milieudefensie v. Shell* was the identity of the defendant. In this case, the defendant was the legislator, while the courts' understanding of harm, causation, and the relevance of cumulative emissions remained largely similar to the previously discussed cases. In this case, the harm was identified as the serious and potentially irreversible consequences of climate change that "can be limited by deep, rapid, and sustained global greenhouse gas emissions reductions," (*Neubauer, et al. v. Germany,* 2018).

The conduct challenged was the failure of the federal legislator and the Federal Government to take suitable and prospectively sufficient measures to stay within the remaining national CO_2 budget, specifically that the Federal Climate Protection Act "lack[ed] sufficient specifications for further emission reductions from 2031 onwards," (*Neubauer, et al. v. Germany,* 2018). With respect to causation, the Court grounded responsibility in the cumulative contribution of national emissions to climate change risks, noting the "nearly linear relationship between the total amount of anthropogenic CO_2 emissions accumulated over time and the global temperature increase," (*Neubauer, et al. v. Germany,* 2018). The Court further explained that "If high emission levels are allowed until 2030, this will lead to increasingly drastic restrictions on freedoms after 2030." such that "The more emissions are permitted today, the more radical the reduction obligations that will be required in the future," (*Neubauer, et al. v. Germany,* 2018) thereby linking the defendant's conduct to future harm to life, freedom, and fundamental rights caused by climate change.

In Juliana v. United States, the discussed harm was the effect of climate change on human life in general, and specifically, causing deaths, shortening human lifespans, damaging property, causing food shortages, and environmental disruptions (*Juliana v. United States, 2020).* As for the challenged action, it concerned the federal government's procedures related to fossil fuels, including permitting, authorization, and subsidizing fossil fuel extraction and consumption, despite the climate risks associated with these actions. Regarding causation, the Ninth Circuit held that the federal government's policies and actions were a substantial factor contributing to the harm under discussion (*Juliana v. United States, 2020).*

C. Results, Challenges and Responses

After analyzing the cases and the relevant international texts, it becomes clear that courts have addressed causation in intergenerational climate litigation by adopting a preventive and precautionary approach. International treaties such as the Rio Declaration and the UNFCCC facilitate this approach by permitting preventive measures and by treating threat and risk as an actionable causal link, even where there is a lack of full scientific certainty, in order to prevent environmental degradation and to allow causation to be established before actual harm occurs. Courts have relied on probabilistic and foreseeable risk, demonstrated in authoritative scientific reports, rather than requiring proof of completed or individualized injury. This approach appears clearly in the *Urgenda* case, where the court grounded its reasoning in IPCC scientific reports.

Courts also addressed the problem of global and cumulative causation. While acknowledging the global nature of the climate problem, to which multiple actors across jurisdictions contribute, courts emphasized that every incremental emission legally matters and forms part of a cumulative harm process, because each contributes to "the total amount of anthropogenic CO_2 emissions accumulated over time and the global temperature increase," (*Neubauer, et al. v. Germany,* 2018). This approach enabled courts to hold legal responsibility across different actors. Courts have therefore held corporate actors (*Milieudefensie*), the state executive branch (*Urgenda*), and the legislative branch (*Neubauer*) accountable where the applicable policy or statutory framework was insufficiently protective.

Constitutional courts, as in *Neubauer*, linked present emissions to future restrictions on fundamental rights and recognized that using more of the carbon budget today will result in "greater restrictions of future freedoms" tomorrow (*Neubauer, et al. v. Germany,* 2018). Courts in different jurisdictions have therefore offered solutions to the causation problem in intergenerational justice by adopting a precautionary approach and by accepting science-based foreseeable risk as the legally sufficient causal link between present acts and omissions and future harm. With these solutions in place, the causation problem in intergenerational justice litigation can be regarded as doctrinally resolved, allowing the analysis to proceed to the separate question of justiciability.

IV. JUSTICIABILITY: THE COURTS COMPETENCY AND AUTHORITY

A. The Concept and the related International and Regional Instruments

Intergenerational justice becomes legally actionable when courts are able to interpret and enforce it, as a right without a remedy is an empty right (Weston, 2012). Without such judicial engagement, environmental duties risk being treated as moral expectations or as a "political question" rather than as legally enforceable obligations. This transition from moral aspiration to legal enforceability occurs when courts acknowledge that "Future generations have a legal right to a clean and healthy environment, and present generations have a corresponding legal obligation to protect it," (Weston, 2012). The decisive justiciability inquiry, therefore, is whether the court is being asked to apply existing legal norms, rather than to formulate policy choices.

The baseline of justiciability in intergenerational environmental litigation is the legal protection of fundamental rights over time (*Neubauer, et al. v. Germany,* 2018). Courts determine whether legally binding duties constitutional, statutory, or treaty-based are being respected (*Urgenda Foundation v. State of the Netherlands, 2015).* Where constitutions or treaties impose obligations explicitly directed to "present and future generations," those obligations furnish a judicially manageable standard (UNECE, 1998). A central justiciability objection is that long-term environmental duties involve complex facts or future impacts; however, courts in different jurisdictions have ruled that neither complexity nor futurity negates justiciability where courts apply legally cognizable obligations and rights (*Urgenda Foundation v. State of the Netherlands, 2015).*

International and regional instruments frame intergenerational justice as a matter of legal duty and legal right. The Stockholm Declaration states that "Humanity bears a solemn responsibility to protect and improve the environment for present and future generations," (Handl, 2012). Similarly, the Rio Declaration provides that "the right to development must be fulfilled so as to equitably meet developmental and environmental needs of present and future generations," (Preparatory, 1992). The Rio Declaration further requires that "Effective access to judicial and administrative proceedings, including redress and remedy, shall be provided," (Preparatory, 1992).

These declarations are not legally binding; however, they have significantly influenced the development of national legal frameworks, with domestic laws built upon their normative principles, thereby providing courts with a foundation for justiciability in intergenerational justice cases (*Urgenda Foundation v. State of the Netherlands, 2015).* Unlike these non-binding declarations, the UNFCCC is a binding convention for its parties following ratification at the state level, and it obliges them to "protect the climate system for the benefit of present and future generations of humankind, on the basis of equity," (United Nations Framework Convention on Climate Change, 2024). Together, these instruments supply courts with authority to review governmental action or omission against legally specified minima.

B. National Case: Justiciability Recognized

Across jurisdictions, courts have affirmed that long-term climate obligations are justiciable even if they involve future impacts, scientific complexity, or affect policy. This is so as long as courts are being asked to apply existing legal obligations, rather than to formulate climate policy. In *Urgenda*, the Court of Appeal held that the State had "breached a duty of care to the Dutch people," (*Urgenda Foundation v. State of the Netherlands, 2015).* and that this breach grounded the court's justiciability to ensure compliance with binding legal obligations.

The Court of Appeal further grounded the justiciability of its decision in Articles 2 and 8 of the European Convention on Human Rights, holding that climate change threatens the protection of human rights and "falls within the scope of the State's positive obligations," which courts are required to review. Accordingly, this form of judicial review does not constitute policy-making, but rather the enforcement of legal obligations derived from both the duty of care and the ECHR (*Urgenda Foundation v. State of the Netherlands, 2015).*

In *Neubauer*, the German Federal Constitutional Court justified its decision on the understanding that the constitution requires the legislature to protect fundamental rights "also in responsibility for future generations," (*Neubauer, et al. v. Germany,* 2018). The Court tied present legislative choices directly to future restrictions on fundamental rights, holding that permitting higher emissions in the present would "lead to increasingly drastic restrictions on freedoms after 2030," and "more radical the reduction obligations that will be required in the future," (*Neubauer, et al. v. Germany,* 2018).

In *Asghar Leghari v. Federation of Pakistan*, the Lahore High Court held that climate change "poses a serious risk to the fundamental rights to life, health, food and an adequate standard of living," (*Leghari v. Federation of Pakistan, 2015).* The court treated governmental failure to implement climate policy as a justiciable violation of these rights guaranteed by the constitution. The Court expressly framed climate governance as a legal obligation, stating that international environmental principles such as "the doctrine of public trust, sustainable development, precautionary principle and intergenerational equity form part of the fundamental rights," (*Leghari v. Federation of Pakistan, 2015).*

In *Future Generations v. Ministry of the Environment*, the Colombian Supreme Court recognized that the Amazon rainforest is a "subject of rights," (*Future Generations v. Ministry of the Environment and Others, 2018)* and held that environmental degradation threatened the fundamental rights of both present and future generations.

C. National Cases: Justiciability Restricted

In *Juliana v. United States*, the youth plaintiffs claimed that they had suffered concrete and particularized injuries as a result of climate change. The U.S. Court of Appeals for the Ninth Circuit accepted this claim and acknowledged that the case raised serious legal concerns. However, the difficulty arose from the nature of the relief sought, as the plaintiffs requested an order requiring the federal government to prepare and implement a national plan to reduce CO_2 emissions, (*Juliana v. United States, 2020)* framing the dispute as one concerning constitutional rights and governmental obligations. The court held that the requested relief was "beyond our constitutional power" (*Juliana v. United States, 2020)* because it would require the judiciary to "order, design, supervise, or implement a comprehensive remedial plan" for national climate policy (*Juliana v. United States, 2020)*. Such relief, the court held, "must be presented to the political branches," (*Juliana v. United States, 2020)* thereby refusing to reach the merits of the constitutional claims. In this posture, the court did not reject the existence of legal duties but instead treated the scope of the requested remedy as non-justiciable under Article III.

These cases show that justiciability challenges in intergenerational climate litigation can be resolved through rights and treaty-based anchors, constitutional and common-law duties, and statutory frameworks. In addition, litigants should frame their claims around judicially manageable obligations and seek actionable remedies, rather than requesting political remedies from the court that lie beyond its power (*Urgenda Foundation v. State of the Netherlands, 2015; Milieudefensie et al. v. Royal Dutch Shell plc*, 2019).

V. CONCLUSION

In this chapter, the concept of intergenerational justice was defined, and the relevant international and domestic legal texts were clarified, alongside highlighting key judicial precedents that established many important details that help overcome any difficulties in achieving the primary objective of this right or duty, which is to pass on the environment to future generations in a condition no worse than that received by the current generation. We deliberately ensured that recommendations and findings were attached to each section so that there is no gap between the issues and the solutions or recommendations. Nevertheless, we summarize the most important recommendations and solutions presented in those sections as follows.

The proposed solutions regarding standing involve defining it broadly through flexible and open-ended language, and allowing public-interest litigation by individuals, groups, and non-governmental organizations. In cases of intergenerational justice, it is essential that standing not be restricted by requirements to prove individualized or immediate harm. Under this approach, the present generation is able to protect the environment in ways that affect future generations. Future generations, by definition, consist of persons who have not yet been born. Accordingly, the only effective way to safeguard their rights at the present time is by enabling the current generation to represent them before courts and to ensure that the environment is preserved in a condition at least comparable to that which it inherited. Judicial precedents, such as *Oposa v. Factoran*, confirm that standing is most effectively achieved when environmental protection is recognized as a fundamental right, since rights-based recognition allows standing to be established through the allegation of a rights violation, without the need to demonstrate any additional personal or individualized interest.

For the causation related challenges, Courts can adopt a preventive and precautionary approach to causation, relying on international instruments such as the Rio Declaration and the UNFCCC, accepting foreseeable and probabilistic risk grounded in authoritative scientific evidence (notably IPCC reports), and linking present emissions to future restrictions on fundamental rights, thereby resolving the causation problem doctrinally and allowing claims to proceed without proof of completed or individualized harm. The challenges of causation could be eased also if the legal system in general acknowledge the global and cumulative nature of climate change, and that every incremental emission legally contributes to cumulative harm, enabling responsibility to be attributed to multiple actors including corporate entities, executive authorities, and legislatures where policies or statutory frameworks are insufficiently protective.

Finally, as demonstrated, intergenerational justice is a matter that can be litigated within environmental law, particularly where courts are asked to enforce constitutional, statutory, or treaty-based obligations. This remains the case despite the long-term nature of environmental duties, scientific complexity, and uncertainty regarding future impacts. In addition to that, and as a recommendation to plaintiffs, to have as successful litigation in this field, require that the remedies sought should be framed around enforceable legal obligations or rights, rather than claiming broad or non-concrete claims that exceed the competence of the court. As happened in *Juliana*, the case was dismissed not as a result of absence of harm or causation, but because the relief sought exceeded the court's authority. Accordingly, the effectiveness of intergenerational climate litigation depends on grounding claims in enforceable legal obligations and seeking remedies that courts are capable of managing in order to protect the environmental rights of present and future generations. From this perspective, intergenerational justice can be understood not as a purely theoretical concept, but as a legally cognizable right and duty.

REFERENCES

African Commission on Human and Peoples' Rights. (2024, August 9). *Social and Economic Rights Action Center (SERAC) and Center for Economic and Social Rights (CESR) / Nigeria - 155/96*. https://achpr.au.int/en/decisions-communications/social-and-economic-rights-action-center-serac-and-center-economic-15596

Allen v. Wright, 468 U.S. 737. (1984). Justia Law. https://supreme.justia.com/cases/federal/us/468/737/

Angstadt, J. M. (2015). Sercuring Access to Justice Through Environmental Courts and Tribunals: A Case in Diversity. *Vt. J. Envtl. L., 17*, 345.

City of Los Angeles v. Lyons, 461 U.S. 95. (1983). Justia Law. https://supreme.justia.com/cases/federal/us/461/95/

Department of Transportation v. Public Citizen, 541 U.S. 752 (2004).

Drigo, A. (2024). Future Generations in Climate Litigation: Early Whispers of an Intergenerational Law? *German Law Journal, 25*(7), 1120–1148. DOI: 10.1017/glj.2024.36

Future Generations v. Ministry of the Environment and Others. (2018). Climatecasechart.com; Sabin. https://www.climatecasechart.com/document/future-generations-v-ministry-of-the-environment-and-others_f076

Garner, B. A. (2019). *Black's Law Dictionary* (11th ed.). Thomson Reuters.

Handl, G. (2012). *Declaration of the United Nations Conference on the Human Environment - Main Page*. Un.org. https://legal.un.org/avl/ha/dunche/dunche.html

Hansungule, M. (2012). The African charter on human and peoples' rights. *The African Union: Legal and institutional framework. A manual on the Pan-African Organization*, 417-453.

India Code. (2015). National Green Tribunal Act, 2010.

Juliana v. United States, 947 F.3d 1159 (9th Cir. 2020).

Leghari v. Federation of Pakistan. (2015). Climatecasechart.com; Sabin. https://www.climatecasechart.com/document/leghari-v-federation-of-pakistan_80c5

Lujan v. Defenders of Wildlife, 504 U.S. 555. (1992). Justia Law. https://supreme.justia.com/cases/federal/us/504/555/

Milieudefensie et al. v. Royal Dutch Shell plc. (2019). Climatecasechart.com; Sabin. https://www.climatecasechart.com/document/milieudefensie-et-al-v-royal-dutch-shell-plc_c3e4

Neubauer, et al. v. Germany. (2018, November 22). Climatecasechart.com; Sabin. https://www.climatecasechart.com/document/neubauer-et-al-v-germany_0a3e

Oposa v. Factoran, Jr., 296 Phil. 694 (1993).

1987. *Philippine Constitution - The LawPhil Project*. (1987). Lawphil.net. https://lawphil.net/consti/cons1987.html

Preparatory, U. N. (1992, April 2). *Rio Declaration on Environment and Development: draft principles /: proposed by the Chairman*. United Nations Digital Library System; UN, https://digitallibrary.un.org/record/141565?v=pdf

Robertson v. Methow Valley Citizens Council, 490 U.S. 332 (1989).

UN. General Assembly (37th sess.: 1982-1983) (1982). World Charter for Nature. United Nations Digital Library. *Digitallibrary.un.org*. https://digitallibrary.un.org/record/39295

UNECE. (1998). *Convention on Access to Information, Public Participation in Decision-making and Access to Justice in Environmental Matters*. United Nations Economic Commission for Europe.

UNHCR. (1975). *Papua New Guinea: Constitution of the Independent State of Papua New Guinea*. Refworld.

United Nations Framework Convention on Climate Change. (2024). ARTICLE 3 PRINCIPLES. Unfccc.int. https://unfccc.int/resource/ccsites/seychell/conven/text/art03.htm

United Nations Treaties. (2018). *Regional Agreement on Access to Information*. Public Participation and Justice in Environmental Matters in Latin America and the Caribbean.

Urgenda Foundation v. State of the Netherlands, Hague District Court (Netherlands, June 24, 2015), ECLI:NL:RBDHA:2015:7196.

U.S. Department of Energy. (2023, June 3). *The National Environmental Policy Act of 1969*. Energy.gov. https://www.energy.gov/nepa/articles/national-environmental-policy-act-1969

Weiss, E. B. (1990). Our rights and obligations to future generations for the environment. *The American Journal of International Law*, *84*(1), 198–207. DOI: 10.2307/2203020

Weiss, E. B. (2008). Climate change, intergenerational equity, and international law. *Vermont Journal of Environmental Law*, *9*(3), 615–627. DOI: 10.2307/vermjenvilaw.9.3.615

Weston, B. H. (2007). Climate change and intergenerational justice: Foundational reflections. *Vt. J. Envtl. L.*, *9*, 375.

Weston, B. H. (2012). The theoretical foundations of intergenerational ecological justice: An overview. *Human Rights Quarterly*, *34*(1), 251–266. DOI: 10.1353/hrq.2012.0003

ENDNOTES

1 S. Afr. Const. 1996 § 24(a)–(b).

2 CONST. (Phil.) art. II, § 16 (1987).

3 42 U.S.C. § 4332(2)(C).

Chapter 12
Jurisdiction for the Climate:
Design, Mandate, and Feasibility of an International Court

Silvio Andrae
http://orcid.org/0000-0002-8586-7812
Independent Researcher, Germany

ABSTRACT

International climate law has experienced rapid normative consolidation, as evidenced by the recognition of binding obligations in recent international jurisprudence for prevention, due diligence, cooperation, and reparation. Despite this progress, institutional mechanisms for systematic enforcement remain inadequate. Existing courts are limited by narrow jurisdiction, restricted standing, and procedural frameworks that are not well-suited for complex, science-intensive climate disputes. This chapter contends that an International Court for the Environment represents the institutional realization of an already mature normative framework, rather than a mere political innovation. It presents a doctrinal and institutional blueprint for climate adjudication, addressing jurisdiction, procedure, evidentiary standards, remedies, compliance mechanisms, and participatory guarantees. By integrating scientific expertise, expanding access to justice, and facilitating structural remedies, such a court could bridge the gap between normative ambition and practical accountability.

DOI: 10.4018/979-8-3373-6635-7.ch012

INTRODUCTION

The Enforcement Deficit In International Climate Law

The global climate crisis has profoundly transformed the normative landscape of international law. Over the past three decades, states have developed an extensive network of treaties, institutions, and policy instruments to mitigate greenhouse gas emissions, facilitate adaptation, and support vulnerable societies. From the United Nations Framework Convention on Climate Change to the Paris Agreement, climate governance has evolved into one of the most densely regulated fields of contemporary international cooperation.

Yet despite this normative expansion, a persistent structural deficit remains. International climate law continues to rely predominantly on cooperative, transparency-based, and facilitative mechanisms. Compliance is encouraged through reporting, peer review, and political dialogue rather than through binding adjudication and enforceable remedies. The Paris Agreement's implementation system, centered on non-adversarial and non-punitive procedures, reflects this orientation. While such mechanisms have contributed to norm diffusion and institutional learning, they have proven insufficient to ensure timely and adequate emissions reductions and effective protection against escalating climate risks.

This enforcement gap is particularly consequential for vulnerable states and affected communities. Small island developing states, indigenous peoples, and populations exposed to extreme weather, sea-level rise, and ecosystem degradation frequently bear the consequences of regulatory failure without access to effective international remedies. In existential situations involving loss of territory, cultural heritage, or livelihood systems, soft compliance mechanisms offer limited protection. The absence of binding decisions, structured reparations, and institutionalized oversight undermines the practical relevance of climate obligations.

Recent developments in international jurisprudence have significantly altered this landscape. International courts have begun to translate previously programmatic climate commitments into legally articulated obligations. Advisory and contentious proceedings before international courts have clarified that climate protection is not merely a matter of political discretion, but engages binding legal obligations. The International Tribunal for the Law of the Sea issued an advisory opinion in 2024 classifying anthropogenic greenhouse gas emissions as marine pollution and imposing stringent duties of prevention and regulation. In 2025, the International Court of Justice articulated a comprehensive framework of climate-related obligations and legal consequences under general international law. In parallel, the Inter-American Court of Human Rights confirmed that state inaction in the face of environmental degradation may constitute a violation of fundamental human rights.

These decisions mark a decisive shift in the legal status of climate governance. They consolidate duties of due diligence, cooperation, precaution, and reparation, and emphasize their collective and *erga omnes* character. Climate change has thus been firmly integrated into the corpus of international responsibility and human rights protection. However, this jurisprudential consolidation has not been accompanied by a corresponding institutional transformation. The most authoritative pronouncements remain largely advisory, sectorally confined, or regionally limited. They clarify the law, but do not provide comprehensive procedural pathways for its systematic enforcement.

As a result, a structural asymmetry has emerged between normative density and institutional capacity. International courts possess increasing interpretative authority in climate matters but lack jurisdictional reach, standing arrangements, evidentiary infrastructure, and compliance mechanisms suited to planetary-scale environmental harm. Climate disputes continue to be fragmented across multiple fora, each operating within mandates that were not designed for cumulative, science-intensive, and long-term regulatory conflicts.

This contribution addresses this institutional deficit through a systematic analysis of the design, mandate, and feasibility of an International Court for the Environment. It proceeds on the premise that establishing specialized climate jurisdiction does not require the creation of new substantive obligations. Rather, it represents the institutional consequence of an already mature normative framework. The central question is therefore not whether international law permits climate adjudication, but how existing legal principles can be translated into an effective and coherent judicial architecture.

The chapter develops this argument in three steps. First, it reconstructs the normative foundations of climate adjudication, focusing on due diligence, precaution, *erga omnes* obligations, and the integration of environmental and human rights law. Second, it analyses recent international jurisprudence and its systemic implications for institutional design. Third, it examines jurisdictional models, procedural regimes, remedial structures, enforcement mechanisms, and participatory guarantees suitable for climate-related disputes.

Methodologically, the contribution adopts a doctrinal and institutional approach. It combines analysis of primary sources, judicial practice, and general principles of international law with comparative insights into existing adjudicatory regimes. Political feasibility considerations are addressed only to the extent they bear on legal design and institutional coherence.

The core thesis advanced is that an International Court for the Environment (ICE) would not fragment international law, but consolidate its emerging climate jurisprudence within a specialized, accessible, and compliance-oriented forum. By integrating scientific expertise, expanding standing, and developing structural remedies, such a court would bridge the gap between normative commitment and practical accountability.

In this sense, the proposal for an environmental court reflects a broader transformation of international law under conditions of planetary risk. As global environmental stability becomes a prerequisite for the enjoyment of fundamental rights and sustainable development, the capacity of legal institutions to enforce collective obligations assumes constitutional significance. Climate adjudication thus emerges not as an exceptional innovation, but as a necessary evolution of the international legal order.

NORMATIVE AND DOCTRINAL FOUNDATIONS OF CLIMATE ADJUDICATION

The legal foundations of international climate adjudication rest on a consolidated body of principles and obligations that have developed within international environmental law, general international law, and human rights law. Contrary to perceptions that climate protection remains primarily a matter of political discretion, contemporary doctrine and jurisprudence demonstrate that states are subject to binding duties whose content is increasingly precise and justiciable.

At the core of international environmental law lies the doctrine of due diligence. States are required to exercise reasonable care in preventing activities within their jurisdiction or control from causing significant environmental harm (Sands et al., 2018; Boyle & Redgwell, 2021; Dupuy & Viñuales, 2018). Due diligence constitutes a conduct-based obligation whose scope depends on available scientific knowledge, technological capacity, and the magnitude of foreseeable risks. In the climate context, this obligation entails the adoption and effective enforcement of regulatory frameworks for emissions reduction, continuous monitoring, environmental impact assessment, and supervision of private actors. As emphasized in the literature, environmental due diligence is inherently dynamic and evolves in response to scientific and normative developments (Dupuy & Viñuales, 2018; Viñuales, 2018).

Closely linked to due diligence is the classical no-harm rule, which prohibits states from causing significant transboundary environmental damage. This principle, firmly established in customary international law, has been extensively analyzed in environmental law doctrine (Sands et al., 2018; Boyle & Redgwell, 2021). In the climate context, the no-harm rule operates under conditions of collective causation. Individual emissions rarely generate identifiable harm in isolation. Responsibility, therefore, arises primarily from the failure to adopt adequate regulatory and preventive measures proportionate to a state's contribution to global emissions and its capacity to act (Mayer, 2018; Brunnée & Toope, 2010). The no-harm principle thus functions as a norm of proportionate prevention rather than of exclusive causal attribution.

A further pillar of climate-related obligations is the duty of international cooperation. Cooperation is not merely a political aspiration under the UNFCCC and the Paris Agreement, but a legally binding requirement grounded in treaty law and customary principles (Bodansky et al., 2017; Brunnée & Toope, 2010). It encompasses duties of information exchange, technical assistance, capacity building, and financial support for vulnerable states. When combined with due diligence, the duty of cooperation forms a legally enforceable framework for shared environmental protection, the violation of which may give rise to international responsibility.

Within this normative structure, the concept of climate justice has acquired increasing legal relevance. Although frequently invoked in political discourse, climate justice is grounded in identifiable legal principles, most notably the principle of common but differentiated responsibilities and respective capabilities (CBDR-RC). CBDR-RC does not negate the universality of climate obligations, but modulates their intensity according to historical responsibility, developmental status, and institutional capacity (Rajamani, 2006; Bodansky et al., 2017). Doctrinally, it functions as a standard of proportionality in the application of due diligence and in the assessment of remedial obligations.

The precautionary principle further reinforces the normative density of climate obligations. Recognized as a general principle of international environmental law, it requires regulatory action in situations of scientific uncertainty where serious or irreversible harm is threatened (Sands et al., 2018; Mayer, 2018). In climate adjudication, precaution has direct implications for standards of proof, interim measures, and evidentiary burdens. It legitimizes judicial intervention based on probabilistic risk assessments and reinforces the preventive orientation of environmental responsibility.

Intergenerational equity constitutes another important interpretative principle. Increasingly reflected in international practice and scholarship, it links environmental protection to the long-term preservation of fundamental rights and ecological systems (Boyd, 2018; Knox, 2018; Sands et al., 2018). Intergenerational considerations justify structural remedies, long-term compliance obligations, and institutional monitoring mechanisms. They also challenge traditional conceptions of standing and temporality in international litigation by extending legal concern beyond present claimants.

As discussed above, the collective character of climate obligations also manifests itself in their erga omnes dimension. Following the doctrine developed in the Barcelona Traction case and its subsequent elaboration in legal scholarship, obligations protecting fundamental interests of the international community generate legal interests for all states (Crawford, 2013; Tams, 2005; Simma, 1994). Climate stability exhibits the defining features of such collective goods, including universality, non-excludability, and systemic vulnerability. Consequently, all states possess a legal interest in compliance with climate-related duties, irrespective of individualized injury (Dupuy, 2002).

Within treaty regimes, particularly under the Paris Agreement, climate obligations may further qualify as obligations *erga omnes partes*. This contractual structure strengthens the doctrinal basis for collective enforcement and broad standing (Bodansky et al., 2017). It provides a legal foundation for public interest-oriented forms of climate litigation, subject to appropriate admissibility safeguards.

The 2025 Advisory Opinion confirms that the normative architecture of climate law cannot be reduced to the Paris Agreement as lex specialis. Instead, the Court adopts a systemic integration approach, insisting that treaty law, customary international law, human rights law, and sectoral environmental regimes must be interpreted harmoniously (Foster, 2025). This interpretative move is doctrinally significant: it prevents the fragmentation of climate obligations and rejects attempts to confine State responsibility to the procedural mechanics of nationally determined contributions.

In particular, the Court elevates due diligence to a unifying standard across treaty and customary law. Rather than treating the preparation of NDCs as a matter of political discretion, the Opinion construes the requirements of 'progression' and 'highest possible ambition' as prescriptive elements subject to stringent due diligence review (Foster, 2025). The normative effect is transformative: ambition is no longer purely political rhetoric but acquires legal density through its linkage to the 1.5°C temperature goal.

Furthermore, the Court clarifies that mitigation obligations under both treaty and customary law possess an *erga omnes* or *erga omnes partes* character, thereby reinforcing the collective legal interest in climate protection (Foster, 2025). This characterization strengthens the normative universality of climate obligations and lays the groundwork for broadened invocation of responsibility beyond directly injured States.

A decisive element in the juridification of climate governance is the integration of environmental and human rights law. The growing body of jurisprudence and scholarship confirms that severe environmental degradation may constitute a violation of fundamental rights, including the rights to life, health, and cultural identity (Boyle, 2012; Knox, 2018; Boyd, 2018; Peel & Osofsky, 2015). This applies not only to domestic conduct, but also to transboundary and extraterritorial effects. States are therefore subject to positive obligations to regulate private actors, prevent foreseeable harm, and provide effective remedies.

This human rights–environment nexus transforms climate protection into a core element of judicially enforceable governance. It reinforces duties of participation, transparency, and access to justice, while legitimizing judicial intervention in areas previously regarded as matters of political discretion.

Taken together, these developments reveal a high degree of normative consolidation. International climate obligations are firmly anchored in due diligence, prevention, cooperation, precaution, proportionality, and human rights protection. They possess *erga omnes* qualities and are increasingly framed in terms of collective interests (Brunnée & Toope, 2010; Bodansky et al., 2017; Viñuales, 2018). What remains structurally deficient is therefore not the existence of legal standards, but the institutional mechanisms required for their systematic enforcement.

Because climate obligations are conduct-based, scientifically contingent, and systemically interconnected, their effective implementation exceeds the functional capacity of traditional bilateral dispute settlement mechanisms. From a doctrinal perspective, the establishment of a specialized environmental jurisdiction would therefore not constitute normative innovation. Rather, it represents the institutional consequence of an already mature legal framework. An ICE would translate dispersed obligations into a coherent system of adjudication capable of addressing complex evidentiary questions, issuing structural remedies, and giving procedural effect to collective environmental interests.

INTERNATIONAL JURISPRUDENCE AND ITS SYSTEMIC IMPLICATIONS

A significant evolution in international jurisprudence has accompanied the consolidation of climate obligations in international law. Over the past decade, and particularly since 2024, international courts have progressively transformed climate-related norms from largely programmatic commitments into legally articulated standards of conduct and responsibility. This jurisprudential development has enhanced normative clarity while simultaneously exposing the structural limitations of existing adjudicatory mechanisms in addressing systemic environmental harm.

A central reference point in this evolution is the 2024 advisory opinion of the International Tribunal for the Law of the Sea on climate change and the marine environment. In this opinion, ITLOS adopted an expansive interpretation of 'pollution of the marine environment' under the United Nations Convention on the Law of the Sea, recognizing anthropogenic greenhouse gas emissions as a form of marine pollution capable of producing deleterious effects on the marine environment (Sands et al., 2018; Boyle & Redgwell, 2021). This qualification entailed far-reaching legal consequences.

First, the Tribunal reaffirmed that states are subject to stringent due diligence obligations to prevent, reduce, and control climate-related marine pollution. These obligations encompass legislative and administrative regulation, continuous monitoring, environmental impact assessment, and oversight of private actors. Second,

ITLOS emphasized that climate-related due diligence must be interpreted dynamically in light of evolving scientific knowledge (Dupuy & Viñuales, 2018; Viñuales, 2018). Third, it clarified that developmental status and capacity affect the modalities of environmental duties, but not their existence.

From a doctrinal perspective, the ITLOS opinion is significant because it integrates climate change into a mature liability regime traditionally associated with maritime pollution. As noted in the literature, this anchoring effect strengthens the normative status of climate obligations by situating them within an established system of prevention and responsibility (Boyle & Redgwell, 2021; Mayer, 2018). At the same time, the Tribunal's mandate remains sectorally confined. Its jurisdiction is limited to marine environments, leaving major dimensions of climate harm beyond its institutional reach.

An even broader systemic impact stems from the 2025 advisory opinion of the International Court of Justice on states' obligations regarding climate change. In this opinion, the Court undertook a comprehensive reconstruction of climate-related duties under customary international law, environmental treaties, and human rights instruments.

The ICJ identified, first, substantive obligations of mitigation and prevention derived from due diligence and no-harm principles (Sands et al., 2018; Dupuy & Viñuales, 2018). These include duties to adopt effective emissions reduction policies, regulate high-risk activities, and integrate climate considerations into national planning. Second, the Court emphasized procedural obligations, such as environmental impact assessment, transparency, public participation, and access to information, reflecting established environmental governance standards (Boyle & Redgwell, 2021; Brunnée & Toope, 2010). Third, it reaffirmed duties of international cooperation and assistance, particularly in relation to vulnerable states, consistent with the Paris regime (Bodansky et al., 2017).

Most importantly, the ICJ elaborated on the legal consequences of breaches of climate obligations. It confirmed that failure to comply with due diligence, precautionary, or cooperative duties may give rise to international responsibility, triggering obligations of cessation, reparation, and guarantees of non-repetition (Crawford, 2013; Dupuy, 2002). Reparation was interpreted broadly to include restitution, compensation, satisfaction, and, where appropriate, environmental restoration. This articulation significantly strengthened the remedial dimension of climate law.

The Court further clarified that climate obligations possess erga omnes and erga omnes partes characteristics, thereby confirming that states have a legal interest in compliance even in the absence of individualized injury (Crawford, 2013; Tams, 2005; Simma, 1994). As emphasized in doctrinal analyses, this finding weakens the traditional bilateralist paradigm of international adjudication and provides a foundation for collective enforcement mechanisms (Dupuy, 2002; Viñuales, 2018).

From a systemic perspective, the ICJ opinion represents a decisive step toward the juridification of climate governance. It translates fragmented treaty commitments and general principles into an integrated structure of rights and duties (Bodansky et al., 2017; Mayer, 2018). Nevertheless, its advisory character limits its operational impact. The Court did not establish procedural pathways for direct enforcement, nor did it address standing beyond the traditional inter-state framework. Normative clarification thus remains institutionally disconnected from systematic adjudication.

Parallel developments have occurred in regional human rights jurisprudence, most prominently in the case law of the Inter-American Court of Human Rights. The IACtHR has consistently held that severe environmental degradation, including climate-related harm, may constitute a violation of fundamental rights, such as the rights to life, health, property, and cultural identity (Boyle, 2012; Knox, 2018; Boyd, 2018).

A distinctive feature of this jurisprudence is its expansive approach to extraterritoriality. The Court has recognized that states may incur responsibility for environmental harm affecting individuals beyond their borders where a sufficient regulatory or causal nexus exists. This approach reinforces the positive obligations of states to regulate private actors and prevent foreseeable harm (Peel & Osofsky, 2015; Knox, 2018).

Doctrinally, regional human rights jurisprudence contributes to climate law in two principal ways. First, it strengthens the rights-based foundations of environmental protection by framing climate inaction as a matter of legal accountability rather than political discretion. Second, it introduces participatory and procedural guarantees—such as consultation, access to justice, and cultural sensitivity—into climate-related adjudication (Boyd, 2018; Boyle, 2012). However, its regional scope and individual-rights orientation limit its capacity to address global emissions attribution and collective loss and damage.

Taken together, the jurisprudence of ITLOS, the ICJ, and the IACtHR reveals a converging pattern. Across different institutional settings, courts have affirmed that climate change engages binding obligations of prevention, regulation, cooperation, and reparation. They have rejected the notion that climate governance lies exclusively within the domain of political discretion. They have acknowledged collective interests and *erga omnes* dimensions, and they have incorporated scientific expertise into legal reasoning (Sands et al., 2018; Viñuales, 2018).

At the same time, this body of case law exhibits structural limitations. All three fora operate within mandates that were not designed for comprehensive climate adjudication. Jurisdiction is either sectoral, advisory, or regionally confined. Standing remains restricted. Procedural frameworks are ill-equipped for large-scale, multiparty, and longitudinal disputes. Monitoring and enforcement mechanisms remain indirect and fragmented (Shany, 2007; Romano et al., 2014).

The advisory character of the most influential opinions accentuates these limitations. Advisory proceedings are indispensable for norm clarification and systemic interpretation, yet they lack binding force, individualized remedies, and institutionalized follow-up (Shany, 2007). Their implementation depends largely on political goodwill, and they provide no procedural platform for affected communities.

From a doctrinal perspective, the current jurisprudential landscape reflects a structural paradox. International courts have articulated increasingly precise and demanding climate obligations, while remaining institutionally unable to ensure their systematic enforcement. This has produced a growing discrepancy between legal ambition and adjudicatory capacity.

This discrepancy has important implications for institutional design. It suggests that incremental jurisprudential development cannot indefinitely compensate for structural constraints embedded in jurisdictional mandates and procedural rules (Koskenniemi, 2006; Romano et al., 2014). The transition from advisory authority to adjudicatory effectiveness requires institutional innovation.

In this context, an ICE would not constitute a departure from established jurisprudence. Rather, it would consolidate and operationalize existing doctrinal achievements. It would transform interpretative authority into enforceable judgments, collective interests into procedural rights, and scientific assessment into structured evidentiary regimes. Recent jurisprudence thus supplies both the normative foundation and the institutional rationale for specialized climate adjudication.

MODELS OF INTERNATIONAL ENVIRONMENTAL JURISDICTION

The institutionalization of climate adjudication must be assessed against the background of existing jurisdictional architectures in international law. Environmental disputes are currently distributed across a fragmented constellation of general courts, sectoral tribunals, trade mechanisms, and regional human rights bodies. This dispersion reflects historical and functional contingencies rather than coherent institutional design. As a result, climate-related litigation is characterized by overlapping competences, divergent procedural standards, and uneven access to justice (Romano et al., 2014; Shany, 2007).

General international courts, most prominently the International Court of Justice, occupy a central position in this landscape. The ICJ possesses broad jurisdiction *ratione materiae* and has developed important doctrines relating to environmental protection, including due diligence, environmental impact assessment, and transboundary harm (Sands et al., 2018; Dupuy & Viñuales, 2018). However, its jurisdiction remains strictly consensual and inter-state in character. Access is limited to states, and contentious jurisdiction depends on explicit acceptance. Moreover, the Court's procedural framework is designed for bilateral disputes and is ill-suited for complex, multi-party, and longitudinal litigation involving cumulative environmental harm (Shany, 2007; Romano et al., 2014).

A similar assessment applies to specialized tribunals such as the International Tribunal for the Law of the Sea. ITLOS has elaborated advanced standards of environmental due diligence under the law of the sea and possesses important expertise in maritime environmental protection (Boyle & Redgwell, 2021). Its compulsory jurisdiction in certain areas constitutes an institutional advantage. Yet its mandate remains confined to disputes arising under UNCLOS. Climate-related harm is addressed only to the extent it affects the marine environment. Terrestrial emissions, atmospheric regulation, and socio-economic impacts largely fall outside its jurisdictional reach (Mayer, 2018; Viñuales, 2018).

Trade-related environmental disputes are primarily adjudicated under the World Trade Organization. WTO jurisprudence has recognized environmental protection as a legitimate public policy objective, notably through general exceptions and interpretative balancing techniques (Sands et al., 2018). Nevertheless, environmental interests remain structurally subordinate to trade liberalization objectives. Climate protection appears as an exception to market access obligations rather than as an autonomous regulatory domain. Consequently, the WTO dispute settlement mechanism cannot serve as a primary forum for climate adjudication (Bodansky et al., 2017).

Regional human rights courts have developed increasingly sophisticated environmental jurisprudence grounded in fundamental rights. The Inter-American Court of Human Rights, in particular, has linked environmental degradation to violations of the rights to life, health, and cultural integrity (Boyle, 2012; Boyd, 2018). These bodies offer direct access for individuals and communities and possess strong remedial mechanisms. However, their jurisdiction is territorially limited and embedded in regional treaty systems. They cannot generate universally binding environmental standards, nor are they institutionally equipped to manage complex global disputes over emissions attribution and burden-sharing (Peel & Osofsky, 2015; Romano et al., 2014).

This fragmented institutional landscape reveals a structural mismatch between the global nature of climate change and the compartmentalized design of international adjudication. Existing courts possess valuable expertise and normative authority, but none combine universal jurisdiction, scientific specialization, broad standing, and comprehensive remedial powers. Against this background, three principal institutional models for environmental adjudication may be distinguished: treaty-based specialized courts, specialized chambers within existing courts, and plurilateral jurisdictional regimes.

The first model consists of independent treaty-based environmental courts. Under this approach, states establish a new judicial body through a multilateral convention that defines its jurisdiction, procedures, and institutional structure. Such a court would possess its own statute, registry, budget, and appointment mechanisms. From a doctrinal perspective, this model offers the highest degree of institutional autonomy and normative coherence. Jurisdiction *ratione materiae*, standing rules, evidentiary standards, and remedial powers can be tailored specifically to environmental and climate disputes (Romano et al., 2014).

Treaty-based courts also benefit from systemic clarity. Their authority is grounded in explicit consent, and their competencies are defined in primary law. This facilitates integration into domestic legal orders and enhances legal certainty (Shany, 2007). Moreover, extended standing provisions, including forms of public interest litigation, may be incorporated without conflicting with pre-existing institutional constraints.

At the same time, this model faces significant legal and practical obstacles. The negotiation and ratification of a comprehensive environmental court treaty requires extensive political coordination and sustained commitment. Fragmented accession may undermine universality and weaken normative authority. Furthermore, treaty-based courts risk initial marginalization if major emitters or economically powerful states remain outside the regime (Bodansky et al., 2017; Romano et al., 2014). From a doctrinal standpoint, however, these challenges concern effectiveness rather than legal permissibility.

The second model relies on establishing specialized chambers within existing judicial institutions, particularly the ICJ or ITLOS. Both courts possess statutory provisions allowing for the formation of chambers to deal with specific categories of cases. A climate or environmental chamber could be created through internal procedural arrangements, supplemented by special agreements conferring jurisdiction.

This model offers several advantages. It builds upon established institutional legitimacy, procedural traditions, and administrative infrastructure. It minimizes transaction costs and avoids complex treaty negotiations. Judgments rendered by such chambers would benefit from the authority of the parent institution and from existing compliance expectations (Romano et al., 2014; Shany, 2007).

However, the chamber model is constrained by the constitutional frameworks of host courts. Standing remains limited to states. Jurisdiction continues to depend on consent. Existing statutes restrict procedural innovation. Extended access for non-state actors, structural remedies, or integrated scientific panels may be difficult to accommodate (Alter, 2014; von Bogdandy & Venzke, 2014). Moreover, specialized chambers remain vulnerable to institutional overload and political influence.

Doctrinally, the chamber model represents an incremental adaptation strategy. It enhances functional specialization without altering the foundational architecture of international adjudication. Its capacity to address systemic enforcement deficits is therefore limited.

The third model consists of plurilateral jurisdictional regimes based on inter se agreements among subsets of states. Under this approach, a coalition of willing states establishes a court or arbitral mechanism with mandatory jurisdiction among its members, combined with accession and extension clauses. Such regimes are well known in trade and investment law and are fully compatible with general international law, provided that the rights of third states are respected (Shany, 2007; Koskenniemi, 2006).

Plurilateral regimes allow for rapid institutional experimentation and gradual expansion. Jurisdictional commitments can be calibrated and extended over time, and innovative procedural mechanisms can be tested without requiring universal consensus. However, they risk reinforcing fragmentation and creating parallel legal orders. Their authority remains limited to participating states, and environmental obligations with *erga omnes* characteristics may be unevenly enforced (Koskenniemi, 2006; Romano et al., 2014).

Against this background, the choice between institutional models must be assessed in light of fundamental principles of international adjudication: consent, universality, legal certainty, and systemic coherence. Treaty-based courts maximize normative autonomy and institutional clarity, but face high entry barriers. Chamber solutions offer institutional continuity, but limited transformative capacity. Plurilateral regimes facilitate experimentation, but risk structural fragmentation.

A two-stage architecture emerges as a legally plausible and systemically balanced approach. In an initial phase, specialized chambers within existing courts may be established through special agreements, allowing for the development of procedural practice and doctrinal refinement. In a subsequent phase, these experiences may be consolidated in a treaty-based independent court with expanded jurisdiction and standing (Romano et al., 2014).

From a constitutional perspective, such a staged approach respects the consensual foundations of international adjudication while enabling gradual institutional evolution. It avoids abrupt jurisdictional ruptures and facilitates doctrinal continuity. Moreover, it reflects broader patterns of institutional development in international law, where specialized regimes evolve from informal coordination to formalized governance structures (Alter, 2014; von Bogdandy & Venzke, 2014).

At this point, a recurring critique of proposals for new international courts must be addressed. Since the early 2000s, international legal scholarship has raised concerns about the growth of judicial bodies. Scholars worry about the potential fragmentation of international law due to overlapping jurisdictions and competing authorities (Koskenniemi, 2006; Alter, 2014). From this perspective, an ICE may seem like another institutional layer in an already complicated landscape.

These concerns are not without foundation. The contemporary international legal order includes an increasing number of specialized tribunals that deal with trade, human rights, investment, and maritime disputes. Critics argue that excessive specialization may produce divergent jurisprudence and encourage the development of partially autonomous legal regimes. Moreover, judicial expansion into politically sensitive domains, such as climate governance, may raise questions of democratic legitimacy and judicial overreach, particularly when courts interpret obligations with significant economic and distributive consequences.

However, the reality of environmental adjudication shows that the main problem is not excessive judicialization but the lack of a clear procedural framework. Environmental disputes are now heard in many forums—such as international courts, sectoral tribunals, trade panels, and human rights bodies. Each follows different rules and has a separate mandate. This setup is due to historical chance, not thoughtful planning.

In this sense, fragmentation already exists. Institutions address climate-related disputes with only partial overlap in mandates. Their procedures were not built for science-heavy environmental litigation. Not having a specialized forum does not prevent fragmentation. Instead, it spreads disputes among institutions that lack a unified framework.

From this perspective, a specialized environmental jurisdiction would not exacerbate fragmentation but could instead contribute to doctrinal consolidation. By concentrating environmental disputes within a coherent procedural framework while maintaining cooperative relations with existing courts, such a forum could serve as a focal point for interpreting international environmental obligations.

Two core principles support this idea: complementarity and subsidiarity. Complementarity means a new environmental court would add to, not replace, current mechanisms. Coordination tools, such as lis pendens rules or institutional agreements, could help avoid conflicts and encourage court dialogue. Subsidiarity means that international courts should act only when local or regional measures fall short of addressing cross-border harm. These principles match trends in international adjudication. Specialized tribunals often arise when regulations become complex. They enable expertise and coherent law within specific fields (Alter, 2014). Environmental governance faces challenges like scientific uncertainty and cross-border harm. Here, specialization is not excess, but a practical answer to the needs of modern international law.

Importantly, none of the models discussed is incompatible with the United Nations Charter or with general principles of international institutional law. The establishment of specialized courts, chambers, or plurilateral tribunals falls squarely within the autonomy of states to organize dispute settlement mechanisms (Shany, 2007). The decisive issue is therefore not legal permissibility, but institutional adequacy.

The comparative analysis demonstrates that existing jurisdictional architectures have reached their functional limits in the climate domain. Sectoral tribunals lack comprehensiveness, general courts lack specialization and access, and regional bodies lack universality. Incremental adaptation cannot fully compensate for these deficits. A dedicated environmental jurisdiction is thus not an institutional luxury, but a doctrinally grounded response to structural mismatches between planetary risk and judicial capacity.

In this sense, models of international environmental jurisdiction must be evaluated not merely in terms of political feasibility but in terms of systemic rationality. The effectiveness of climate adjudication depends on the alignment between normative density and institutional form. Where obligations are collective, science-dependent, and preventive, adjudicatory structures must be correspondingly specialized, accessible, and coherent. An ICE would represent the institutional crystallization of this requirement.

The following sections, therefore, move from institutional design to the procedural and substantive mechanisms through which such a jurisdiction could operate.

JURISDICTION AND STANDING BEFORE AN INTERNATIONAL COURT FOR THE ENVIRONMENT

The effectiveness of any system of international adjudication depends fundamentally on the scope of its jurisdiction and the configuration of standing. In the climate context, these issues acquire particular significance, as environmental harm is typically collective, cumulative, and transboundary, while affected interests extend far beyond traditional inter-state relations. A central challenge for an International Court for the Environment, therefore, lies in reconciling established principles of consent and sovereignty with the need for accessible and comprehensive legal protection.

Ratione Personae: Subjects of Proceedings

Under classical international law, access to international courts is predominantly reserved for states. This paradigm continues to shape the jurisdiction of the International Court of Justice and the International Tribunal for the Law of the Sea, whose contentious procedures remain strictly inter-state in character (Romano et al., 2014; Shany, 2007). While this model reflects the consensual foundations of the international legal order, it is increasingly misaligned with the structural realities of climate-related harm.

An ICE should recognize three categories of potential parties: states, international organizations, and non-state actors.

States remain the primary bearers of climate obligations and the principal subjects of international responsibility (Crawford, 2013). Their standing is therefore indispensable. Inter-state proceedings may concern insufficient mitigation policies, failures of regulatory oversight, breaches of cooperation obligations, or responsibility for transboundary damage. Retaining full state access ensures continuity with established jurisdictional traditions and reinforces institutional legitimacy.

International organizations may also be granted limited standing, particularly where they exercise regulatory, supervisory, or financial functions in environmental and developmental governance. Regional organizations, development banks, and treaty bodies may possess institutional interests in compliance with climate-related obligations (Bodansky et al., 2017). Allowing such entities restricted access may enhance systemic coherence and facilitate coordination between judicial and administrative regimes.

The most innovative and contested element concerns the participation of non-state actors, including indigenous communities, non-governmental organizations, and affected groups. From a doctrinal perspective, their inclusion reflects the gradual erosion of strict state centrism in international law, particularly in environmental and human rights regimes (Boyd, 2018; Knox, 2018). Excluding such actors would perpetuate an enforcement deficit, given that states may lack political incentives to initiate proceedings against major emitters.

At the same time, unrestricted access would risk procedural overload and politicization. A balanced model, therefore, requires carefully designed admissibility filters, including representativeness, substantial impact, public-interest relevance, and the exhaustion or futility of domestic remedies (Peel & Osofsky, 2015; Shany, 2007). Such criteria do not undermine sovereignty, but operationalize the collective-interest character of climate obligations within a legally controlled framework.

Ratione Materiae: Subject-Matter Jurisdiction

The subject-matter jurisdiction of an ICE must reflect the multidimensional character of climate-related disputes. A narrowly defined competence limited to treaty compliance would fail to capture the systemic nature of climate harm. Conversely, overly expansive jurisdiction risks institutional overlap and fragmentation.

At a minimum, *ratione materiae* jurisdiction should encompass four categories.

First, failure to act in violation of international climate obligations constitutes a central category. This includes inadequate mitigation measures, deficient regulatory frameworks, insufficient monitoring systems, and failures to integrate scientific standards into policymaking (Mayer, 2018; Dupuy & Viñuales, 2018). In climate law, responsibility often arises from persistent omissions rather than discrete acts.

Second, transboundary environmental damage forms a traditional and indispensable jurisdictional domain. Loss and damage associated with sea-level rise, desertification, biodiversity collapse, and extreme weather events frequently transcend national borders. International law has long recognized that such harm may engage responsibility where preventive obligations are breached (Sands et al., 2018; Boyle & Redgwell, 2021).

Third, violations of procedural environmental rights should fall within the court's competence. Access to information, public participation, and access to justice constitute core elements of contemporary environmental governance (Boyle, 2012; Knox, 2018). Their denial undermines substantive climate protection and justifies judicial intervention.

Fourth, extraterritorial and indirect harm requires explicit jurisdictional recognition. This includes support for carbon-intensive activities abroad, export of environmental risks, and transnational supply-chain effects. Climate obligations cannot be confined to territorial boundaries when regulatory and economic activities produce foreseeable global consequences (Viñuales, 2018; Peel & Osofsky, 2015).

To prevent jurisdictional conflicts, delimitation clauses should establish principles of complementarity. The ICE would not function as an appellate body for trade or human rights courts, but as a specialized forum for climate-related dimensions of disputes (Koskenniemi, 2006; Shany, 2007).

Admissibility and Public Interest Litigation

Expanded standing necessitates a refined system of admissibility criteria. Without such mechanisms, collective enforcement risks degenerating into opportunistic or politically motivated litigation.

A central element is the requirement of sufficient interest. Applicants must demonstrate that the alleged violation affects protected collective interests in a substantial and non-hypothetical manner (Tams, 2005; Crawford, 2013). In the case of non-state actors, this may be established through geographical proximity, functional representation, or demonstrable expertise.

Representativeness constitutes a further safeguard. Organizations and community groups must possess legitimate internal governance structures and a genuine connection to the interests invoked. This requirement prevents instrumentalization of climate litigation by unaccountable actors (Peel & Osofsky, 2015).

The principle of exhaustion of domestic remedies, adapted to environmental realities, also plays a role. Where effective national procedures exist, international adjudication should remain subsidiary. However, exhaustion should not be required where remedies are unavailable, ineffective, or unduly prolonged, particularly in situations of urgent or irreversible damage (Knox, 2018; Boyd, 2018).

Abuse-of-process doctrines provide an additional layer of control. Manifestly unfounded, duplicative, or vexatious applications may be dismissed at an early stage. Such mechanisms preserve judicial resources and protect institutional integrity (Shany, 2007).

Within this framework, forms of international public interest litigation become doctrinally defensible. They reflect the *erga omnes* character of climate obligations and translate collective interests into procedural rights without abandoning legal discipline (Simma, 1994; Dupuy, 2002).

To ensure procedural manageability, these admissibility requirements may be supplemented by a substantial environmental impact threshold. Applicants should demonstrate that the alleged conduct produces or risks producing significant harm to the climate system, biodiversity, or ecological integrity. Proof of completed damage is not demanded at the admissibility stage. This requirement ensures that proceedings address genuine environmental risks rather than abstract policy disagreements.

In procedural terms, admissibility may be assessed through preliminary review mechanisms. Dedicated chambers of the court may evaluate standing, representativeness, and admissibility criteria before the merits stage begins. Such preliminary scrutiny is widely used in international adjudication. It filters manifestly unfounded claims while preserving meaningful access to justice (Romano et al., 2014).

Properly designed admissibility filters do not restrict environmental litigation; they structure it. They enable courts to address systemic environmental harm. These filters also prevent procedural overload. In this sense, admissibility mechanisms are an essential institutional complement to expanded standing in international environmental adjudication.

Relationship to Other Courts and Jurisdictional Coordination

The expansion of jurisdiction and standing inevitably raises questions of institutional interaction. Climate disputes frequently intersect with trade law, investment law, human rights law, and maritime law. Without clear coordination rules, parallel proceedings and inconsistent judgments may undermine legal certainty.

A first mechanism is the lis pendens principle. Where substantially identical disputes between the same parties are pending before another tribunal, the ICE should defer or suspend proceedings (Shany, 2007; Romano et al., 2014).

Second, the doctrine of res judicata ensures the finality of judgments. Matters definitively resolved by competent courts should not be reopened, unless materially new circumstances justify reconsideration (Koskenniemi, 2006).

Third, consultation and referral mechanisms may facilitate doctrinal coherence. Questions concerning individual human rights, trade compatibility, or maritime regulation may be addressed in cooperation with specialized bodies. Conversely, other courts may seek interpretative guidance on climate-specific issues.

Fourth, institutional agreements, including memoranda of understanding, can structure information exchange and mutual recognition of findings. Such arrangements reflect the principle of systemic integration under Article 31(3)(c) of the Vienna Convention on the Law of Treaties (Koskenniemi, 2006).

Jurisdiction, Consent, and Collective Interests

A recurring objection to expanded jurisdiction and standing concerns the principle of state consent. International adjudication remains, in principle, voluntary. No court may exercise compulsory jurisdiction without an adequate legal basis (Shany, 2007).

This objection must, however, be contextualized. Consent may be expressed through treaties, optional clauses, special agreements, or inter se arrangements. Moreover, by ratifying environmental and human rights conventions, states already accept extensive supervisory and reporting mechanisms (Bodansky et al., 2017; Knox, 2018). An ICE would reorganize existing commitments rather than impose qualitatively new constraints.

Furthermore, the *erga omnes* character of climate obligations modifies traditional consent logic. Where duties protect fundamental collective interests, the legitimacy of collective enforcement is enhanced (Crawford, 2013; Tams, 2005). The transition from bilateralism to community-oriented adjudication reflects an evolutionary development of international law (Simma, 1994).

Doctrinal Assessment

From a doctrinal standpoint, a jurisdictional regime combining broad subject-matter competence with differentiated standing represents a necessary adaptation to systemic environmental risk. Limiting access to states alone would perpetuate enforcement deficits. Uncontrolled access would undermine judicial stability. A calibrated model, grounded in admissibility criteria and coordination mechanisms, reconciles effectiveness with legal certainty.

The proposed jurisdictional architecture translates the collective, preventive, and science-dependent nature of climate obligations into procedural form. It enables the enforcement of erga omnes interests, integrates non-state participation, and preserves coherence with existing institutions. In doing so, it moves international adjudication beyond the bilateral paradigm without abandoning its foundational principles.

Jurisdiction and standing thus become instruments for operationalizing collective responsibility. They transform diffuse normative commitments into legally actionable claims and enable international courts to respond effectively to cumulative and transboundary environmental harm. An ICE would not merely extend existing models, but reconfigure them in light of the structural features of climate governance.

PROCEDURAL LAW AND EVIDENCE IN CLIMATE ADJUDICATION

The following section examines the procedural architecture required for climate adjudication, including sources of procedural law, scientific integration, standards of proof, interim measures, and multi-party proceedings. The procedural architecture of an ICE constitutes the decisive interface between normative ambition and judicial effectiveness. Climate-related disputes differ fundamentally from classical inter-state conflicts. They are characterized by cumulative causation, probabilistic harm, scientific complexity, long temporal horizons, and asymmetric access to information. A procedural regime designed for bilateral, factually discrete disputes cannot adequately address such features. The institutional credibility of an environmental court, therefore, depends on its capacity to integrate scientific knowledge, manage evidentiary uncertainty, and structure complex, multi-actor proceedings within a coherent framework of due process.

Sources of Procedural Law

The procedural law of an ICE would be derived from multiple normative layers. At its foundation stands the constitutive statute, which defines jurisdiction, composition, and core procedural guarantees. Complementary rules of court regulate pleadings, evidence, interim measures, participation of third parties, and enforcement procedures. These internal sources are supplemented by general principles of international procedural law, including equality of arms, audi alteram partem, impartiality, and reasoned decision-making (Romano et al., 2014; Shany, 2007).

Beyond these traditional sources, climate adjudication requires procedural adaptation. As emphasized in the literature on international adjudication, evidentiary regimes must evolve where disputes involve systemic risk and complex factual matrices (Alter, 2014; von Bogdandy & Venzke, 2014). Procedural flexibility thus becomes a functional necessity rather than an exception.

A central procedural innovation of the Advisory Opinion lies in its treatment of science as an integral component of legal reasoning. The Court does not merely reference scientific findings; it embeds Intergovernmental Panel on Climate Change assessments into the determination of the applicable standard of conduct. In doing so, it implicitly recognizes that due diligence in climate matters is inseparable from the evolving state of scientific knowledge (Foster, 2025).

This integration has significant evidentiary consequences. The Court emphasizes that States must actively acquire and assess scientific information and adjust regulatory measures in light of new knowledge. Due diligence thus becomes dynamic rather than static, and the evidentiary benchmark is calibrated to contemporary climate science (Foster, 2025).

From a procedural perspective, this approach reduces the space for epistemic uncertainty as a defensive strategy. The precautionary principle, coupled with stringent due diligence, shifts the burden of proof towards demonstrating an adequate regulatory response rather than disputing climate risk as such. An international environmental court would need to institutionalize this science-law interface through specialized chambers, expert panels, and structured evidentiary procedures.

Scientific Integration and Expert Governance

Scientific knowledge is constitutive of climate adjudication. Questions concerning emissions trajectories, tipping points, biodiversity loss, and public health impacts cannot be resolved without specialized expertise. Contemporary environmental law doctrine therefore emphasizes the central role of science in regulatory and judicial decision-making (Sands et al., 2018; Dupuy & Viñuales, 2018).

An ICE should institutionalize scientific integration through several mechanisms. First, it should adopt the principle of 'best available science' as a guiding evidentiary standard, requiring systematic engagement with authoritative scientific assessments, including those of the Intergovernmental Panel on Climate Change (IPCC) and Intergovernmental Science-Policy Platform on Biodiversity and Ecosystem Services (IPBES) (Mayer, 2018; Viñuales, 2018).

Second, the court should be empowered to appoint independent experts or establish permanent scientific panels. Court-appointed experts enhance neutrality and coherence, reducing dependence on party-appointed testimony and mitigating strategic manipulation of scientific uncertainty (Romano et al., 2014).

Third, regulated participation of amici curiae from scientific institutions and research networks may enrich the evidentiary record. Such participation must be subject to procedural safeguards to preserve balance and judicial independence (Shany, 2007).

The adjudication of environmental and climate disputes presents a distinctive procedural challenge: legal determination is inseparable from complex and evolving scientific knowledge. Climate change litigation involves questions of atmospheric modeling, attribution science, ecological thresholds, and long-term risk assessment. Traditional evidentiary frameworks developed for bilateral disputes between states are often ill-equipped to handle such technically dense, interdisciplinary material (Sands et al., 2018; Foster, 2025).

An effective environmental jurisdiction must therefore institutionalize mechanisms for systematically integrating scientific expertise. This integration cannot be limited to the occasional appointment of experts under general procedural rules. Instead, it requires a structured architecture that ensures both epistemic reliability and procedural fairness.

A first component of this architecture concerns the use of court-appointed independent experts. International courts already possess the authority to appoint experts, yet this power has historically been exercised sparingly. In climate adjudication, however, expert evidence is not ancillary but central to the determination of due diligence standards and causation. A specialized environmental court could therefore maintain a standing roster of independent experts covering relevant disciplines such as climate science, oceanography, biodiversity science, environmental economics, and public health. Experts drawn from this roster could assist the court in interpreting complex technical evidence and in assessing competing scientific claims presented by the parties.

Second, the court may establish permanent scientific advisory panels to support judicial deliberation. Such panels would function similarly to technical advisory bodies in other international regimes, providing methodological guidance on interpreting scientific consensus and uncertainty. Their role would not be to determine legal outcomes but to assist the court in evaluating the evidentiary weight of competing scientific arguments. Importantly, advisory panels should operate under strict transparency requirements to ensure the legitimacy of their input.

A third procedural mechanism concerns the structured use of party-appointed experts and scientific submissions. In many environmental disputes, states and other litigants rely on extensive scientific reports prepared by research institutions or independent experts. While such submissions can enrich judicial understanding, they may also produce adversarial scientific contestation. Procedural rules should therefore establish clear standards for the admissibility and evaluation of scientific evidence, including disclosure obligations, peer-review requirements, and opportunities for cross-examination.

In addition to these mechanisms, the court may recognize authoritative scientific assessments as relevant evidentiary benchmarks. Reports produced by international scientific bodies—most prominently the IPCC and the IPBES—constitute comprehensive syntheses of current scientific knowledge. While such reports lack formal legal authority, their methodological rigor and consensus-based procedures render them particularly relevant for assessing the state of scientific knowledge in environmental litigation (Foster, 2025).

The recognition of such assessments has important implications for evidentiary standards. Where a broad scientific consensus exists, litigants challenging that consensus may bear a heightened burden of justification. Conversely, where significant uncertainty persists, the precautionary principle may justify judicial intervention even in the absence of complete scientific certainty. In this way, scientific integration interacts directly with fundamental principles of international environmental law.

Finally, procedural design must ensure that integrating scientific expertise does not undermine judicial independence. Experts and advisory panels should assist the court, not replace its adjudicatory function. Decisions regarding legal responsibility, causation, and remedies remain matters of judicial determination. The role of science is to inform the evidentiary assessment underlying these determinations.

Institutionalizing scientific expertise in this manner serves several purposes. It enhances the epistemic credibility of judicial decisions, reduces the risk of inconsistent scientific interpretation across cases, and facilitates the development of coherent evidentiary standards for environmental adjudication. Moreover, it strengthens the legitimacy of judicial intervention in technically complex policy domains by demonstrating that legal reasoning is grounded in robust scientific understanding.

In this respect, the integration of scientific expertise should not be regarded as an exceptional feature of environmental adjudication but as a structural requirement of contemporary international governance. As environmental disputes increasingly involve complex interactions between legal norms and scientific knowledge, international courts must develop institutional mechanisms to bridge this epistemic divide.

Standards of Proof and Causation

Causation represents one of the most contested issues in climate litigation. Traditional international responsibility doctrine often requires proof of a sufficiently direct causal nexus between wrongful conduct and injury (Crawford, 2013). In climate cases, however, harm results from the cumulative effect of multiple actors over extended periods.

A rigid insistence on direct causation would render climate obligations largely unenforceable. Doctrinal scholarship, therefore, supports differentiated standards adapted to collective harm (Mayer, 2018; Viñuales, 2018).

First, in cases of systemic environmental degradation, the 'substantial contribution' standard provides an appropriate threshold. Where regulatory failure materially contributes to harmful outcomes, responsibility may arise despite the presence of multiple causal factors.

Second, in precautionary contexts, the standard of 'material increase in risk' may suffice. Where state conduct significantly elevates the probability of serious or irreversible harm, judicial intervention is justified even absent full causal certainty (Sands et al., 2018).

Third, proportional attribution mechanisms may be employed in reparation contexts. Allocation of responsibility based on emissions shares, economic capacity, or risk contribution reflects evolving doctrines of shared responsibility (Dupuy & Viñuales, 2018).

The burden of proof must also be calibrated. In situations of pronounced informational asymmetry—such as undisclosed emissions data or opaque regulatory processes—the court may shift or relax evidentiary burdens. This approach reflects general principles of procedural fairness and prevents strategic non-disclosure (Shany, 2007; Peel & Osofsky, 2015).

Interim Measures and Urgency

Climate-related harm is frequently irreversible. Ecosystem collapse, territorial loss, and cultural destruction cannot be fully remedied ex post. The availability of provisional measures is therefore central to effective adjudication.

International courts traditionally possess authority to order interim measures where rights are at risk of irreparable harm (Romano et al., 2014). In the climate context, this authority must be exercised in light of precautionary obligations. Demonstration of plausibility and serious risk should suffice, without requiring conclusive proof of responsibility (Sands et al., 2018; Mayer, 2018).

Procedural acceleration mechanisms—such as expedited hearings and shortened submission deadlines—are equally necessary. Sunset clauses and periodic review ensure proportionality and adaptability.

Structural Proceedings and Multi-Party Litigation

Climate disputes frequently involve multiple respondents, numerous affected communities, and long-term compliance obligations. Bilateral litigation models are inadequate for managing such complexity.

The procedural framework should therefore accommodate structural proceedings, including consolidated cases, representative actions, and phased adjudication separating liability determination from remedial planning (Alter, 2014; Romano et al., 2014). Such mechanisms enhance coherence and reduce fragmentation.

Structural proceedings may require periodic reporting, independent auditing, and judicial supervision of compliance plans. While courts must avoid direct administrative control, ongoing oversight remains compatible with judicial functions where systemic violations have been established (von Bogdandy & Venzke, 2014).

Transparency, Participation, and Knowledge Justice

Procedural legitimacy depends not only on technical accuracy but also on transparency and inclusiveness. Public hearings, accessible documentation, and reasoned judgments enhance accountability and trust (Shany, 2007).

Given the technical complexity of climate litigation, transparency also requires intelligibility. Publication of scientific summaries and explanatory materials enables meaningful public scrutiny.

At the same time, procedural rules must protect sensitive ecological, cultural, and security-related information. Confidentiality regimes and protective measures are therefore indispensable.

Participation rights for indigenous communities and affected groups require procedural accommodation, including language access, culturally appropriate evidentiary practices, and recognition of traditional knowledge systems (Boyd, 2018; Knox, 2018). Integrating plural epistemologies enhances contextual accuracy without undermining scientific rigor.

Procedural Fairness and Judicial Restraint

Despite its innovative elements, the climate procedure must remain anchored in core guarantees of due process. Equality of arms, impartial adjudication, access to legal representation, and reasoned judgments are non-negotiable components of judicial legitimacy (Romano et al., 2014).

Expanded evidentiary flexibility must not result in arbitrary reasoning. Courts must maintain clear standards for relevance, reliability, and probative value. Scientific evidence informs the scope of due diligence, but does not replace normative evaluation.

Judicial restraint remains essential. Courts must distinguish legal obligations from policy preferences and respect institutional boundaries. Procedural innovation must therefore be accompanied by principled self-limitation (von Bogdandy & Venzke, 2014).

Doctrinal Assessment

From a doctrinal perspective, procedural law in climate adjudication reflects an adaptation of international litigation to conditions of planetary risk. It reconciles probabilistic harm with legal responsibility, integrates science without surrendering judicial autonomy, and accommodates collective interests within structured proceedings.

These developments do not represent a departure from international legal tradition. International courts have long relied on expert evidence, interim measures, and flexible standards of proof. Climate adjudication extends these techniques to a domain where uncertainty and complexity are systemic rather than exceptional (Sands et al., 2018; Viñuales, 2018).

Procedural law thus becomes a substantive instrument. Standards of proof shape prevention obligations, evidentiary rules influence risk allocation, and participation rights reinforce human rights integration. An ICE would institutionalize a model of science-informed, precautionary, and participatory justice capable of translating diffuse climate obligations into operational legal decisions.

SUBSTANTIVE REMEDIES AND LEGAL CONSEQUENCES

The effectiveness of international climate adjudication ultimately depends on the availability and credibility of legal remedies. Normative clarification and procedural sophistication remain insufficient if judicial determinations cannot be translated into concrete obligations capable of altering harmful practices and repairing environmental and social damage. In the climate context, remedies must address cumulative harm, long-term risks, and structural regulatory failures. This requires a doctrinal reconfiguration of traditional concepts of responsibility and reparation in international law.

Legal Consequences of Climate-Related Breaches

Under the general law of state responsibility, internationally wrongful acts entail the obligation to cease the wrongful conduct, to offer appropriate assurances and guarantees of non-repetition, and to make full reparation for injury caused (Crawford, 2013). These principles, codified in the Articles on Responsibility of States for Internationally Wrongful Acts, apply equally to environmental and climate-related violations.

Climate breaches rarely arise from isolated acts, but from persistent regulatory omissions and failures to align national policies with scientific risk thresholds. Such violations are often continuous in character. Consequently, cessation obligations assume particular importance. States may be required to revise legislation, strengthen enforcement mechanisms, or adopt comprehensive mitigation strategies (Dupuy & Viñuales, 2018; Mayer, 2018).

Assurances and guarantees of non-repetition complement cessation by stabilizing compliance over time. In climate cases, these guarantees may involve institutional reforms, transparency mechanisms, and permanent monitoring arrangements. Their function is preventive rather than punitive and reflects the forward-looking orientation of environmental responsibility (Boyle & Redgwell, 2021).

About legal consequences, the Advisory Opinion unequivocally affirms that breaches of climate obligations entail international responsibility under the law of State responsibility. By explicitly linking climate obligations to the framework of ARSIWA, the Court confirms that cessation, assurances of non-repetition, and reparation are in principle available remedies (Foster, 2025).

Of particular importance is the Court's discussion of causation. While acknowledging the cumulative and diffuse nature of greenhouse gas emissions, the Opinion sketches a pathway for attributing responsibility when a State fails to exercise due diligence in regulating private actors (Foster, 2025). The emphasis shifts from identifying a singular causal contribution to assessing regulatory omission within a shared global harm structure.

Moreover, the recognition of *erga omnes* obligations implies that the invocation of responsibility is not limited to specially injured States. This structural expansion of standing has direct implications for remedial design in a prospective International Court for the Environment. Remedies may need to include structural injunctions, forward-looking compliance orders, and supervisory mechanisms capable of addressing systemic mitigation deficits rather than isolated wrongful acts.

Declaratory and Injunctive Relief

Declaratory judgments constitute the most basic form of judicial relief. By authoritatively determining the existence and scope of climate obligations, they contribute to legal certainty and normative consolidation (Sands et al., 2018). In international practice, declaratory relief often exerts significant influence on domestic courts and regulatory authorities.

However, declaratory relief alone is insufficient in the face of persistent environmental harm. Injunctive orders represent a more intrusive, yet increasingly necessary, remedial instrument. Such orders may require states to suspend environmentally harmful projects, revise regulatory frameworks, or implement defined mitigation measures within prescribed timeframes.

Although international law has traditionally approached injunctive relief with caution, environmental jurisprudence has gradually accepted forward-looking, performance-oriented remedies (Romano et al., 2014; Shany, 2007). Where preventive obligations are breached and irreparable harm is imminent, injunctions function as extensions of the duty of cessation.

To preserve proportionality and respect regulatory autonomy, injunctive measures should be framed in outcome-oriented terms, allowing states discretion in choosing implementation modalities (von Bogdandy & Venzke, 2014).

Remedial and Restorative Measures

Beyond cessation and injunctions, climate adjudication requires mechanisms for remediation and ecological restoration. Environmental harm frequently affects ecosystems, livelihoods, and cultural practices in ways that monetary compensation alone cannot adequately address.

Restorative remedies may include habitat rehabilitation, reforestation programs, coastal protection measures, or investments in climate adaptation infrastructure. In doctrinal terms, such measures constitute forms of restitution in kind adapted to environmental contexts (Dupuy & Viñuales, 2018; Sands et al., 2018).

Where full restoration is impossible, courts may order substitute or compensatory measures that mitigate residual harm. This reflects the principle of full reparation, interpreted flexibly in light of ecological realities (Crawford, 2013).

Structured remediation plans represent a particularly important instrument. These plans combine judicial determination of obligations with phased implementation schedules, performance indicators, and reporting duties. They enable courts to address systemic failures without engaging in continuous administration (Alter, 2014).

Compensation, Loss, and Damage

Monetary compensation remains an essential component of climate remedies, particularly in cases involving economic loss, displacement, health impacts, and property destruction. Climate-related damage often disproportionately affects vulnerable populations, raising questions of distributive justice (Boyd, 2018; Knox, 2018).

Traditional compensation doctrine requires a causal link between wrongful conduct and quantifiable harm. As discussed in Section 6, climate causation necessitates flexible attribution standards. Proportional liability and risk-based allocation mechanisms allow courts to operationalize compensation without demanding unattainable precision (Mayer, 2018; Viñuales, 2018).

Loss and damage associated with irreversible harm, such as loss of territory, cultural heritage, or biodiversity, pose special challenges. In such cases, compensation serves symbolic, distributive, and preventive functions. It acknowledges responsibility, supports affected communities, and incentivizes compliance (Bodansky et al., 2017). Compensation schemes may be administered through dedicated funds or trust mechanisms supervised by the court or affiliated bodies. Such institutional arrangements enhance transparency and ensure effective distribution of resources (Dupuy, 2002).

One of the central doctrinal challenges in climate adjudication concerns the attribution of responsibility for cumulative, globally distributed environmental harm. Unlike many traditional transboundary disputes, climate change does not result from a single identifiable act or a discrete bilateral interaction. Rather, it arises from the aggregated emissions of numerous state and non-state actors over extended periods. This structural feature complicates the application of classical causation models in international law, which have often relied on relatively direct causal links between wrongful conduct and identifiable injury (Crawford, 2013).

International environmental law has increasingly recognized that such traditional causation models are insufficient for addressing complex ecological harm. Contemporary scholarship, therefore, emphasizes more flexible attribution frameworks capable of accommodating cumulative and probabilistic forms of causation (Peel & Osofsky, 2015; Sands et al., 2018). The law of state responsibility itself does not require a single exclusive cause of harm; it requires that a wrongful act make a sufficient contribution to the injury in question (Crawford, 2013). This doctrinal flexibility provides an entry point for developing proportionate responsibility models adapted to climate change.

Under a proportionate responsibility approach, liability is not contingent on demonstrating that a single state caused a specific climate impact in isolation. Instead, responsibility may arise where a state's conduct has contributed materially to a broader harmful process and where that conduct falls below the standard of due diligence required under international law. The relevant inquiry, therefore, shifts from exclusive causation to contributory responsibility within a structure of cumulative harm.

This approach aligns with emerging scientific methods of climate attribution. Advances in climate modeling increasingly enable researchers to estimate the relative contributions of specific emission sources to observed climatic changes and extreme weather events. While such attribution studies do not produce precise causal determinations in the traditional legal sense, they can identify statistically significant relationships between emissions trajectories and environmental impacts. In judicial settings, these findings may support proportional allocation of responsibility among multiple contributors.

From a doctrinal perspective, proportionate responsibility may operate in two complementary ways. First, it may inform breach determinations by evaluating whether a state's regulatory framework and mitigation efforts satisfy the due diligence obligations under international climate law. Where states fail to adopt reasonable measures to limit emissions or prevent foreseeable harm, responsibility may arise regardless of whether their emissions alone caused a specific climatic event.

Second, proportionate responsibility may guide reparations allocation. Where multiple states contribute to climate harm, compensation or restoration obligations may be apportioned according to relative contributions to the harmful process. Such a proportional allocation reflects principles of fairness and equity while avoiding the evidentiary impossibility of identifying a single responsible actor in a cumulative-emissions system.

Crucially, proportionate responsibility does not remove the need for a causal link. Instead, it reshapes causation into probabilistic and systemic terms. Courts must still decide whether a state's actions made a meaningful contribution to the harm and whether the damage falls within the risks the breached obligation should have prevented.

A specialized environmental court would be particularly well-positioned to operationalize this model. By integrating scientific expertise, structured evidentiary procedures, and flexible remedial mechanisms, such a court could develop jurisprudence capable of addressing the distinctive causal structures of climate harm. Instead of treating the complexity of climate attribution as an obstacle to legal responsibility, adjudication would recognize cumulative contribution as a legally relevant basis for allocating responsibility.

In sum, proportionate responsibility provides a doctrinal bridge between the collective nature of climate change and the individualized framework of state responsibility. In so doing, it allows international adjudication to address systemic environmental harm while remaining anchored in established principles of international law.

Collective and Structural Remedies

One of the distinctive features of climate remedies is their emphasis on institutional transformation. Guarantees of non-repetition extend beyond formal assurances to encompass concrete governance reforms.

These may include the establishment of independent regulatory authorities, mandatory climate impact assessments, enhanced disclosure obligations, and strengthened domestic-level judicial review mechanisms (Brunnée & Toope, 2010; Bodansky et al., 2017). Courts may also require integrating climate considerations into budgetary planning and infrastructure development.

From a doctrinal perspective, such measures operationalize due diligence and precaution by embedding them in institutional practice. While they approach the boundary between adjudication and administration, they remain justified where systemic regulatory failures have been judicially established (von Bogdandy & Venzke, 2014).

Guarantees of Non-Repetition and Institutional Reform

Given the collective nature of climate harm, remedies often require collective and structural dimensions. Individualized relief is insufficient where violations affect entire ecosystems or populations.

Collective remedies may involve regional adaptation programs, shared mitigation initiatives, or cross-border environmental management frameworks. Structural remedies address underlying governance deficits, such as fragmented regulatory regimes or inadequate enforcement capacities (Alter, 2014; Romano et al., 2014).

Comparative jurisprudence indicates that such remedies are increasingly employed in contexts of mass harm and systemic rights violations. Climate adjudication represents a paradigmatic case for its cautious expansion (Boyd, 2018; Peel & Osofsky, 2015).

Enforcement-Oriented Design of Remedies

Remedies must be designed with enforceability in mind. Abstract obligations lacking operational content are unlikely to produce behavioral change. Effective remedies therefore, combine legal clarity with institutional feasibility.

Key elements include measurable benchmarks, timelines, verification procedures, and consequences for non-compliance (Shany, 2007). Graduated remedial structures may enhance effectiveness by linking escalating obligations to persistent non-compliance.

Embedding such mechanisms within judicial orders strengthens the connection between adjudication and compliance architecture (Romano et al., 2014).

Doctrinal Assessment

From a doctrinal standpoint, climate remedies reflect an evolution from episodic reparation toward structural governance correction. Traditional international responsibility focused on bilateral injury and retrospective compensation. Climate adjudication requires forward-looking, collective, and adaptive remedies.

This evolution does not undermine foundational principles of state responsibility. Cessation, reparation, and guarantees of non-repetition remain central. What changes is their operationalization in response to systemic risk and scientific uncertainty (Crawford, 2013; Viñuales, 2018).

An ICE would thus function not merely as an arbiter of past wrongdoing, but as an architect of compliance-oriented legal frameworks. Its remedial practice would integrate prevention, restoration, and institutional reform within a coherent doctrine of responsibility.

In this sense, substantive remedies constitute the normative bridge between legal obligation and environmental protection. They transform abstract duties into actionable standards and translate judicial authority into tangible ecological and social outcomes. Without such remedial capacity, climate adjudication would remain symbolically significant but practically marginal.

ENFORCEMENT AND COMPLIANCE MECHANISMS

The practical relevance of international adjudication depends on the existence of reliable mechanisms for implementing judicial decisions. Without effective enforcement structures, even the most sophisticated jurisdictional and remedial regimes risk remaining largely symbolic. This challenge is particularly acute in climate governance, where compliance requires long-term regulatory transformation, sustained investment, and politically sensitive policy adjustments. An ICE must therefore be embedded in a multilayered compliance architecture capable of translating judicial determinations into durable behavioral change.

Domestic Implementation and Legal Effect

he primary locus of enforcement in international law remains the domestic legal order. International judgments generally lack direct coercive force and depend on national authorities for execution. This decentralization reflects the consensual and non-hierarchical structure of the international legal system (Shany, 2007; Romano et al., 2014).

For an International Court for the Environment, domestic implementation should be institutionalized through explicit treaty obligations requiring states to recognize and give effect to judgments within their internal legal systems. Such obligations may include the designation of competent authorities, the establishment of expedited execution procedures, and the availability of judicial review mechanisms for compliance disputes.

Recognition regimes comparable to those governing international arbitral awards provide a useful model. By limiting grounds for non-recognition to narrowly defined public policy exceptions, legal certainty and uniformity can be enhanced (Alter, 2014). This approach respects constitutional autonomy while strengthening transnational judicial authority.

Domestic implementation also reinforces vertical integration between international and national environmental governance. It facilitates judicial dialogue, promotes regulatory learning, and enhances the legitimacy of international oversight (von Bogdandy & Venzke, 2014).

Institutional Supervision and Monitoring

Effective compliance in climate cases requires continuous supervision. One-off judgments are insufficient where remedial obligations involve emissions trajectories, adaptation planning, or institutional reform.

An ICE should therefore be empowered to establish compliance committees or monitoring chambers to review implementation reports, conduct follow-up hearings, and issue supplementary determinations (Romano et al., 2014). These bodies would operate under judicial authority and remain bound by legal standards rather than political discretion.

Periodic reporting obligations constitute a central component of this system. States subject to remedial orders may be required to submit detailed progress reports supported by independent audits and scientific assessments (Sands et al., 2018; Dupuy & Viñuales, 2018). Failure to comply with reporting duties may itself constitute a procedural violation.

Public transparency further reinforces monitoring. Publication of compliance data, implementation benchmarks, and evaluative summaries enhances accountability and enables civil society participation (Shany, 2007). At the same time, confidentiality safeguards must protect sensitive ecological and security-related information.

Financial Conditionality and Legal Incentives

Contemporary international governance increasingly relies on conditionality mechanisms linking legal compliance to access to financial resources. In the climate domain, such instruments are particularly relevant given the central roles of development finance, climate funds, and export credit agencies.

Judicially recognized compliance obligations may be integrated into the operational policies of multilateral development banks and climate finance mechanisms. Conditioning funding on the implementation of court-approved remediation plans transforms economic incentives into enforcement tools (Bodansky et al., 2017; Brunnée & Toope, 2010).

From a doctrinal perspective, such conditionality must respect principles of proportionality, non-discrimination, and due process. Financial restrictions cannot substitute for judicial determination. Rather, they serve as auxiliary mechanisms that support legally established obligations (Shany, 2007).

When properly designed, conditionality enhances coherence between judicial, financial, and regulatory regimes. It reduces free-rider incentives and aligns developmental assistance with environmental responsibility (Viñuales, 2018).

Countermeasures and Collective Response

Under the law of state responsibility, injured states may, under certain conditions, adopt countermeasures in response to internationally wrongful acts (Crawford, 2013). In the climate context, however, traditional bilateral countermeasures are of limited utility. Harm is diffuse, and retaliation risks undermining cooperative frameworks.

Nevertheless, collective response mechanisms may complement judicial enforcement in cases of persistent non-compliance. Coordinated measures—such as suspending treaty privileges, withdrawing technical cooperation, or imposing targeted restrictions—may be justified where a state systematically refuses to implement binding judgments (Tams, 2005; Dupuy, 2002).

The role of an ICE in this context would be primarily declaratory and supervisory. By authoritatively determining non-compliance, the court provides the legal predicate for collective action. Political bodies may then act within established legal parameters. This separation preserves judicial independence while enhancing systemic responsiveness.

Interaction with Multilateral Compliance Regimes

Climate governance already features extensive compliance mechanisms, most notably under the UNFCCC and the Paris Agreement. These regimes emphasize facilitation, transparency, and capacity building rather than sanctions (Bodansky et al., 2017).

An ICE must be designed to complement, rather than undermine, these structures. Judicial enforcement should address serious or persistent violations that exceed the scope of facilitative mechanisms. Conversely, findings of compliance committees may inform judicial assessment of diligence and good faith (Brunnée & Toope, 2010).

Institutional coordination may be formalized through memoranda of understanding, information-sharing agreements, and joint review procedures. Such arrangements prevent duplication and foster systemic integration (Koskenniemi, 2006).

Doctrinally, this reflects the principle of subsidiarity. Judicial intervention operates as a backstop where cooperative governance fails to ensure minimum compliance.

Reputational Sanctions and Public Accountability

Reputation remains a powerful enforcement mechanism in international relations. Judicial condemnation carries normative stigma and may affect diplomatic standing, investment attractiveness, and domestic legitimacy.

An ICE can reinforce reputational effects through reasoned judgments, public compliance assessments, and transparent publication practices (Shany, 2007).

Compliance rankings and performance indicators, while not legally binding, may enhance peer pressure and public scrutiny.

These mechanisms are particularly effective in democratic systems and in states dependent on international financing and trade. They transform judicial findings into broader governance signals (Alter, 2014).

Escalation Mechanisms and Graduated Enforcement

Effective compliance systems require graduated responses to non-compliance. Immediate resort to coercive measures is rarely proportionate or sustainable. Instead, enforcement should follow an escalation logic.

Initial stages may focus on clarification, technical assistance, and extended deadlines. Persistent non-compliance may trigger intensified monitoring, financial conditionality, and public reporting. Only in cases of systematic defiance should collective countermeasures be considered (Romano et al., 2014; Shany, 2007).

Embedding such escalation mechanisms in procedural rules enhances predictability, fairness, and legitimacy. It also incentivizes early cooperation and reduces adversarial polarization.

Jurisdictional Coordination and Trade Law Interfaces

The development of a specialized environmental jurisdiction also raises questions about its interaction with other international dispute settlement regimes, most notably the World Trade Organization's (WTO) dispute settlement system. Climate-related regulatory measures—such as carbon border adjustment mechanisms, environmental subsidies, or restrictions on carbon-intensive products—may simultaneously implicate environmental obligations and international trade rules. Consequently, the possibility of overlapping jurisdiction must be addressed in the institutional design.

Tensions between trade liberalization and environmental protection are not new at the normative level. WTO jurisprudence has repeatedly recognized environmental protection as a legitimate public policy objective, particularly through the environmental exceptions in Article XX of the General Agreement on Tariffs and Trade (GATT) (Sands et al., 2018). Panels have acknowledged that states retain regulatory autonomy for environmental objectives, provided their measures are not applied in a discriminatory or protectionist manner.

The institutional mandates of the WTO dispute settlement system and a prospective International Court for the Environment differ fundamentally. WTO adjudication focuses on whether national measures align with trade obligations. An environmental court would instead assess compliance with international environmental duties. These include obligations of due diligence, prevention, and cooperation.

Even when the same regulatory measure is at issue, respective tribunals address distinct legal questions.

Potential jurisdictional overlaps can be managed through established coordination mechanisms in international adjudication. The doctrine of lis pendens is one mechanism. Under this doctrine, proceedings may be suspended if the same dispute is already pending before another competent tribunal. Another mechanism is complementarity clauses. These clauses clarify that environmental adjudication does not invalidate trade rulings but instead addresses the environmental legality of the underlying conduct.

The principle of systemic integration requires international tribunals to interpret legal obligations in harmony with other relevant rules. Trade and environmental norms should be understood as parts of a broader legal framework, not as competing regimes (Koskenniemi, 2006).

Institutional dialogue between adjudicatory bodies also contributes to coherence. Courts often consider jurisprudence from other legal regimes when interpreting related obligations. Such cross-referencing promotes doctrinal consistency and reduces conflicting interpretations.

Accordingly, the existence of parallel jurisdiction does not necessarily produce fragmentation. When accompanied by appropriate coordination mechanisms, it may instead enhance the international legal system's capacity to address complex regulatory challenges. An International Court for the Environment would complement existing dispute settlement systems by providing a forum specifically designed to adjudicate environmental obligations while leaving trade law enforcement to established institutions.

Doctrinal Assessment

From a doctrinal perspective, enforcement in climate adjudication reflects the hybrid character of contemporary international law. It combines decentralized implementation, institutional supervision, economic incentives, reputational pressure, and collective response mechanisms. Rather, it corresponds to the structural realities of a non-hierarchical legal order confronting global public goods problems. Coercion is limited, but coordination is extensive (Koskenniemi, 2006; Viñuales, 2018).

An ICE would function as the normative and procedural hub of this architecture. By providing authoritative determinations, structured monitoring, and legally grounded escalation pathways, it would transform diffuse compliance expectations into an integrated enforcement regime.

In this sense, enforcement mechanisms complete the juridification of climate governance. They ensure that legal obligations do not remain aspirational but become operational components of global environmental regulation. Judicial authority thus becomes embedded in a broader system of accountability capable of of sustaining long-term compliance and reinforcing the operational effectiveness of international environmental law.

RIGHTS OF PARTICIPATION AND PROCEDURAL JUSTICE

The legitimacy of an ICE depends not only on the substantive correctness of its judgments or the effectiveness of its enforcement mechanisms, but also on the fairness and inclusiveness of its procedures. Climate adjudication concerns collective goods, intergenerational interests, and vulnerable populations whose voices are frequently marginalized in international decision-making. A credible judicial architecture must therefore embed participation, transparency, and procedural equality at its core. Procedural justice is not an ancillary feature of climate adjudication; it is constitutive of its normative authority.

Access to Justice in Environmental Matters

Access to justice constitutes a foundational principle of contemporary environmental governance. Procedural environmental rights—access to information, participation in decision-making, and access to judicial remedies—are increasingly recognized as indispensable components of effective environmental protection. At the international level, these guarantees have evolved from soft-law commitments into enforceable standards.

An ICE must reflect this evolution. Access to the court, as discussed in Section 5, should not be confined to states. Where climate harm affects identifiable communities or collective interests, affected groups must be able to initiate or meaningfully participate in proceedings, subject to admissibility safeguards. Without such access, the enforcement of climate obligations remains structurally dependent on state initiative, which political or economic considerations may compromise.

Access to justice also entails affordability and practical accessibility. Provisions for legal aid, reduced procedural fees, and digital participation mechanisms are essential to prevent socio-economic exclusion. In climate disputes, where vulnerable communities often lack resources, procedural design must compensate for structural inequalities.

Participation of Indigenous Peoples and Affected Communities

Climate change disproportionately impacts indigenous peoples, small island populations, and communities whose livelihoods are closely linked to fragile ecosystems. International law increasingly recognizes the distinct rights of such groups, including rights to cultural integrity, land, and self-determination.

Procedural participation before an ICE must therefore accommodate culturally specific forms of representation and knowledge. Mechanisms may include collective representation, culturally appropriate evidentiary standards, and respect for traditional knowledge systems. Where disputes concern territories or cultural heritage, the principles of free, prior, and informed consent may inform procedural practice.

Participation is not limited to formal standing. Even where communities are not parties to proceedings, structured opportunities for written submissions, expert testimony, and amicus curiae participation should be guaranteed. This enhances factual accuracy and reinforces the court's responsiveness to the lived realities of environmental harm.

Transparency and Public Reason

Transparency is a cornerstone of procedural legitimacy. Climate adjudication addresses matters of global public concern; its proceedings should therefore, as a rule, be public. Hearings, written submissions, and judgments should be accessible, subject to narrowly tailored confidentiality exceptions.

Publication of scientific summaries accompanying judgments enhances comprehensibility and public trust. Given the technical complexity of climate litigation, transparency requires not only formal openness but also intelligibility. Reasoned judgments must articulate the evidentiary basis and normative reasoning underpinning conclusions.

At the same time, transparency must be balanced against the protection of sensitive information. Disclosure of precise locations of vulnerable ecosystems, sacred sites, or critical infrastructure may entail security risks. The procedural framework should provide for confidential submissions and protective measures where justified.

Equality of Arms and Protection Against Procedural Abuse

Climate litigation frequently involves asymmetries of power and resources. States and corporate actors often possess extensive legal teams and technical expertise, whereas affected communities may face significant constraints. Ensuring equality of arms requires active judicial management.

The court should possess the authority to appoint independent experts, to request clarifications, and to structure proceedings in a manner that prevents dominance by resource-rich parties. Legal aid mechanisms and pro bono support frameworks help balance disparities.

Protection against procedural abuse is equally important. Strategic lawsuits against public participation (SLAPPs) or dilatory tactics may undermine effective adjudication. The procedural rules should allow for the early dismissal of manifestly abusive claims and sanction conduct intended to intimidate or obstruct.

Intergenerational Representation

Climate disputes inherently implicate the interests of future generations. Traditional procedural law, however, is oriented toward present claimants and tangible injury. Bridging this temporal gap requires innovative yet doctrinally grounded solutions.

One approach involves recognizing collective or public trusteeship claims, where states or designated representatives act on behalf of future generations. Another mechanism may involve the appointment of guardians ad litem or special representatives to articulate long-term environmental interests in proceedings.

While such measures challenge conventional standing doctrines, they are consistent with the recognition of intergenerational equity as an interpretative principle. Procedural accommodation of future interests strengthens the court's capacity to address irreversible harm and long-term risk.

Knowledge, Justice, and Epistemic Pluralism

Climate adjudication relies heavily on scientific expertise. Yet exclusive reliance on technocratic knowledge may marginalize experiential and traditional forms of understanding. Procedural justice, therefore, includes epistemic justice.

An ICE may establish advisory panels that incorporate diverse forms of expertise, including the natural and social sciences and indigenous knowledge systems. Evidentiary rules should allow for culturally grounded testimony and community-based observations, subject to reliability and relevance standards.

Epistemic pluralism does not dilute scientific rigor; it broadens the evidentiary landscape and enhances contextual accuracy. Courts remain responsible for evaluating the probative value of evidence, but should avoid rigid hierarchies that systematically exclude non-Western or non-academic knowledge traditions.

Political Feasibility and the Problem of Partial Participation

Proposals for new international courts often face skepticism about their political feasibility. In the context of climate governance, such concerns are particularly pronounced. Major emitting states may be reluctant to accept binding jurisdiction in an area with far-reaching implications for national energy policy, industrial regulation, and economic competitiveness. The possibility that key actors remain outside a new jurisdiction, therefore, raises an important question: can an International Court for the Environment function effectively under conditions of partial participation?

This concern is neither hypothetical nor unique to environmental law. Many international judicial institutions have emerged under conditions of incomplete participation. International investment arbitration, regional human rights courts, and trade dispute settlement regimes all developed gradually, expanding their jurisdiction through successive accessions rather than universal adoption from the outset (Alter, 2014). The effectiveness of these institutions has depended less on immediate universality than on establishing credible legal frameworks capable of generating normative influence and institutional learning.

From a legal perspective, partial participation does not render the establishment of a specialized environmental court impermissible. International law allows states to create dispute settlement mechanisms through treaties or inter se agreements, provided that these mechanisms do not affect the rights of non-participating states (Shany, 2007). Willing states can establish a jurisdictional regime that binds only its members while remaining open to future accessions.

Building on this understanding, a plurilateral institutional pathway may provide a realistic starting point for environmental adjudication. Under such an approach, a coalition of climate-vulnerable and climate-ambitious states could establish an International Court for the Environment with mandatory jurisdiction among participating members. Over time, accession clauses and jurisdictional extensions could allow additional states to join the regime. This gradual, evolutionary pattern has characterized the development of several international legal institutions.

Importantly, the influence of an international court extends beyond its formal jurisdiction. Judicial decisions may generate interpretative authority, shape expectations of lawful conduct, and influence domestic and international legal practice beyond the circle of formally bound parties. The jurisprudence of regional human rights courts, for example, has frequently influenced global legal debates despite limited geographical jurisdiction (Alter, 2014). A specialized environmental court could similarly contribute to the progressive development of international environmental law even before achieving universal participation.

Despite these potential benefits, it must be acknowledged that the absence of major emitters would raise practical challenges, particularly regarding the enforcement of judgments and the effectiveness of remedial orders. To address this, institutional design must incorporate mechanisms that can operate under conditions of asymmetrical participation. Several complementary strategies may mitigate this risk.

First, jurisdiction may be structured to focus initially on disputes among participating states and on transboundary environmental harm occurring within their territories or jurisdictions. Even without universal membership, such cases may clarify legal standards and develop jurisprudence relevant to global climate governance.

Second, the court's jurisdiction could be linked to cooperative mechanisms within existing international regimes. For instance, states participating in climate finance initiatives, loss-and-damage facilities, or environmental cooperation agreements could agree to submit disputes arising from these arrangements to the court's jurisdiction. Such jurisdictional clauses would expand the practical relevance of the institution while remaining compatible with the consensual foundations of international adjudication.

Third, the court may gradually increase its institutional legitimacy by issuing credible jurisprudence. When courts demonstrate procedural fairness, scientific competence, and doctrinal coherence, they often draw broader participation over time. The history of international adjudication shows that institutional authority frequently grows through practice rather than immediate universal acceptance (Romano et al., 2014).

Taken together, these considerations suggest that political feasibility should not be understood as a binary condition determining whether a specialized environmental jurisdiction is possible or impossible. Instead, feasibility must be assessed in terms of institutional adaptability and evolutionary potential. A court designed with flexible accession mechanisms, cooperative jurisdictional arrangements, and procedural credibility may gradually expand its influence even in the absence of immediate universal participation.

Ultimately, the challenge of non-participation reflects a broader structural feature of international law. Collective action problems often delay the rapid creation of comprehensive governance mechanisms. Yet, coalitions of willing actors have repeatedly developed institutions that reshape international legal practice. In this light, one should understand the establishment of an International Court for the Environment as a pragmatic step toward progressively institutionalizing climate accountability, rather than merely an unrealistic attempt to achieve immediate universality.

Doctrinal Integration

The incorporation of participation and procedural justice principles into climate adjudication reflects broader developments in international law. Human rights jurisprudence, environmental treaty practice, and global governance standards increasingly emphasize procedural guarantees as prerequisites for substantive compliance.

An ICE would synthesize these strands. It would transform participatory rights from peripheral consultative mechanisms into enforceable procedural standards within the adjudication process. This integration reinforces both legitimacy and effectiveness. Decisions reached through inclusive and transparent processes are more likely to command compliance and societal acceptance.

From a doctrinal standpoint, the expansion of participation does not undermine sovereignty. States remain central actors, but their regulatory autonomy operates within a framework of accountability to affected populations and the international community. Procedural justice thus functions as a mediating principle between collective environmental protection and state authority.

In conclusion, rights of participation and procedural justice are not supplementary features of environmental adjudication. They are essential structural components that align judicial authority with the democratic, human rights–based, and science-informed character of contemporary climate governance. By embedding these guarantees in its procedural design, an ICE would enhance both the normative coherence and the practical legitimacy of global climate adjudication.

CONCLUSION: FROM NORMATIVE CONSOLIDATION TO INSTITUTIONAL CAPACITY

This chapter has demonstrated that the juridification of climate governance has reached an advanced stage. Contrary to persistent narratives portraying climate protection as a predominantly political or aspirational domain, contemporary international law contains a dense and coherent body of binding obligations. Duties of due diligence, prevention, cooperation, and precaution, reinforced by human rights jurisprudence and *erga omnes* structures, have transformed climate governance into a matter of legal responsibility.

Recent international jurisprudence confirms this transformation. The advisory opinions of international courts have clarified the content of climate obligations, articulated their collective character, and affirmed the legal consequences of noncompliance. These developments mark a decisive shift from programmatic commitments toward normative determinacy. At the same time, they reveal a structural asymmetry between legal ambition and institutional capacity. Existing courts possess the authority to interpret climate law, but lack the procedural architecture and jurisdictional reach required for systematic enforcement.

The analysis of jurisdiction, procedure, remedies, and compliance mechanisms illustrates that this deficit is not primarily normative, but institutional. Climate obligations are scientifically contingent, cumulatively generated, and structurally embedded in regulatory systems. Their enforcement requires specialized evidentiary regimes, differentiated standards of causation, mechanisms for collective standing, and long-term monitoring structures. Traditional bilateral dispute settlement models are ill-equipped to accommodate these features.

An ICE would address this structural mismatch. It would consolidate dispersed normative standards into a coherent adjudicatory framework capable of translating abstract duties into operational legal obligations. By integrating scientific expertise, expanding access to justice, and developing compliance-oriented remedies, such a court would institutionalize the preventive and collective logic of contemporary environmental law.

Importantly, the establishment of an ICE would not constitute a rupture with existing international legal structures. It would represent an evolutionary adaptation of adjudicatory institutions to conditions of planetary risk. Specialized jurisdiction, participatory procedures, and structural remedies are already present, albeit in fragmented form, across different regimes. Their systematic integration responds to the principle of systemic coherence rather than fragmenting international law.

Concerns relating to sovereignty, consent, and institutional proliferation must therefore be assessed in light of existing commitments. States have already accepted extensive environmental and human rights obligations, including supervisory and reporting mechanisms. An ICE would reorganize these commitments within a judicial framework governed by due process and legal certainty. Collective enforcement of *erga omnes* interests reflects an established trajectory of international law toward community-oriented responsibility.

From a broader perspective, the institutionalization of climate adjudication carries implications for the future of the international legal order. It signals a shift from reactive dispute settlement toward anticipatory governance, from episodic responsibility toward structural compliance, and from purely inter-state accountability toward inclusive legal protection. Climate courts do not replace political negotiation or regulatory experimentation. They provide the juridical infrastructure that enables such processes to operate within legally defined boundaries.

The effectiveness of international climate law ultimately depends on its capacity to convert normative consensus into durable practice. Advisory authority and voluntary compliance mechanisms have reached their functional limits. The next stage of legal development lies in constructing institutions capable of sustaining long-term accountability under conditions of scientific uncertainty and global interdependence.

An ICE would not offer a panacea for climate change. It would, however, supply the missing judicial pillar of global climate governance. By anchoring climate justice in enforceable rights, reasoned adjudication, and monitored implementation, international law would be capable of responding to one of the defining challenges of the twenty-first century.

REFERENCES

Alter, K. J. (2014). *The New Terrain of International Law: Courts, Politics, Rights*. Princeton University Press.

Bodansky, D., Rajamani, L., & Werksman, J. (2017). *International Climate Change Law*. Oxford University Press. DOI: 10.1093/law/9780199664290.001.0001

Boyd, D. R. (2018). *The Human Right to a Healthy Environment*. UBC Press.

Boyle, A. (2012). Human rights and the environment: Where next? *European Journal of International Law*, *23*(3), 613–642. DOI: 10.1093/ejil/chs054

Boyle, A., & Redgwell, C. (2021). *Birnie, Boyle, and Redgwell's International Law and the Environment* (4th ed.). Oxford University Press. DOI: 10.1093/he/9780199594016.001.0001

Brunnée, J., & Toope, S. J. (2010). *Legitimacy and Legality in International Law*. Cambridge University Press. DOI: 10.1017/CBO9780511781261

Crawford, J. (2013). *State Responsibility: The General Part*. Cambridge University Press. DOI: 10.1017/CBO9781139033060

Dupuy, P.-M. (2002). A general stocktaking of the connections between the multilateral dimension of obligations and codification of the law of responsibility. *European Journal of International Law*, *13*(5), 1053–1069. DOI: 10.1093/ejil/13.5.1053

Dupuy, P.-M., & Viñuales, J. E. (2018). *International Environmental Law* (2nd ed.). Cambridge University Press. DOI: 10.1017/9781108399821

Foster, C. E. (2025). The 2025 International Court of Justice Advisory Opinion on Obligations of States in respect of climate change. *The International and Comparative Law Quarterly*, *74*(4), 1–35. DOI: 10.1017/S0020589325101292

Knox, J. H. (2018). Report of the Special Rapporteur on human rights and the environment. UN Doc. A/73/188. https://www.ohchr.org/en/documents/thematic-reports/a73188-report-special-rapporteur-issue-human-rights-obligations-relating

Koskenniemi, M. (2006). Fragmentation of international law: Difficulties arising from the diversification and expansion of international law. ILC. https://legal.un.org/ilc/documentation/english/a_cn4_l682.pdf

Mayer, B. (2018). *The International Law on Climate Change*. Cambridge University Press. DOI: 10.1017/9781108304368

Peel, J., & Osofsky, H. M. (2015). *Climate Change Litigation: Regulatory Pathways to Cleaner Energy*. Cambridge University Press. DOI: 10.1017/CBO9781139565851

Rajamani, L. (2006). *Differential Treatment in International Environmental Law*. Oxford University Press. DOI: 10.1093/acprof:oso/9780199280704.001.0001

Romano, C., Alter, K. J., & Shany, Y. (2014). *The Oxford Handbook of International Adjudication*. Oxford University Press.

Sands, P., Peel, J., Fabra, A., & MacKenzie, R. (2018). *Principles of International Environmental Law* (4th ed.). Cambridge University Press. DOI: 10.1017/9781108355728

Shany, Y. (2007). *Regulating Jurisdictional Relations Between National and International Courts*. Oxford University Press. DOI: 10.1093/acprof:oso/9780199211791.001.0001

Simma, B. (1994). From bilateralism to community interest in international law. *Recueil des cours, 250*, 217–384.

Tams, C. J. (2005). *Enforcing Obligations Erga Omnes in International Law*. Cambridge University Press. DOI: 10.1017/CBO9780511494116

Viñuales, J. E. (2018). The Organisation of the Anthropocene. In *Our Hands?* Cambridge University Press.

von Bogdandy, A., & Venzke, I. (2014). *In Whose Name? A Public Law Theory of International Adjudication*. Oxford University Press. DOI: 10.1093/acprof:oso/9780198717461.001.0001

Chapter 13

Policy Reform, Legal Innovation, and the Nexus of ESG and Environmental Law:

Toward a Just and Sustainable Climate Economy

Oindrilla Ghosh
http://orcid.org/0000-0001-9609-0781
Christ University, India

Binod Kumar
http://orcid.org/0009-0002-9721-1367
Radha Govind University, India

ABSTRACT

Policy reform and legal innovation are interdependent for a climate-resilient future. ESG frameworks align corporate duties with societal goals, but require robust environmental law to avoid greenwashing. A just climate economy demands accessible, participatory rulemaking, transparent metrics, and enforceable standards that scale from local to levels. Innovations include performance-based regulations, climate risk disclosure, and dynamic permitting that update with science. Legal tools such as citizen lawsuits, right-to-information, and climate justice considerations channel accountability toward vulnerable communities. Integrating ESG with environmental justice requires harmonizing fiduciary duties with public interest, ensuring minority and frontline voices input into policy design. Cross-border cooperation, standard-

DOI: 10.4018/979-8-3373-6635-7.ch013

ized reporting, and adaptive governance enable timely responses to climate shocks. State capacity, funding for transition, and clear decarbonization pathways reduce uncertainty for investors and workers alike, fostering a just, regenerative economy.

INTRODUCTION

Policy reform, legal innovation, and the integration of environmental, social, and governance (ESG) considerations into law are mutually reinforcing engines driving toward a just and sustainable climate economy. The contemporary challenge, climate disruption compounded by inequality, demands a framework where policy design aligns with enforceable norms, innovative regulatory tools, and a governance ethos that foregrounds equity and long-term resilience (Fan & Wang, 2024). This chapter outlines the core ideas, aims, and guiding questions that animate the field: How can policy reform translate evolving scientific knowledge and public values into effective rules and incentives? In what ways can legal innovation expand the toolkit beyond traditional regulation to catalyze rapid decarbonization, adaptation, and responsible stewardship of natural capital? And how can the ESG construct operationalize accountability, transparency, and risk management in both public and private sectors to create a just, inclusive climate economy?

Policy reform in this space seeks to reconcile competing imperatives: ambitious climate outcomes, cost-effectiveness, and distributive justice. It requires rethinking the design of emissions trading schemes, carbon pricing, performance standards, and subsidies to ensure they are not only efficient but also fair. Reform debates increasingly emphasize mechanism design that drives measurable results while avoiding perverse incentives, such as leakage or regressive fiscal impacts. They also stress the need to harmonize sectoral policies, energy, transportation, industry, land use, with macroeconomic objectives, workers' livelihoods, and communities disproportionately affected by climate risk (Adamowicz, 2022). A just climate economy demands that policy reforms incorporate explicit distributive analyses, labor and Indigenous rights considerations, and guarantees for communities facing environmental burdens. The policy entrepreneur must navigate a landscape of federal, state, and local authorities, international agreements, and private sector voices, crafting coherent strategies that scale up ambition without sacrificing political feasibility (Söderholm, 2020).

Legal innovation refers to the creation and adaptation of rules, institutions, and processes that better align law with climate realities. This includes new forms of governance, such as outcome-based regulations, performance-based standards, and adaptive rules that can adjust as conditions evolve. It also encompasses novel rights-based approaches, ambient environmental rights, climate justice claims, and public trust doctrines, that empower communities to participate in decision-making and seek redress for harm (Batrancea et al., 2021). Legal innovation leverages data, science, and technology to improve compliance, enforcement, and accountability. It envisions integrated climate policy that spans environmental, social, and economic dimensions, promoting cross-border cooperation, financeable green infrastructure, and resilient supply chains. Infusing ESG into law goes beyond corporate disclosure to embed material ESG risks into regulatory decision-making, financial markets, procurement, and risk management frameworks (Hariyani et al., 2024). This requires clearer standards, independent verification, and robust governance structures to prevent greenwashing and ensure credible commitments translate into real-world outcomes.

The nexus of ESG and environmental law thus sits at the intersection of ethics, economics, and governance. ESG signals a shift from narrow environmental protection to a holistic approach that values social capital, governance integrity, and long-term stability. In practice, ESG integration can drive capital toward sustainable projects, incentivize corporate behavior aligned with climate goals, and raise the standard of corporate accountability (Yang et al., 2023). Environmental law provides the binding scaffolding, rules, liabilities, and remedies that give ESG commitments teeth and ensure consistency across markets and jurisdictions. The result is a climate economy that is not only low-carbon but also socially inclusive, regionally resilient, and procedurally legitimate (Plachkov, 2024). Achieving this vision requires ongoing experimentation, transparent evaluation, and a willingness to recalibrate in light of new evidence and lived experience. The future of policy reform and legal innovation lies in adaptive, participatory, and justice-centered frameworks that mobilize science, finance, and civil society toward a shared horizon: a just and sustainable climate economy.

CLARIFICATION OF CENTRAL ARGUMENT AND SCOPE

This chapter advances the argument that effective climate governance requires a carefully integrated framework that aligns policy reform, legal innovation, ESG governance, and technological transformation within a justice-oriented paradigm. While these domains are often discussed in parallel, their interaction is neither seamless nor inherently complementary. The central research question guiding this chapter is: how can legal and policy systems reconcile efficiency-driven environmental governance mechanisms with equity-based climate justice objectives, while incorporating emerging technologies as tools of institutional transformation? The chapter adopts a normative-analytical approach, combining theoretical evaluation with policy-oriented insights to assess the limitations of existing frameworks and propose a more coherent model for a sustainable climate economy.

At present, many regulatory approaches rely on fragmented interventions, where ESG operates as a voluntary or semi-regulated corporate mechanism, legal innovation evolves unevenly across jurisdictions, and technological tools such as artificial intelligence and immersive platforms are introduced without sufficient legal grounding. This fragmentation risks producing governance systems that are technologically advanced yet socially exclusionary, or economically efficient yet distributively unjust. Accordingly, the chapter seeks to reposition these elements within a unified analytical structure, where technological innovation is understood not as an independent driver but as an instrument embedded within legal and policy frameworks.

The argument proceeds by first identifying the structural limitations of prevailing ESG and environmental law regimes, particularly their reliance on market-based logic and their vulnerability to strategic compliance. It then examines the role of emerging technologies in reshaping governance processes, highlighting both their transformative potential and their regulatory challenges. Finally, the chapter proposes an integrated framework that combines enforceable legal standards, adaptive policy mechanisms, and equity-oriented principles to ensure that climate governance is both effective and just.

To enhance conceptual clarity, the chapter is structured as follows: the first section develops the theoretical framework underpinning the analysis; the second examines the role of technological innovation in legal systems; the third evaluates case studies to illustrate practical applications and limitations; and the final section synthesizes these insights into policy recommendations. By clearly articulating its central thesis and analytical trajectory, the chapter aims to move beyond descriptive synthesis and contribute a focused and rigorous perspective on the intersection of law, technology, and climate governance.

HISTORICAL EVOLUTION

The historical arc of policy reform, legal innovation, and ESG-aligned environmental law traces a trajectory from traditional command-and-control regulation to a more nuanced, integrative, and dynamic climate governance paradigm. Early environmental law emerged in modest form, protecting air and water quality, regulating point-source pollution, and establishing liability regimes (Lwamba et al., 2022). These measures laid foundational norms, but they often struggled with scale, speed, and distributive impact. As climate science clarified the urgent need to decarbonize energy systems, policy reform began to incorporate market-based tools, performance standards, and sector-specific mandates. The shift from purely prescriptive rules to incentives and flexible pathways reflected lessons from economic efficiency theories and political acceptability (Shandas & Hellman, 2022). Emissions trading experiments, renewable portfolio standards, and clean energy subsidies became common instruments, designed to convert environmental imperatives into investable markets and bankable projects.

The late 20th and early 21st centuries witnessed a methodological expansion in environmental law and policy. Global instruments, such as the Kyoto Protocol and later the Paris Agreement, reframed climate action as a shared, multilateral obligation, emphasizing Nationally Determined Contributions (NDCs) and cross-border cooperation. At the same time, domestic reform advanced in layers: federal standards and state-level innovations in the United States; similar multi-tier approaches in the European Union, with its emissions trading system and the European Green Deal; and rising attention to climate risk disclosure in financial markets (Mottershead et al., 2021). These developments underscored a crucial insight: legal rules alone cannot realize systemic decarbonization without robust financial, technological, and social ecosystems that support transition.

A pivotal evolution has been the formal embrace of ESG frameworks within both public policy and corporate governance. ESG considerations emerged from ethical investing, risk management, and sustainability reporting, gradually hardening into regulatory expectations and standardized disclosures (Hill et al., 2020). Governments began requiring or endorsing climate-related financial risk disclosures, supply chain transparency, and sustainability reporting for large organizations. This alignment of corporate behavior with environmental and social objectives has accelerated private-sector investment in decarbonization, resilience, and sustainable innovation. It also illuminated governance gaps board accountability, executive incentives, and risk governance that ESG-oriented reforms aim to address (Sharif et al., 2024).

Technological and social innovations further reshaped the landscape. Digitalization, data analytics, and advanced modeling improved measurement, verification, and traceability of emissions and environmental impacts. Decentralized energy systems, electrification of transport, and nature-based solutions offered practical pathways to reduce emissions while supporting communities (Raihan, 2023). The social dimension environmental justice, just transition for workers, and inclusive decision-making moved from peripheral concern to central policy design, reflecting a broader understanding that climate action must be equitable to be durable.

Today, the nexus of ESG and environmental law embodies a transformative governance paradigm: rules that are smarter and more adaptive, backed by transparent disclosure and accountable governance. Legal innovations such as performance-based standards, market mechanisms with social safeguards, and public-trust or rights-based claims coexist with ESG disclosures, climate risk reporting, and sustainable finance frameworks (Chou et al., 2023). The aim is a climate economy characterized by credible commitments, measurable progress, and resilience to shocks, while ensuring that the benefits and burdens of the transition are distributed justly. As global finance, technology, and civil society converge, the historical evolution points toward a more integrated, proactive, and justice-centered approach to environmental governance (Bolton et al., 2020). The path ahead requires continuous experimentation, vigilant oversight, and a shared commitment to a sustainable and inclusive future

THEORETICAL FRAMEWORKS

The frameworks outlines core theoretical lenses that structure thinking about policy reform, legal innovation, and the intersection of ESG and environmental law within a just and sustainable climate economy. The aim is to map how concepts from political theory, law, economics, and environment-and-society studies illuminate pathways to reform, guide institutional experimentation, and foreground equity, precaution, and resilience.

Justice, Capabilities, and Distributive Equity - Rawlsian and capabilities-inspired frameworks emphasize fairness in the distribution of climate-related costs and benefits. Policy reform becomes legitimate when it promises to enhance the real freedoms and capabilities of the least advantaged, not merely aggregate efficiency. Application to environmental law: design standards, access to remedies, and procedural safeguards should address vulnerable communities disproportionately bearing climate burdens. This frame pushes for universal social protections, just transition mechanisms for workers, and participatory governance structures in ESG decision-making. Implication for reform: inclusive stakeholder engagement, targeted subsidies or transfers, and transparent accountability to ensure reforms don't privilege elites or externalize risk (Onu et al., 2023).

Environmental Law as Adaptive Governance - Adaptive governance treats environmental law as dynamic, experiment-oriented, and capable of learning under uncertainty and changing ecological and economic conditions. Key ideas include nested institutions, polycentric governance, rule of law that supports experimentation, and horizon-scanning for emerging risks (e.g., tipping points, climate migration). Implication for reform: institutions should be designed to adjust policies as data streams evolve, with sunset clauses, built-in review cycles, and learning loops that incorporate scientific uncertainty and local knowledge.

Systemic Risk and Precautionary Principle - In the climate realm, precautionary and risk-management frameworks prioritize preventing irreversible harm in the face of scientific uncertainty. Theoretical stance: policies should err on the side of safety, even when full certainty is unavailable, particularly for high-stakes environmental impacts. Implication for reform: preemptive regulation, robust environmental impact assessments, and frameworks that require cost-of-inaction analyses alongside cost-of-action estimates (D'amato & Korhonen, 2021).

Law and Economics plus Distributional Externalities - While standard cost–benefit analysis has limits, integrating distributional weights and non-market values within a law-and-economics lens helps internalize health, biodiversity, and intergenerational costs (Kumar et al., 2023). Implication for reform: design of incentives (carbon pricing, subsidies for clean tech) should explicitly consider distributional effects and co-benefits to public health and welfare.

Critical and Feminist Legal Theories - These perspectives highlight power asymmetries, gendered impacts of climate policy, and the role of care economies. Implication for reform: ensure that climate policy embeds care responsibilities, recognizes informal labor, and strengthens community resilience, particularly for marginalized groups.

Governance, Institutions, and Legitimacy - The legitimacy of reform depends on trust, legitimacy of institutions, and credible enforcement. Implication for reform: transparency, participatory rulemaking, and credible enforcement mechanisms are central to successful policy reform and sustainable ESG governance (Li et al., 2022).

Integrative Frameworks for ESG - A synthesis framework aligns environmental outcomes with social justice and governance quality, ensuring that ESG metrics drive policy and legal innovation toward systemic sustainability (Amirova et al., 2022). Implication for reform: harmonize environmental, social, and governance standards across sectors; foster cross-institutional data sharing and interoperable reporting.

USE OF AI AND ML IN POLICY REFORMS

AI and ML can accelerate policy reform by informing evidence-based decisions, identifying inequities, forecasting outcomes, and enabling participatory governance.

Evidence Synthesis and Forecasting - AI tools can process vast data streams, from climate models, economic indicators, to social media sentiment to forecast policy outcomes, stress-test scenarios, and identify potential unintended consequences (Shulimova et al., 2024). Applications: predictive modeling for emissions trajectories, optimization of subsidy programs, and scenario planning under different regulatory regimes.

Equity and Inclusion Analytics - ML can detect distributional effects of policies by mapping who benefits or bears costs, enabling targeted interventions to protect vulnerable groups. Applications: monitoring environmental justice indicators, identifying disparities in access to clean energy, and guiding reform to prioritize marginalized communities.

Policy Design and Optimization - Optimization algorithms and causal inference methods help craft policy mixes that achieve multiple objectives (emissions reductions, job creation, health gains) with minimal trade-offs (Savchenko & Borodina, 2020). Applications: design of carbon pricing with compensation mechanisms, allocation of clean energy subsidies, and tailored regulatory frameworks for different sectors.

Public Participation and Deliberative AI - AI-enabled platforms can support broad, inclusive engagement by summarizing complex policy trade-offs, translating technical jargon, and capturing stakeholder preferences at scale. Applications: interactive dashboards, deliberative forums, and AI-assisted public comment analysis to inform reform directions.

Monitoring, Compliance, and Verification - Computer vision, sensor networks, and ML-based anomaly detection improve compliance monitoring, environmental reporting, and enforcement efficiency (Dwivedi et al., 2022). Applications: real-time emissions monitoring, supply-chain traceability, and detection of greenwashing.

Governance, Bias Mitigation and Transparency - Responsible AI governance is essential to prevent opacity, bias, and manipulation in policy processes. Measures: explainable AI, audit trails, algorithmic impact assessments, data provenance, and public disclosure of model assumptions and limitations.

Challenges and Safeguards - Data quality, representativeness, and accountability are critical. There is a risk of automation bias, where overreliance on models obscures contextual knowledge (Olsen et al., 2022). Safeguards: human-in-the-loop processes, robust validation, ethical guidelines, and alignment with constitutional rights and democratic norms.

LEGAL INNOVATION THROUGH AI AND ML

AI/ML reshape legal practice, enforcement, and access to justice, enabling faster analysis, better risk assessment, and more transparent decision-making.

Rule of Law and Procedural Fairness - AI tools for risk-based sentencing, legal research, and case outcome prediction raise questions about due process, accountability, and bias (Swain & Karimu, 2020). Safeguards: algorithmic transparency, human oversight, and rights-centered governance.

Access to Justice and Legal Tech - AI-powered chatbots, document automation, and predictive analytics can make legal services more affordable and accessible. Equity focus: ensure that low-income and marginalized groups can benefit, with safeguards against misuse or infringement of rights.

Intellectual Property and Data Governance - AI-generated outputs raise IP questions; robust data governance policies ensure data rights, consent, and fair use in legal tech ecosystems (Stroebel & Wurgler, 2021).

ROLE OF AR, VR AND MR

AR (Augmented Reality): Enhancing Perception, Practice, and Policy

Augmented Reality (AR) overlays digital information onto the real world, creating a hybrid experience that augments human perception. In the context of climate action, environmental law, and a just climate economy, AR holds several transformative roles.

First, AR can revolutionize environmental monitoring and compliance (Hong et al., 2020). Field technicians can wear AR headsets that display real-time sensor data, regulatory thresholds, and historical trends overlaid on the physical site. This boosts accuracy, speeds up inspections, and reduces human error. For example, in pollution control or hazardous-wac situations, AR guides workers through stepwise procedures, flags deviations from permitted limits, and records geo-tagged evidence for audits and enforcement. This directly supports policy implementation by making compliance more visible, auditable, and scalable (Bhutta et al., 2022).

Second, AR enhances public participation and environmental justice. By projecting interactive overlays onto neighborhoods or ecosystems e.g., visualizing flood risk, heat island effects, or proposed infrastructure impacts community members can better understand potential outcomes and participate in hearings (Yeow & Ng, 2021). This democratizes environmental decision-making, helping to center vulnerable communities in policy reform and project planning. AR can also translate complex environmental data into accessible formats for non-experts, supporting informed consent and shared decision-making.

Third, AR as a tool for legal professionals and policymakers. In litigation or regulatory rulemaking, AR can visualize scenarios, show causal links, and simulate policy outcomes (Monk & Perkins, 2020). Attorneys can present immersive demonstrations of environmental harms, while regulators can stress-test compliance regimes by walking through enforcement workflows in situ. For training, prosecutors, judges, and compliance officers can engage with realistic, interactive case simulations that reflect current environmental statutes, executive orders, and international standards (Glanemann et al., 2020). Such experiential learning lowers the barrier to understanding sophisticated environmental law and strengthens the practical implementation of reform.

Fourth, AR supports design, planning, and sustainable infrastructure. Urban planners and engineers can overlay climate projections, green infrastructure designs, and resilience metrics onto real sites, facilitating adaptive planning. Visualizing future scenarios on-site helps stakeholders weigh trade-offs, optimize land use, and accelerate approvals, contributing to a more timely and participatory transition to a low-carbon urban fabric.

Finally, privacy, data governance, and equity considerations must guide AR deployment. Data overlays can reveal sensitive information about individuals, communities, or critical infrastructure (Wang & Ma, 2021). Clear governance around data ownership, consent, access controls, and bias mitigation is essential to ensure AR-driven environmental action remains fair and respectful of rights.

VR (Virtual Reality): Immersive Experiments in Climate Risk and Justice

Virtual Reality creates entirely simulated environments, enabling deep immersion without dependence on the physical world. In climate policy, law, and a sustainable climate economy, VR offers unique capabilities.

First, VR enables safe, scalable risk and resilience simulations. With immersive environments, policymakers and planners can model extreme weather, sea-level rise, or wildfire scenarios, testing adaptation strategies, emergency responses, and infrastructure designs (Casady et al., 2024). This experiential understanding can shape more robust building codes, zoning, and disaster-management policies that reflect lived realities. VR simulations help translate abstract climate data into tangible implications, facilitating consensus-building and prioritization of investments in resilience.

Second, VR supports education and capacity-building across sectors. Lawyers, judges, and regulators can practice applying environmental laws in lifelike courtroom or regulatory scenarios (Gao et al., 2023). Students and professionals can experience the consequences of different policy choices, deepening comprehension of legal nuance, compliance implications, and enforcement challenges. This elevates the quality of legal innovation and policy reform by nurturing a workforce fluent in climate law and ethics.

Third, VR fosters stakeholder engagement and environmental justice. Immersive experiences can reveal disproportionate climate impacts on marginalized communities, enabling empathetic understanding and more inclusive deliberations (Vassileva, 2022). By sharing a VR-enabled experience with diverse audiences, advocates can build coalitions around equitable climate strategies, ultimately informing policy reform with a justice-centered lens.

Fourth, VR accelerates design innovation for sustainable infrastructure. Engineers and architects can prototype and visualize resilient systems sea walls, floodproof buildings, green corridors in an immersive 3D space, reducing costs and accelerating iterative testing before construction (Celestin et al., 2024). This capitalizes on rapid experimentation to identify low-carbon, high-benefit solutions .

Fifth, ethical and logistical considerations matter. VR interventions must be accessible, culturally sensitive, and mindful of potential misrepresentations of data. Costs, technical capacity, and maintenance define where VR delivers the most value, especially in resource-constrained jurisdictions (Shahzad et al., 2023).

MR (Mixed Reality): Integrating Digital and Physical Worlds for Smart Climate Solutions

Mixed Reality (MR) blends AR and VR to create seamless integrations of digital content with the physical environment. In climate leadership and environmental law, MR unlocks synergistic capabilities.

First, MR supports on-site decision-making with real-time data overlays and immersive context. Field engineers can see regulatory constraints, performance metrics, and model projections anchored to the exact location of a project. This enhances accuracy, speeds approvals, and ensures consistency with environmental standards (Zhu et al., 2023).

Second, MR improves collaboration across disciplines and geographies. Teams can co-create in shared mixed-reality spaces, combining legal analysis, engineering design, and policy considerations in a single immersive workflow. Stakeholders from regulators to local communities can interact with the same MR scenario, promoting transparency and trust in reform processes (Abdullah et al., 2023).

Third, MR enhances enforcement and compliance administration. Inspectors can access dynamic policy workflows, checklists, and evidence streams while in the field, creating auditable trails that strengthen accountability and reduce disputes about compliance status.

Fourth, MR enables adaptive infrastructure management. Operators can monitor performance, receive live alerts, and adjust operations with contextual guidance embedded in the environment (Licastro & Sergi, 2021). This supports a climate-smart, responsive economy where systems learn from ongoing data.

Fifth, governance and equity considerations apply. MR deployments must safeguard privacy, ensure equitable access to technology, and avoid reinforcing existing inequities through biased data overlays or unequal benefits (Nazari & Musilek, 2023).

CHALLENGES FOR A JUST AND SUSTAINABLE CLIMATE ECONOMY

Bringing together policy reform, innovative legal frameworks, and ESG integration within environmental law is essential for a just and sustainable climate economy. Yet this convergence faces multifaceted challenges across governance, markets, technology, and society. Identifying and addressing these obstacles is critical to designing effective, durable, and equitable solutions.

Political fragmentation and short-termism: Policy cycles driven by elections can undermine long-horizon climate reforms, delaying investments in low-carbon transitions and resilience (Dwivedi et al., 2022). Regulatory coherence vs. complexity: Aligning sectoral policies (energy, transport, land use, finance) to avoid conflicts is difficult in federated or plural legal systems, leading to regulatory gaps or overlaps. Trade-offs and distributional impacts: Reform often shifts costs and benefits, raising concerns about equity for workers, communities dependent on high-carbon industries, and vulnerable populations. Designing just transition mechanisms is essential but politically arduous (Muth, 2023). Speed of reform vs. precaution: Rapid reforms risk unintended consequences, while slow reforms miss climate windows. Achieving calibrated, evidence-based pacing is challenging. Public participation and legitimacy: Ensuring inclusive, transparent rulemaking that meaningfully incorporates diverse voices, including marginalized communities, remains uneven and resource-intensive. Data gaps and analytical capacity: Policymakers need robust, timely data to design, monitor, and adjust reforms. Data availability, quality, and interoperability vary widely (Schröder, 2020).

Regulatory burden and innovation tension: Overly prescriptive rules can stifle innovation; dynamic, performance-based, or outcomes-based approaches require advanced measurement and enforcement regimes. Liability, accountability, and ambiguity: Rapidly evolving technologies and business models create uncertainties about responsibility for harms, data use, and enforcement pathways (Balezentis et al., 2024). Cross-border and harmonization challenges: Global supply chains, multinational finance, and cross-jurisdictional operations complicate the application of ESG and environmental laws, risking regulatory discrepancies. Access to justice and capacity constraints: Courts, regulators, and compliance teams in many jurisdictions lack the resources to adjudicate novel disputes, leading to enforcement gaps (Bluedorn et al., 2023). Standardization and comparability: Diverse ESG metrics and environmental indicators hinder consistent reporting, verification, and investor confidence. Standardization efforts must balance rigor with practicality. Public-private collaboration: Achieving effective innovation often relies on pilots, sandboxes, or co-regulation, which require trust, clear guardrails, and risk-sharing mechanisms (van Niekerk, 2024).

Greenwashing peril: Inconsistent or vague ESG disclosures allow superficial sustainability claims, undermining credibility and investor protection. Measurement, verification, and disclosure: Developing universal, verifiable metrics for environmental impact, social outcomes, and governance quality is technically and institutionally challenging (Fan & Wang, 2024). Social equity and environmental justice: Ensuring ESG practices genuinely advance just outcomes not merely cosmetic improvements requires robust inclusion, impact assessments, and accountability mechanisms. Corporate governance vs. public interest: Tensions arise when corporate incentives prioritize short-term financial gains over long-term climate resilience and community well-being (Adamowicz, 2022). Financing structure and risk: Aligning finance with climate goals (sectoral decarbonization, resilience investments) demands risk disclosures, climate-tied fiduciary duties, and suitable incentives, which can be costly for smaller entities. Regulatory capture and credibility: Well-resourced actors may shape ESG norms and enforcement in ways that favor incumbents, marginalizing marginalized groups or newer entrants (Söderholm, 2020).

Pathways Toward a Just and Sustainable Climate Economy - Build interoperable standards and shared data ecosystems to improve transparency and comparability. Use phased, performance-based regulation with clear milestones and sunset clauses to maintain accountability while allowing innovation. Strengthen capacity-building, public participation, and co-design with affected communities to ensure legitimacy and equity (Batrancea et al., 2021). Develop robust verification, auditing, and independent oversight to deter greenwashing and build market trust. Align finance with climate outcomes through enhanced disclosures, risk taxonomies, and incentives that reward just transitions and resilience. Foster global cooperation to harmonize core ESG metrics, while respecting local contexts and development needs.

CASE STUDIES

Policy Reform, Green Infrastructure, and ESG-Driven Municipal Finance

Context and problem

Cities globally face aging infrastructure, flood and heat risks, and inequitable access to services. Traditional public finance often failed to align with climate resilience and social equity. Municipal budgets relied on volatile revenue streams, while capital projects overlooked long-term externalities. In many regions, fragmented governance and limited access to private capital hindered scalable climate action, producing a gap between aspirational climate goals and feasible implementation (Hariyani et al., 2024). ESG reporting began to shape investor expectations, but practical integration into municipal policy lagged.

Policy response

A consortium of megacities and regional authorities launched a policy reform package centered on climate-aligned budgeting, performance-based capital allocation, and mandatory resilience standards for new projects. The reform required climate risk disclosures for municipal investments and established a pipeline of green infrastructure projects transit, flood defenses, urban cooling, and water security prioritized by social vulnerability indices (Yang et al., 2023). An accompanying public procurement reform tightened environmental criteria, mandated lifecycle cost analysis, and incentivized local job creation. Cities also created blended finance facilities (green bonds, concessional loans, and guarantees) to de-risk projects with long horizons.

Legal innovation

Legislation codified climate-budget alignment, embedding ESG criteria into municipal finance law. The instruments created a legally binding requirement to quantify externalities, assign ownership of climate risk in project documents, and ensure equity-weighted project evaluation. A new form of municipal climate obligation contract was introduced, linking debt service to realized climate benefits, such as reductions in heat exposure hours or flood damage avoided (Plachkov, 2024). Public-private partnerships adopted standardized ESG dashboards, with independent verification and cross-jurisdictional interoperability. Courts recognized climate risk disclosure as a fiduciary duty in public finance, enabling citizen suits for noncompliance.

Outcomes and impacts

Early pilots demonstrated improved resilience and cost efficiency. Reduced energy costs, lower flood losses, and cooler urban zones produced tangible quality-of-life gains for vulnerable residents. Access to private capital improved as ESG metrics provided a transparent risk-return framework, expanding the investor base to sovereign wealth funds and green banks (Lwamba et al., 2022). Job creation in construction and maintenance supported a just transition by prioritizing equitable labor standards and local hiring. Cross-border knowledge sharing accelerated replication in neighboring regions.

Challenges and lessons

Data gaps impeded full lifecycle analyses, particularly for indirect climate benefits. Political cycles threatened long-term commitments; reform required durable governance arrangements. There were tensions between short-term fiscal consolidation and long-horizon climate investments. Equity concerns necessitated ongoing refinement of the vulnerability index to avoid marginalization (Shandas & Hellman, 2022). Certification and standardization across jurisdictions remained a work in progress, raising concerns about greenwashing.

Implications for climate justice

The case demonstrates that integrating ESG into municipal finance can democratize climate investments, align risk with long-term benefits, and deliver immediate social resilience. When climate justice is embedded in budgeting, procurement, and legal duties, marginalized communities gain protection against climate shocks (Mottershead et al., 2021). The approach also provides a scalable template for other regions seeking to fuse policy reform with legal innovation to finance a just, sustainable climate economy.

National Carbon Markets and Legal Innovation for Fair Transition

Context and problem

National carbon markets emerged to price emissions and incentivize decarbonization. However, uneven sectoral coverage, auction design, and allocation methods often exacerbated inequality, affecting low-income households and carbon-intensive workers (Hill et al., 2020). In some jurisdictions, opportunities to leverage market mechanisms for a just transition were constrained by legal uncertainty, weak enforcement, and limited distributional protections.

Policy response

A comprehensive reform package expanded coverage to transport, industry, and power while introducing a gradual price trajectory with revenue recycling to protect vulnerable groups. The reform integrated a social climate fund to fund retraining, energy assistance, and community resilience projects in regions with high dependence on fossil fuels (Sharif et al., 2024). It also created a just-transition framework, mandating stakeholder consultation and transition plans for major emitters and regions.

Legal innovation

Legislation established a centralized market operator with robust governance safeguards, transparent auction rules, and independent monitoring. It codified distributional impact assessments for all rule changes and mandated a minimum share of allowance revenues to be earmarked for affected workers and communities (Raihan, 2023). Courts recognized non-compliance with just-transition duties as a matter of public interest, enabling citizen petitions and expedited remedies. Arbitration provisions for cross-border compliance facilitated international participation and avoided trade-related disputes.

Outcomes and impacts

Emissions declined in covered sectors while revenue funds mitigated the regressive effects of policy shocks. Workers benefitted from retraining programs, and communities gained investment in diversification and energy efficiency. Market liquidity improved due to standardized rules and predictable price signals, attracting international traders and investors (Chou et al., 2023). A broader ESG investor community emerged, recognizing the market as a credible channel for green capital.

Challenges and lessons

Distributional safeguards struggle against path dependency and political economy constraints. Ensuring robust governance to prevent leakage, gaming, and corruption requires continuous strengthening. Technical capacity to monitor and verify emissions across sectors remains essential. Generating credible, transparent data for social impact reporting is critical to maintain trust in the market.

Implications for climate justice

A well-designed carbon market with strong social mechanisms can align market incentives with equity and resilience goals. By reinvesting proceeds into affected workers and communities, the approach underpins a fair transition and reinforces the legitimacy of climate action as a human-rights issue (Bolton et al., 2020).

Regulatory Sandboxes for Clean Tech and Environmental Innovation

Context and problem

Regulations often lag behind fast-moving clean technologies, impeding deployment of carbon capture, energy storage, and advanced manufacturing. Regulators faced a tension between ensuring safety, environmental protection, and enabling rapid innovation. Startups and incumbents alike struggled to navigate uncertain approval paths, creating gaps in the climate economy.

Policy response

A national program introduced regulatory sandboxes for clean tech pilots, combining temporary relaxations of certain rules with stringent monitoring and exit criteria. The program allowed pilots to test new tech in controlled environments, with predefined metrics for environmental impact, safety, and labor standards (Onu et al., 2023). Sandboxes were coupled with a fast-track permitting track and a modular approvals process designed for scalable deployment.

Legal innovation

Legislation created a safe harbor framework for select innovations, clarifying liability allocation in pilot contexts and establishing standardized compliance templates. Environmental and energy agencies issued adaptive rules that could be updated in response to real-world data from sandbox pilots. Public interest standards were incorporated to ensure community engagement, equity safeguards, and disproportionate impact mitigation. Multistakeholder boards provided guidance and oversight for sandbox projects, ensuring alignment with climate justice principles.

Outcomes and impacts

Pilot outcomes included reduced costs for demonstration-scale deployments, accelerated permitting, and improved data sharing among regulators and innovators. Companies gained confidence to scale promising technologies, attracting funding and partnerships. Environmental safeguard learning informed permanent rules and better risk assessment practices (D'amato & Korhonen, 2021). Communities saw early-stage benefits in job opportunities and local investment.

Challenges and lessons

Safeguarding environmental integrity while allowing experimentation required rigorous metrics and independent verification. Balancing speed with accountability remained a primary tension. Ensuring inclusive community engagement in pilot design and sharing outcomes was essential to avoid exclusion. Coordinating across agencies and jurisdictions tested governance capacity (Kumar et al., 2023).

Implications for climate justice

Sandbox approaches can democratize access to new technologies by reducing barriers and enabling co-created solutions with affected communities. When designed with robust safeguards, transparency, and equitable benefit-sharing, these programs advance a just climate transition while expanding the toolkit of environmental law for rapid innovation.

ESG Disclosure Regulation and Corporate Legal Obligations

Context and problem

Rising investor demand for robust ESG data coincided with concerns over inconsistent reporting, greenwashing, and material risk misalignment. Corporate legal duties around climate-related financial risk disclosure lagged behind market expectations, creating information gaps that hinder prudent decision-making and capital allocation toward sustainable outcomes.

Policy response

A cross-cutting legal reform established mandatory, standardized ESG disclosure aligned with climate risk. The policy mandated board-level accountability for ESG strategy, integrated climate risk into financial reporting, and required scenario analysis for transition and physical risk (Li et al., 2022). A dedicated regulator introduced a tiered enforcement regime and standardized templates to ensure comparability and reliability across sectors.

Legal innovation

Augmented fiduciary duties elevated the responsibility of directors to consider climate-related risk as part of enterprise risk management. A new liability regime held misstatements or omissions in ESG disclosures to account, with penalties proportional to impact. Data integrity requirements spurred the creation of third-party assurance standards and cross-border cooperation to harmonize reporting frameworks (Amirova et al., 2022). Jurisdictional cooperation facilitated mutual recognition of disclosures, reducing compliance burdens for multinational companies.

Outcomes and impacts

Improved disclosure quality, comparability, and investor confidence. Capital allocation aligned more closely with decarbonization trajectories, increasing funding for climate innovation and low-carbon assets. Directors gained clearer expectations around climate governance, and regulators gained better visibility into systemic risk. Stakeholders, including local communities and workers, benefited from better information about company strategies and impacts.

Challenges and lessons

Heterogeneity of business models remains a barrier to perfect comparability. Assurance costs and limited availability of specialized auditors could burden smaller firms. Ensuring the framework remains adaptable to rapid climate tech evolution poses ongoing regulatory challenges (Shulimova et al., 2024). Balancing transparency with proprietary concerns requires careful design.

Implications for climate justice

Transparent ESG disclosures enable investors and communities to scrutinize corporate practices, directing capital toward equitable transitions. Strong governance and accountability mechanisms ensure that environmental and social risk considerations are embedded in corporate strategy, supporting just, sustainable outcomes across the economy (Savchenko & Borodina, 2020).

Environmental Justice in Climate Adaptation Policy

Context and problem

Adaptation policy often prioritized infrastructure resilience without adequately addressing social equity. Marginalized communities faced higher vulnerability, limited access to adaptation benefits, and weakened trust in governance. Environmental justice lacked a cohesive framework linking adaptation investments to human rights and social equity.

Policy response

A national adaptation plan integrated a justice-focused framework, prioritizing vulnerable communities, participatory planning, and the right to safe, clean air, water, and housing. The plan mandated explicit equity criteria in project selection, targeted investments in heat resilience, flood defenses, and affordable housing, and required ongoing community engagement to co-design risks and solutions (Dwivedi et al., 2022). It also established a climate justice fund to support local, community-led adaptation ventures.

Legal innovation

Legislation codified climate justice as a guiding principle for adaptation policy, including anti-displacement and just compensation provisions. The policy enabled community-benefit agreements with developers and ensured access to legal assistance for affected residents (Olsen et al., 2022). A new regulatory mechanism empowered local governments to enforce equity targets in adaptation projects and provided safe harbor for citizen enforcement when plans fell short of commitments.

Outcomes and impacts

Adaptation investments targeted at vulnerable communities reduced exposure to heat islands and flood risks. Improved access to cooling centers, resilient housing, and green space enhanced health and well-being. Community ownership of projects increased trust in governance and empowered residents to influence decision-making. Data collection on equity outcomes improved, enabling better future policy design.

Challenges and lessons

Funding gaps persisted in high-need areas, and capacity constraints in communities slowed implementation. Ensuring meaningful participation required sustained engagement, translation of technical information, and long-term funding commitments (Swain & Karimu, 2020). Measuring justice outcomes remained complex, demanding robust indicators beyond traditional metrics.

Implications for climate justice

Embedding environmental justice in adaptation policy reframes resilience as a human-rights issue. When equity is a core criterion for investment and governance, climate adaptation can deliver both protective benefits and social empowerment, creating a fairer climate-resilient economy (Stroebel & Wurgler, 2021).

Energy Market Reforms for Clean Power and Just Access

Context and problem

Energy transitions faced affordability and access challenges, with low-income households disproportionately bearing energy costs and exposure to pollution. Market designs often favored large incumbents, hindering distributed energy resources (DERs) and community-led clean power. Regulatory obstacles limited innovation in storage, demand response, and microgrids.

Policy response

A reform package introduced dynamic pricing to reflect real-time costs, with targeted rebates and social tariffs to protect vulnerable consumers. It expanded DER integration, simplified interconnection standards, and promoted community solar, storage, and microgrid pilots (Hong et al., 2020). A capacity market evolved to value resilience and flexibility, incentivizing distributed resources alongside traditional generation.

Legal innovation

Legislation redefined grid access rights, established clear cost allocation for grid modernization, and formalized community ownership models for local clean energy projects. A framework for performance-based incentives linked subsidies to reliability, emission reductions, and equity goals. Regulatory sandboxes supported new business models, while a robust complaint mechanism protected consumers.

Outcomes and impacts

Retail prices stabilized, and access to clean energy expanded for lower-income households, with reduced pollution exposure. DER adoption growth accelerated, enabling local resilience and job creation in installation and maintenance (Bhutta et al., 2022). Utilities shifted toward cleaner generation portfolios, boosting investor confidence in the energy transition.

Challenges and lessons

Ensuring fair cost sharing among customers and safeguarding grid reliability remained critical. Technical barriers delayed some DER projects, and regulatory lag threatened to slow deployment. Equitable access required ongoing monitoring to prevent new forms of energy poverty (Yeow & Ng, 2021).

Implications for climate justice

Energy reforms that center affordability and community ownership advance energy justice. When policies incentivize distributed, low-emission resources with direct consumer protections, vulnerable populations gain reliable, clean power and economic opportunities.

Urban Planning and Green Mobility for Inclusive Climate Action

Context and problem

Urban areas are key nodes of emissions, but growth often outpaces sustainable planning. Transportation and housing policies frequently excluded low-income residents, leading to pollution hotspots and inequitable access to mobility and services. Cities needed integrated planning to align climate goals with social equity.

Policy response

An integrated urban planning package linked zoning reform, affordable housing, and public transit investments with climate targets. It prioritized dense, transit-oriented development, non-motorized transport infrastructure, and electrification of public fleets (Monk & Perkins, 2020). A mobility fund supported subsidized transit for low-income residents, bike-share programs, and safe routes to schools.

Legal innovation

A comprehensive land-use statute embedded climate resilience, social equity, and environmental justice into all planning decisions. It required inclusive stakeholder engagement, community benefit agreements, and mandatory environmental impact assessments for major developments. The legal framework established enforceable urban heat mitigation standards and penalties for noncompliance.

Outcomes and impacts

Efficient transit options reduced car dependency, lowering emissions and improving air quality. Affordable housing near transit corridors increased mobility and reduced commute times. Green spaces diminished heat islands, improving health outcomes for vulnerable populations (Glanemann et al., 2020). Local employment opportunities expanded through construction and maintenance work.

Challenges and lessons

Gentrification risk required strong affordability safeguards. Financing complex, long-horizon urban projects demanded blended finance and private sector participation. Data integration across agencies posed governance hurdles.

Implications for climate justice

Inclusive urban planning aligns climate action with social equity, ensuring that all residents can participate in and benefit from a cleaner, healthier, and more resilient city.

IMPLEMENTATION OF REGULATORY FRAMEWORKS

Regulatory frameworks are the backbone of translating policy reform into measurable climate outcomes. The successful implementation of policy reform, coupled with legal innovation, requires coherent governance, precise mandates, and adaptive enforcement mechanisms that align private incentives with public goods. In the nexus of ESG and environmental law, regulators must harmonize disclosure, performance standards, and accountability to propel a just and sustainable climate economy.

Policy reform begins with clear objectives: decarbonization trajectories, resilience against extreme weather, and inclusive participation for workers and communities disproportionately affected by transition. Designing regulations that are technology-agnostic where possible, yet performance-based where necessary, allows firms to innovate while meeting public goals (Wang & Ma, 2021). This balance reduces regulatory risk and accelerates investment in low carbon assets, nature-based solutions, and circular business models.

Legal innovation translates policy aims into enforceable rights and duties. From smart compliance regimes to dynamic rulebooks, regulators should leverage digital platforms for real-time reporting and data sharing, while protecting privacy and competitive integrity. Impact assessments, sandboxing for new approaches, and sunset clauses prevent regulatory drift. International cooperation, mutual recognition, and alignment with human rights standards extend the reach and legitimacy of domestic rules, supporting cross-border supply chains and climate finance flows (Casady et al., 2024).

The ESG–environmental law interface presents both opportunity and challenge. Mandates on disclosure create the information infrastructure needed to compare performance across firms, sectors, and jurisdictions. When disclosures are standardized, verified, and linked to credible benchmarks, capital markets price climate risk more efficiently and direct capital toward sustainable activity (Gao et al., 2023). At the same time, environmental law must ensure that ESG disclosures do not become mere optics. Regulators should require verifiable, material disclosures that reflect scope, impact, and governance factors, including board accountability for climate strategy, remuneration linked to decarbonization, and credible methane and fugitive emission reporting (Vassileva, 2022).

Enforcement is pivotal to legitimacy. A tiered approach, combining warnings, technical guidance, and meaningful penalties, signals seriousness while preserving the ability to correct course. Compliance assistance, industry-specific technical standards, and collaborative enforcement programs with civil society can expand regulatory reach without stifling innovation. Parity between public and private actors matters: whistleblower protections, third-party audits, and open data access bolster trust and fairness (Celestin et al., 2024).

Equity considerations must permeate every regulatory choice. Just transition principles require that regulations protect workers, communities dependent on high-emission sectors, indigenous lands, and vulnerable populations. Revenue recycling, retraining programs, and accelerated permitting for clean energy projects in disadvantaged regions help ensure that the climate economy does not reproduce inequality (Shahzad et al., 2023). Public procurement, land use planning, and urban design should embed climate resilience, biodiversity protection, and social inclusion as core criteria.

Measurement and learning are the implementation's lifeblood. Regulators should institutionalize periodic review, impact evaluation, and cross-jurisdictional learning. Metrics for policy reform success include emissions reductions, job quality, energy access, and biodiversity restoration, all tracked with transparent dashboards (Zhu et al., 2023). Finally, the climate economy will mature only when policy reform, legal innovation, and ESG integration converge into a coherent, participatory, and just governance architecture. This vision requires continuous iteration, robust institutions, and a shared commitment to planetary boundaries and human dignity for all.

CONCLUSION

Policy reform, legal innovation, and the ESG–environmental law nexus must be understood as an integrated project for climate justice and sustainable development. Regulatory frameworks that align environmental objectives with social equity require clear legitimacy, credible monitoring, and adaptive governance that can absorb technological change and market transformation. A just climate economy depends on transparent rulemaking that incorporates affected communities, workers, and businesses in collaborative design so that regulations do not merely constrain but enable sustainable investment and innovation. The implementation challenge is not only technical but political: translating high-level commitments into actionable standards, predictable enforcement, and timely reviews. To achieve this, reform agendas should emphasize coherence across agencies, alignment of fiscal incentives with green outcomes, and the integration of lifecycle accounting into reporting obligations. Legal innovation will flourish where courts, legislators, and regulators share a common language for risk, precaution, and resilience, including clear pathways for compliance, remediation, and protean dispute resolution. ESG criteria must be anchored in robust metrics that capture externalities, transition risks, and social impacts, while avoiding performative greenwashing. Regulatory instruments ranging from carbon pricing and product standards to procurement rules and monetary policy signals should be designed with a strong focus on distributive effects, targeting the most exposed communities and small- and medium-sized enterprises. In addition, institutions must cultivate data systems, interoperability, and open access to information so that stakeholders can assess progress and hold actors accountable. The climate economy cannot be sustainable without predictable timelines, credible enforcement, and a culture of continual improvement. Finally, international cooperation, knowledge sharing, and mutual recognition of standards can reduce fragmentation and raise global ambition, while respecting local contexts. A comprehensive, just, and forward-looking regulatory architecture can catalyze investment, protect ecosystems, and promote human well-being while keeping pace with innovation.

Suggestion

To translate theory into practice, policymakers should establish iterative policy cycles that pair performance-based standards with ambitious timelines, pilot zones, and sunset clauses. This approach encourages experimentation while ensuring accountability and revision in light of new evidence. Regulatory design should embed equity impact assessments, ensuring that low-income communities, workers, and marginalized groups benefit from climate actions rather than bearing disproportionate burdens. Assistive mechanisms such as transitional aid, retraining programs, and targeted financing can mitigate displacement effects and support a just transition. ESG governance requires binding, standardized disclosures, standardized verification, and interoperability of data across sectors and jurisdictions. Regulators should leverage digital technologies, open data platforms, and common taxonomies to improve comparability, reduce compliance costs, and enhance public trust. Sanctions, incentives, and incentive alignment must be calibrated to avoid unintended consequences, such as carbon leakage or regressive pricing. Cross-border cooperation should harmonize core standards while respecting local contexts, permitting mutual recognition of credible ESG reporting and environmental impact assessments. Capacity-building in public institutions, judiciary, and regulatory agencies is essential to sustain rigorous enforcement and adaptive governance. Finally, stakeholder engagement should be continuous, inclusive, and calibrated to different expertise levels, enabling citizens, businesses, and civil society to contribute meaningfully to reform pathways and to monitor progress toward shared climate justice goals.

Future Research Direction

Future research should map the regulatory landscape across policy reform, legal innovation, and ESG–environmental law to identify gaps, tensions, and opportunities for synergy. Comparative studies can illuminate how different governance models handle distributed impacts, inform criteria for legitimacy, and reveal best practices in adaptive regulation. Methodological innovations in governance science such as dynamic risk assessment, evidence synthesis, and scenario planning can strengthen decision-making under uncertainty and accelerate responsible experimentation. Investigations should examine the timelines, incentives, and enforcement mechanisms that maximize compliance while maintaining social equity, as well as the governance architecture required to scale pilot programs into nationwide reforms. The nexus of ESG reporting and environmental law calls for rigorous validation of disclosure metrics, third-party assurance standards, and the interoperability of data standards across jurisdictions. Research should also analyze the political economy of regulatory reform, including interest group dynamics, path dependence, and the distributional consequences of climate policy. Finally, scholars ought to develop policy instruments that integrate climate objectives with biodiversity preservation, water security, and health outcomes, pursuing systemic solutions rather than narrow sectoral fixes. By advancing empirical knowledge, normative analysis, and practical guidance, future work can help design regulatory frameworks that are just, flexible, and effective in building a sustainable climate economy.

REFERENCES

Abdullah, A., Saraswat, S., & Talib, F. (2023). Barriers and strategies for sustainable manufacturing implementation in SMEs: A hybrid fuzzy AHP-TOPSIS framework. *Sustainable Manufacturing and Service Economics*, *2*, 100012. DOI: 10.1016/j.smse.2023.100012

Adamowicz, M. (2022). Green deal, green growth and green economy as a means of support for attaining the sustainable development goals. *Sustainability (Basel)*, *14*(10), 5901. DOI: 10.3390/su14105901

Amirova, E. F., Gavrilyeva, N. K., Romanishina, T. S., & Asfandiarova, R. A. (2022). On the problem of the development of 'sustainable' agriculture in modern economic realities. *Siberian Journal of Life Sciences and Agriculture*, *14*(3), 392–406. DOI: 10.12731/2658-6649-2022-14-3-392-406

Balezentis, T., Streimikiene, D., Stankuniene, G., & Shobande, O. A. (2024). Willingness to pay for climate change mitigation measures in households: Bundling up renewable energy, energy efficiency, and renovation. *Sustainable Development (Bradford)*, *32*(3), 2385–2402. DOI: 10.1002/sd.2784

Batrancea, L., Pop, M. C., Rathnaswamy, M. M., Batrancea, I., & Rus, M. I. (2021). An empirical investigation on the transition process toward a green economy. *Sustainability (Basel)*, *13*(23), 13151. DOI: 10.3390/su132313151

Bhutta, U. S., Tariq, A., Farrukh, M., Raza, A., & Iqbal, M. K. (2022). Green bonds for sustainable development: Review of literature on development and impact of green bonds. *Technological Forecasting and Social Change*, *175*, 121378. DOI: 10.1016/j.techfore.2021.121378

Bluedorn, J., Hansen, N. J., Noureldin, D., Shibata, I., & Tavares, M. M. (2023). Transitioning to a greener labor market: Cross-country evidence from microdata. *Energy Economics*, *126*, 106836. DOI: 10.1016/j.eneco.2023.106836

Bolton, P., Musca, X., & Samama, F. (2020). Global Public-Private Investment Partnerships: A Financing Innovation with Positive Social Impact. *The Bank of America Journal of Applied Corporate Finance*, *32*(2), 31–41. DOI: 10.1111/jacf.12403

Casady, C. B., Cepparulo, A., & Giuriato, L. (2024). Public-private partnerships for low-carbon, climate-resilient infrastructure: Insights from the literature. *Journal of Cleaner Production*, *470*, 143338. DOI: 10.1016/j.jclepro.2024.143338

Celestin, M., Vasuki, M., Sujatha, S., & Kumar, A. D. (2024). Implementing Green Technologies to Reduce Environmental Impact: Economic and Competitive Advantages of Eco-Friendly Practices. *International Journal of Scientific Research and Modern Education*, *9*(2), 33–39.

Chou, C. H., Ngo, S. L., & Tran, P. P. (2023). Renewable energy integration for sustainable economic growth: Insights and challenges via bibliometric analysis. *Sustainability (Basel)*, *15*(20), 15030. DOI: 10.3390/su152015030

D'amato, D., & Korhonen, J. (2021). Integrating the green economy, circular economy and bioeconomy in a strategic sustainability framework. *Ecological Economics*, *188*, 107143. DOI: 10.1016/j.ecolecon.2021.107143

Dwivedi, Y. K., Hughes, L., Kar, A. K., Baabdullah, A. M., Grover, P., Abbas, R., Andreini, D., Abumoghli, I., Barlette, Y., Bunker, D., Chandra Kruse, L., Constantiou, I., Davison, R. M., De', R., Dubey, R., Fenby-Taylor, H., Gupta, B., He, W., Kodama, M., & Wade, M. (2022). Climate change and COP26: Are digital technologies and information management part of the problem or the solution? An editorial reflection and call to action. *International Journal of Information Management*, *63*, 102456. DOI: 10.1016/j.ijinfomgt.2021.102456

Fan, L., & Wang, D. (2024). Natural resource efficiency and green economy: Key takeaways on clean energy, globalization, and innovations in BRICS countries. *Resources Policy*, *88*, 104382. DOI: 10.1016/j.resourpol.2023.104382

Gao, B., Ozturk, I., & Ullah, S. (2023). A new framework to the green economy: Asymmetric role of public-private partnership investment on environment in selected Asian economies. *Economic research-. Ekonomska Istrazivanja*, *36*(1), 1960–1971. DOI: 10.1080/1331677X.2022.2094441

Glanemann, N., Willner, S. N., & Levermann, A. (2020). Paris Climate Agreement passes the cost-benefit test. *Nature Communications*, *11*(1), 110. DOI: 10.1038/s41467-019-13961-1 PMID: 31988294

Hariyani, D., Hariyani, P., Mishra, S., & Sharma, M. K. (2024). A literature review on green supply chain management for sustainable sourcing and distribution. *Waste Management Bulletin*.

Hill, R., Adem, Ç., Alangui, W. V., Molnár, Z., Aumeeruddy-Thomas, Y., Bridgewater, P., Tengö, M., Thaman, R., Adou Yao, C. Y., Berkes, F., Carino, J., Carneiro da Cunha, M., Diaw, M. C., Díaz, S., Figueroa, V. E., Fisher, J., Hardison, P., Ichikawa, K., Kariuki, P., & Xue, D. (2020). Working with indigenous, local and scientific knowledge in assessments of nature and nature's linkages with people. *Current Opinion in Environmental Sustainability*, *43*, 8–20. DOI: 10.1016/j.cosust.2019.12.006

Hong, H., Karolyi, G. A., & Scheinkman, J. A. (2020). Climate finance. *Review of Financial Studies, 33*(3), 1011–1023. DOI: 10.1093/rfs/hhz146

Kumar, S., Darshna, A., & Ranjan, D. (2023). A review of literature on the integration of green energy and circular economy. *Heliyon, 9*(11), e21091. DOI: 10.1016/j.heliyon.2023.e21091 PMID: 37954368

Li, J., Song, G., Cai, M., Bian, J., & Mohammed, B. S. (2022). Green environment and circular economy: A state-of-the-art analysis. *Sustainable Energy Technologies and Assessments, 52*, 102106. DOI: 10.1016/j.seta.2022.102106

Licastro, A., & Sergi, B. S. (2021). Drivers and barriers to a green economy. A review of selected balkan countries. *Cleaner Engineering and Technology, 4*, 100228. DOI: 10.1016/j.clet.2021.100228

Lwamba, E., Shisler, S., Ridlehoover, W., Kupfer, M., Tshabalala, N., Nduku, P., Langer, L., Grant, S., Sonnenfeld, A., Anda, D., Eyers, J., & Snilstveit, B. (2022). Strengthening women's empowerment and gender equality in fragile contexts towards peaceful and inclusive societies: A systematic review and meta-analysis. *Campbell Systematic Reviews, 18*(1), e1214. DOI: 10.1002/cl2.1214 PMID: 36913184

Monk, A., & Perkins, R. (2020). What explains the emergence and diffusion of green bonds? *Energy Policy, 145*, 111641. DOI: 10.1016/j.enpol.2020.111641

Mottershead, D., Watkins, E., Gore, T., Andersen, M. S., Pedersen, A. B., Whiteoak, K., & Chewpreecha, U. (2021). *Green taxation and other economic instruments. Internalizing environmental costs to make the polluter pay*. European Commission.

Muth, D. (2023). Pathways to stringent carbon pricing: Configurations of political economy conditions and revenue recycling strategies. A comparison of thirty national level policies. *Ecological Economics, 214*, 107995. DOI: 10.1016/j.ecolecon.2023.107995

Nazari, Z., & Musilek, P. (2023). Impact of digital transformation on the energy sector: A review. *Algorithms, 16*(4), 211. DOI: 10.3390/a16040211

Olsen, K. H., Kolenda, M., & Konrad, S. (2022). *Promoting transformational change through carbon markets. Strengthening the transformational impact of carbon market cooperation under Article 6 of the Paris Agreement. Final report* (No. UBA-FB--000837/ENG). Umweltbundesamt (UBA).

Onu, P., Pradhan, A., & Mbohwa, C. (2023). The potential of industry 4.0 for renewable energy and materials development–The case of multinational energy companies. *Heliyon, 9*(10), e20547. DOI: 10.1016/j.heliyon.2023.e20547 PMID: 37822636

Plachkov, D. (2024). *GREEN ECONOMY: CONCEPT AND PRINCIPLES. Scientific WORKS of the Union of Scientists in Bulgaria-Plovdiv*. Series A. Social Sciences, Art & Culture.

Raihan, A. (2023). The dynamic nexus between economic growth, renewable energy use, urbanization, industrialization, tourism, agricultural productivity, forest area, and carbon dioxide emissions in the Philippines. *Energy Nexus*, *9*, 100180. DOI: 10.1016/j.nexus.2023.100180

Savchenko, A. B., & Borodina, T. L. (2020). Green and digital economy for sustainable development of urban areas. *Regional Research of Russia*, *10*(4), 583–592. DOI: 10.1134/S2079970520040097

Schröder, P. (2020). *Promoting a just transition to an inclusive circular economy*. Royal Institute of International Affairs.

Shahzad, K., Abdul, D., Umar, M., Safi, A., Maqsood, S., Baseer, A., & Lu, B. (2023). Analysis of obstacles to adoption of solar energy in emerging economies using spherical fuzzy AHP decision support system: A case of Pakistan. *Energy Reports*, *10*, 381–395. DOI: 10.1016/j.egyr.2023.06.015

Shandas, V., & Hellman, D. (2022). Toward an Equitable Distribution of Urban Green Spaces for People and Landscapes: An Opportunity for Portland's Green Grid. In *Green Infrastructure and Climate Change Adaptation: Function, Implementation and Governance* (pp. 289–301). Springer Nature Singapore. DOI: 10.1007/978-981-16-6791-6_18

Sharif, A., Bashir, U., Mehmood, S., Cheong, C. W., & Bashir, M. F. (2024). Exploring the impact of green technology, renewable energy and globalization towards environmental sustainability in the top ecological impacted countries. *Geoscience Frontiers*, *15*(6), 101895. DOI: 10.1016/j.gsf.2024.101895

Shulimova, M., Tekeeva, H., & Baysaeva, M. (2024). The role of green economy in achieving sustainable development goals: prospects and challenges for business. *Reliability: Theory & Applications, 19*(SI 6 (81)), 1615-1622.

Söderholm, P. (2020). The green economy transition: The challenges of technological change for sustainability. *Sustainable Earth*, *3*(1), 6. DOI: 10.1186/s42055-020-00029-y

Stroebel, J., & Wurgler, J. (2021). What do you think about climate finance? *Journal of Financial Economics*, *142*(2), 487–498. DOI: 10.1016/j.jfineco.2021.08.004

Swain, R. B., & Karimu, A. (2020). Renewable electricity and sustainable development goals in the EU. *World Development*, *125*, 104693. DOI: 10.1016/j.worlddev.2019.104693

Takacs, A. (2023). The positive effects of green technology investments on growth expectations. *Technology in Society*, *75*, 102407. DOI: 10.1016/j.techsoc.2023.102407

van Niekerk, A. J. (2024). Economic Inclusion: Green Finance and the SDGs. *Sustainability (Basel)*, *16*(3), 1128. DOI: 10.3390/su16031128

Vassileva, A. G. (2022). Green Public-Private Partnerships (PPPs) as an instrument for sustainable development. *Journal of World economy: Transformations & transitions, 2*(5), 2.

Wang, N., & Ma, M. (2021). Public–private partnership as a tool for sustainable development–What literatures say? *Sustainable Development (Bradford)*, *29*(1), 243–258. DOI: 10.1002/sd.2127

Yang, M., Chen, L., Wang, J., Msigwa, G., Osman, A. I., Fawzy, S., Rooney, D. W., & Yap, P. S. (2023). Circular economy strategies for combating climate change and other environmental issues. *Environmental Chemistry Letters*, *21*(1), 55–80. DOI: 10.1007/s10311-022-01499-6

Yeow, K. E., & Ng, S. H. (2021). The impact of green bonds on corporate environmental and financial performance. *Managerial Finance*, *47*(10), 1486–1510. DOI: 10.1108/MF-09-2020-0481

Zhu, Y., Zhang, H., Siddik, A. B., Zheng, Y., & Sobhani, F. A. (2023). Understanding corporate green competitive advantage through green technology adoption and green dynamic capabilities: Does green product innovation matter? *Systems*, *11*(9), 461. DOI: 10.3390/systems11090461

Chapter 14

Reframing Sustainable Architecture Through Climate Justice Design, Governance, and Equity in the Built Environment

Ahmed AlQodsi
http://orcid.org/0009-0005-9143-9418
Modern University for Technology and Information, Egypt

Mohammad Ibrahim Hesso
Modern University for Technology and Information, Egypt

Enass Salama
Modern University for Technology and Information, Egypt

Noha Mohamed Ezz Eldein
Modern University for Technology and Information, Egypt

Mariam Osama Shaker
Modern University for Technology and Information, Egypt

ABSTRACT

Climate change increasingly affects the built environment, placing greater pressure on cities and communities, particularly in regions experiencing rapid urbanization and environmental stress. This chapter examines the role of sustainable architecture in addressing climate challenges through design strategies that promote environmental

DOI: 10.4018/979-8-3373-6635-7.ch014

responsibility, resilience, and social equity. Focusing on architectural practice in the Egyptian context, the chapter explores climate-responsive design approaches such as passive cooling, energy-efficient building systems, sustainable materials, and context-sensitive urban design. It highlights how sustainable architectural solutions can reduce environmental impact, improve thermal comfort, and enhance the quality of life for diverse communities. The chapter emphasizes that architecture plays a crucial role in shaping resilient and sustainable built environments capable of responding to climate change while supporting more balanced and inclusive urban development.

INTRODUCTION

Climate change has profoundly transformed the intellectual landscape within which architecture operates. What was once approached primarily as an environmental or technical concern centered on energy efficiency, material optimization, and carbon reduction has evolved into a multidimensional challenge that intersects with questions of justice, governance, and socio-spatial equity. The accelerating frequency of heatwaves, water scarcity, extreme weather events, and ecological degradation has revealed that climate impacts are not evenly distributed. Instead, they disproportionately affect communities already facing economic marginalization, infrastructural deficits, and limited political representation. In this context, climate change emerges not only as an environmental crisis but as a justice crisis embedded within the spatial fabric of cities. The built environment lies at the center of this transformation. Buildings and urban systems are responsible for a significant share of global greenhouse gas emissions, yet they also determine patterns of exposure, adaptation, and protection. The design of streets influences airflow and heat retention; material choices shape thermal performance; infrastructure planning affects access to water, energy, and mobility. These spatial decisions structure daily life and mediate environmental risk. Architecture therefore functions as both a contributor to climate change and a critical agent in shaping resilience. This dual role places architectural practice within broader debates on climate justice and environmental governance. Climate justice challenges the assumption that sustainability is inherently equitable. While environmental law and policy reforms establish regulatory frameworks for emission reduction and resilience planning, the translation of these frameworks into lived realities depends on how architectural and urban strategies are implemented. A green building that remains financially inaccessible does little to advance distributive equity. Similarly, climate-adaptive infrastructure concentrated in affluent districts may inadvertently reinforce spatial inequality. Thus, the effectiveness of environmental policy is inseparable from spatial and material execution.

Architecture operates as a mediating interface between normative commitments and everyday experience, transforming abstract sustainability goals into tangible environments that either mitigate or reproduce inequality.

In rapidly urbanizing regions such as Egypt, the urgency of this intersection becomes particularly pronounced. Climatic intensification, rising temperatures, and urban heat island effects coincide with rapid demographic growth and uneven urban development. Informal settlements, energy-intensive construction practices, and disparities in infrastructure access reveal how environmental vulnerability is spatially differentiated. At the same time, Egypt's architectural heritage contains climate-responsive knowledge rooted in vernacular design traditions—knowledge that offers valuable insights for contemporary adaptation. This duality of vulnerability and opportunity underscores the need for an integrated framework that aligns sustainable design with justice-oriented planning.

This chapter argues that sustainable architecture can serve as a structural pathway toward climate justice when reoriented through an ethical and governance-based lens. It first reexamines the conceptual transition from conventional sustainability to justice-oriented design, interrogating the social limits of purely technical environmental performance. It then situates this theoretical reflection within the Egyptian context, analysing how climatic stress intersects with socio-economic inequality to shape uneven adaptive capacity. Building on this contextual analysis, the chapter explores how climate justice can be operationalized through architectural strategies passive environmental systems, material selection, urban configuration, and affordability-sensitive design. Finally, it addresses the necessity of adaptive governance and intergenerational responsibility, emphasizing that long-term justice requires institutional continuity, measurable equity indicators, and flexible built environments capable of evolving alongside climatic and social change.

By integrating conceptual critique, contextual analysis, operational design strategies, and governance mechanisms, this chapter positions architecture not as a reactive response to climate change but as an active instrument of socio-environmental transformation. Sustainable architecture, when embedded within policy reform and guided by distributive ethics, becomes more than a mitigation tool; it becomes a platform for reshaping urban development toward resilience, inclusivity, and shared environmental security. In doing so, the built environment emerges as a decisive arena in which climate justice is not merely debated, but materially enacted.

To advance this argument in a structured manner, the chapter unfolds across four interrelated sections. The first section critically reframes the relationship between sustainability and climate justice, questioning the neutrality of green design and repositioning architecture within an ethical and distributive framework. The second section situates this discussion within the Egyptian urban condition, examining how climatic pressures intersect with socio-economic inequality to produce differentiated patterns of vulnerability and resilience. The third section translates these conceptual and contextual insights into operational design strategies, identifying concrete architectural mechanisms through which climate justice can be spatially enacted. The final section addresses the governance dimension, emphasizing adaptive regulatory frameworks and intergenerational responsibility as essential conditions for sustaining equitable climate-responsive development over time.

1. FROM SUSTAINABILITY TO CLIMATE JUSTICE: RETHINKING THE ETHICAL ROLE OF ARCHITECTURE

For decades, sustainability in architecture has been framed primarily as a technical objective—reducing energy consumption(Taylor, Jokela, Laine, & Ferguson, 2025), lowering carbon emissions, and improving material efficiency. While these goals remain essential, they are no longer sufficient (Bennett, 2024). Climate change has revealed that environmental performance alone does not guarantee justice. Buildings may be energy-efficient yet socially exclusionary; cities may adopt green technologies while marginalized communities continue to bear the greatest environmental burdens(Abraham, Ganyaglo, Andoh, Asamoah, & Koomson, 2025). Climate justice shifts the focus from environmental metrics to human consequences. It asks: Who benefits from sustainable design? Who remains exposed to heat stress, flooding, or poor air quality? Who can afford climate-resilient housing? In rapidly transforming urban contexts, architectural decisions shape access to shade(Cole, 2020), ventilation, infrastructure, and public space. These are not merely design choices; they are decisions that structure vulnerability and resilience. Architecture, therefore, operates within an ethical terrain(Pelsmakers & Brown, 2025). Designers influence patterns of social equity through spatial organization, material selection, and technological integration. Climate-responsive strategies such as passive cooling, adaptive façades, and climate-sensitive urban morphology must be understood not only as environmental tools but as instruments of distributive justice(Vogt, Ritson, & Blanc, 2010). In this sense, sustainable architecture becomes a mediator between environmental responsibility and social fairness, translating climate policy into lived spatial reality.

1.1 Beyond Technical Sustainability: Questioning the Neutrality of Green Design

For several decades, sustainable architecture has been largely framed within a technical paradigm(Vogt et al., 2010). Design discourse emphasized measurable indicators energy performance, carbon emissions, material life cycles, and certification systems such as LEED or BREEAM. This technocratic orientation has undoubtedly contributed to lowering environmental impact and advancing innovation in building systems(White, 2020). However, it has also cultivated an implicit assumption: that sustainability is inherently beneficial and socially neutral. Yet sustainability is never neutral. Environmental strategies are embedded within political economies, land markets, and power structures(Fu, Hao, Wu, Xu, & Zuo, 2026). High performance green buildings frequently emerge in affluent urban districts where investment capital is concentrated. Eco-districts often become sites of rising property values and, in some cases, green gentrification, displacing lower-income populations(Araújo, 2026). Thus, while the building performs efficiently from an environmental perspective, its social consequences may contradict the very ethos of sustainability. The dominance of technological solutions can also obscure the uneven geography of vulnerability. Climate change does not impact all communities equally. Exposure to extreme heat, inadequate ventilation, flooding, or infrastructural fragility is closely tied to socio-economic status and urban form. When sustainability discourse prioritizes innovation without addressing distribution, it risks reinforcing structural inequality under an environmentally progressive narrative(Gondi, 2026). This realization compels a conceptual shift. Sustainability must be evaluated not only by how much energy a building saves(Subrahmanyam, 2026), but by whom it protects and whom it excludes. The question moves from "How efficient is this building?" to "For whom is this environment resilient?" Architectural practice, therefore, must interrogate its own complicity within broader systems of inequality and reconsider the social implications of its environmental interventions.

1.2 Repositioning Architecture within the Framework of Climate Justice

Climate justice introduces an ethical lens that reorients architectural responsibility. Rather than viewing climate adaptation and mitigation solely as environmental imperatives(Bennett, 2024), climate justice situates them within questions of equity, rights, and distributive fairness. It recognizes that climate risks are socially produced and spatially distributed, and that the built environment mediates these patterns. Architecture is not a passive backdrop to climate change; it actively shapes exposure and resilience(Comelli et al., 2026). The spatial configuration of neighbourhood's influences airflow and shading. Building materials determine thermal absorption. Urban density affects heat island intensity. Access to green infrastructure alters public health outcomes(Gonzalez, 2021). Each of these architectural decisions has justice implications. Repositioning architecture within climate justice requires expanding the scope of design evaluation. It demands integrating vulnerability mapping, community participation, and socio-economic assessment into architectural processes. Design becomes a form of governance—an instrument through which environmental commitments are operationalized and through which public policy is spatially enacted. This repositioning also challenges the conventional separation between environmental law and architectural practice. Policies may mandate emission reductions or resilience planning, but their tangible realization occurs through material and spatial design choices(Alqodsi & Abdallah, 2024a). Architecture thus operates as a bridge between normative frameworks and lived environments(Alqodsi, 2025). It translates legal aspirations into thermal comfort, structural safety, and accessible public space. Ultimately, climate justice reframes sustainable architecture as an ethical project. It asks architects to consider not only environmental performance but also spatial equity, affordability, and long-term adaptability(Alqodsi & Abdallah, 2024b). By embedding justice principles into design methodologies, architecture can move beyond reactive adaptation and contribute to transformative urban change aligning environmental responsibility with social inclusion in the face of accelerating climate uncertainty.

2. DESIGNING RESILIENCE IN UNEQUAL CONTEXTS: THE EGYPTIAN URBAN CONDITION

The Egyptian built environment presents a compelling illustration of the intersection between climate stress and socio-spatial inequality(Eltinay, 2019). Rising temperatures, urban heat island effects, and energy-intensive construction practices coincide with rapid urban expansion and informal development. In such contexts, vulnerability is unevenly distributed. Informal settlements and low-income communities often lack adequate thermal insulation, green spaces, and reliable infrastructure, amplifying exposure to climate-related risks. Designing for resilience in Egypt requires moving beyond imported technological models toward context-sensitive solutions grounded in local climate, culture, and economic realities. Passive cooling techniques rooted in vernacular traditions such as courtyard typologies, shading devices, and natural ventilation systems offer adaptive potential when reinterpreted through contemporary architectural practice(Hafiz, 2025). Similarly, sustainable material strategies and compact urban planning can reduce environmental impact while improving accessibility and affordability. Resilience, however, must not be reduced to structural durability alone(Esmail & Afifi, 2026). It encompasses social stability, economic viability, and equitable access to environmental comfort. Architectural interventions that lower energy demand, enhance thermal performance, and integrate public space can contribute to more inclusive urban futures. In this way, sustainable architecture becomes a strategic component of climate justice addressing environmental risk while simultaneously promoting spatial equity within the evolving Egyptian city.

2.1 Climate Vulnerability and Urban Inequality in the Egyptian Context

Egypt's urban landscape is shaped by a complex convergence of climatic pressure, demographic expansion, and socio-economic disparity. Rising temperatures, intensified heatwaves, water stress, and urban heat island effects increasingly affect Egyptian cities, particularly Cairo and rapidly expanding peri-urban areas. Yet climate exposure is not evenly distributed. Vulnerability is spatially patterned, often concentrated in informal settlements and lower-income districts where infrastructure deficits intersect with environmental stress(Mahmoud, 2017).

In many dense urban areas, limited green coverage, narrow streets, inadequate ventilation corridors, and heat-absorbing construction materials amplify thermal discomfort. Buildings frequently rely on energy-intensive mechanical cooling systems, increasing both financial burden and carbon emissions(Soliman & Soliman, 2022). For households with limited income, the cost of artificial cooling becomes unsustainable, creating a cycle in which environmental stress reinforces economic vulnerability.

Urban expansion patterns further complicate resilience. Rapid development in desert fringes has often prioritized speed and economic return over climatic adaptation. Standardized building typologies—frequently imported or replicated without climatic calibration fail to respond adequately to local environmental conditions(Elbahrawy & Gammaz, 2026). This disconnect between design practice and environmental reality deepens inequality by producing spaces that are climatically inefficient and socially stratified.

Climate justice in Egypt therefore requires acknowledging that resilience is not merely a matter of engineering durability; it is a matter of spatial equity. The capacity to withstand climate stress is linked to housing quality, infrastructure access, and urban planning decisions. Without deliberate corrective strategies, climate adaptation efforts risk reinforcing pre-existing inequalities within the built environment.

2.2 Vernacular Intelligence and Contemporary Reinterpretation

While contemporary urban expansion often overlooks climatic specificity, Egypt possesses a rich architectural heritage rooted in environmental responsiveness. Traditional courtyard houses, mashrabiya screens, wind catchers (malqaf), thick masonry walls, and shaded narrow streets historically functioned as passive climatic regulators. These design strategies moderated temperature, enhanced airflow, and reduced reliance on mechanical systems long before the concept of sustainability emerged as a global discourse.

However, modernization and globalization gradually marginalized vernacular intelligence in favor of standardized construction models associated with modernization and economic growth. Glass façades, exposed concrete surfaces, and sealed building envelopes became symbols of progress, often disregarding thermal performance within the local climate. This shift reflects not only aesthetic preference but also deeper socio-economic transformations influencing construction markets and regulatory frameworks.

Re-engaging with vernacular strategies does not imply romanticizing the past. Rather, it involves critically reinterpreting indigenous climatic knowledge within contemporary technological and social contexts. Hybrid design approaches—combining passive cooling systems with modern energy-efficient technologies—can produce adaptable, low-energy buildings suited to Egypt's environmental conditions. Sustainable materials sourced locally can reduce embodied carbon while supporting local economies.

This reinterpretation aligns with climate justice principles by promoting affordability, accessibility, and environmental responsiveness. When passive strategies reduce energy demand, they lower operational costs for households, benefiting economically vulnerable populations. Climate-responsive design thus becomes not only an environmental intervention but a socially redistributive strategy that enhances everyday comfort without exacerbating economic inequality.

2.3 Integrating Design, Policy Reform, and Social Equity

Achieving climate justice through architecture requires alignment between design innovation and policy reform. Environmental law and urban regulations establish the framework within which architectural practice operates(Guo, 2026). Building codes, energy standards, zoning policies, and housing regulations shape material choices, spatial density, and infrastructural integration. Without supportive regulatory environments, even well-intentioned sustainable design remains fragmented.

In Egypt, policy initiatives addressing energy efficiency and green development are gradually expanding, yet gaps remain in enforcement, affordability mechanisms, and integration with social housing programs(Alhuzail & Majumdar, 2026). Climate justice demands that sustainable architectural standards extend beyond high-end developments and become embedded within mainstream housing policy. Public housing projects, urban renewal programs, and infrastructure investments must incorporate climate-responsive design as a core principle rather than an optional enhancement.

Community participation further strengthens this integration. Residents possess experiential knowledge of local climatic challenges and social dynamics. Participatory design processes can reveal micro-level vulnerabilities such as shading needs, ventilation patterns, or public space usage that standardized planning may overlook. Embedding community input into architectural and planning decisions enhances legitimacy, cultural relevance, and long-term sustainability(Tandon & Nag, 2026).

Ultimately, architecture functions as a mediating layer between environmental policy aspirations and lived spatial realities. Through coordinated regulatory reform, inclusive design methodologies, and context sensitive innovation, resilient built environments can emerge as instruments of climate justice. In the Egyptian context, this integration offers an opportunity not only to mitigate environmental risk but to reshape urban development toward greater equity, sustainability, and social cohesion.

3. OPERATIONALIZING CLIMATE JUSTICE IN ARCHITECTURAL DESIGN

Climate justice, while conceptually compelling, must ultimately materialize through spatial decisions, construction systems, and urban configurations. Translating justice principles into architectural practice requires moving beyond abstract commitments toward measurable design strategies that redistribute environmental comfort, reduce vulnerability(Klinsky & Mavrogianni, 2020), and expand equitable access to resilient infrastructure. Operationalizing climate justice within the built environment involves integrating environmental performance with social inclusion, ensuring that adaptive design is not confined to elite developments but embedded within mainstream architectural production(Hamstead, 2023). This section identifies two primary operational dimensions through which architecture can concretely advance climate justice in rapidly urbanizing contexts.

3.1 Climate-Responsive Design as a Tool for Spatial Equity

Climate-responsive design strategies such as passive cooling systems, optimized orientation, shading devices, high performance envelopes, and natural ventilation are often evaluated through energy metrics. However, their justice dimension lies in their capacity to democratize environmental comfort(Ayo-Odifiri, Ebekozien, & Aigbavboa, 2025).

In hot climates such as Egypt's, thermal comfort becomes a matter of socio-economic survival. Households lacking financial capacity for continuous mechanical cooling are disproportionately exposed to heat stress. By integrating passive cooling principles at the design stage, architects reduce dependence on energy-intensive systems and lower operational costs(Li, Zou, Tian, & Ying, 2022). This directly benefits lower-income residents and enhances long term affordability.

Urban-scale interventions further reinforce spatial equity. Street orientation, tree canopy integration, permeable surfaces, and shaded public spaces can significantly reduce heat island effects(Nag, 2026). When such strategies are implemented in underserved districts rather than exclusively in commercial or high-income zones, architecture becomes an instrument of distributive justice.

Importantly, climate-responsive design must be incorporated early in project development rather than appended as an afterthought. Justice-oriented architecture requires systemic integration of climatic analysis, microclimate simulation, and community-based needs assessment. In this framework, environmental performance becomes inseparable from social performance, and resilience becomes a shared urban asset rather than a private privilege.

3.2 Institutional Mechanisms and Regulatory Integration

Design innovation alone cannot sustain climate justice without institutional reinforcement. Architectural practice operates within regulatory, economic, and professional systems that influence feasibility and scale (Elrefaie, 2026). Building codes, urban planning regulations, energy standards, and procurement policies determine whether climate-responsive solutions remain isolated experiments or evolve into standardized practice(Bampou, 2017). Integrating justice principles into policy reform involves revising building regulations to mandate minimum thermal performance standards, incentivizing passive design adoption, and aligning public housing policies with sustainability criteria. Financial instruments such as subsidies for energy-efficient materials or tax incentives for climate-adaptive construction can expand accessibility beyond high-income markets(Al-shawabkeh, Shouaib, Al Qudah, & Aldabbas, 2025). Professional education and interdisciplinary collaboration also play critical roles(Othman & Youssef, 2021). Architects must engage with urban planners, environmental engineers, policymakers, and community organizations to ensure that design decisions correspond with legal frameworks and social priorities(Alshawabkeh, 2024). Institutional coordination strengthens accountability and bridges the gap between environmental commitments and on-the-ground implementation(Al Shawabkeh, 2023).

By embedding climate-responsive design within regulatory systems and professional practice, architecture transitions from isolated sustainable interventions to structurally supported justice-oriented development. In this way, climate justice becomes not merely an ethical aspiration but an operational paradigm shaping the future of the built environment.

4. MEASURING IMPACT AND ENSURING LONG-TERM CLIMATE JUSTICE IN THE BUILT ENVIRONMENT

While climate-responsive design strategies and regulatory integration provide essential foundations for resilient development, a critical question remains: how can the long-term justice impact of architectural interventions be assessed and sustained? Climate justice within the built environment cannot rely solely on initial design intentions or policy declarations; it requires continuous evaluation, adaptive governance, and measurable indicators that link environmental performance with social equity outcomes(Amin & Zaher, 2026).

Architectural interventions often prioritize quantifiable environmental metrics such as energy consumption reduction, carbon emissions, and material efficiency. However, justice-oriented architecture demands expanded evaluation criteria. Thermal comfort distribution, affordability of energy use, access to shaded public space, infrastructure reliability, and reduced exposure to environmental hazards must become integral components of performance assessment(Ofosu-Asare, 2026). These indicators reveal whether resilience strategies genuinely serve vulnerable populations or primarily enhance already privileged environments.

Long-term climate justice also depends on adaptability. Climate conditions evolve, urban populations shift, and socio-economic pressures intensify. Buildings and urban systems must therefore be designed with flexible frameworks that allow retrofitting, modification, and technological upgrading without imposing excessive financial burdens(Al Ali, Al Shawabkeh, & Alqodsi, 2026). Modular systems, adaptable façades, and infrastructure-ready designs increase the lifespan and social value of architectural interventions(Abdelhamid, 2026).

Equally important is governance continuity. Justice-centered design must be supported by monitoring institutions capable of enforcing building standards, evaluating urban microclimates, and tracking socio-spatial disparities. Without institutional oversight, progressive policies risk becoming symbolic rather than transformative. Collaboration between municipal authorities, academic institutions, and professional bodies can generate data-driven assessments that refine future architectural strategies(Dahiyat, 2007).

Finally, climate justice requires intergenerational responsibility. The built environment constructed today will shape environmental exposure for decades. Architectural decisions must therefore anticipate long-term climate projections and demographic change. Prioritizing durable materials(Jadalhaq, 2019), passive design foundations, and inclusive urban infrastructure contributes not only to immediate resilience but to sustained equity across generations.

In this context, sustainable architecture evolves from a design methodology into a long-term justice infrastructure. It becomes a systemic mechanism through which environmental responsibility, social inclusion, and policy reform converge. By embedding measurable equity indicators, adaptive capacity, and governance accountability into architectural practice, the built environment can function as a durable platform for climate justice rather than a temporary mitigation response.

4.1 Expanding Performance Metrics: From Environmental Efficiency to Justice Indicators

Evaluating sustainable architecture has traditionally relied on quantifiable environmental indicators such as energy reduction, carbon footprint, and material efficiency. While these metrics remain essential, they are insufficient for assessing whether architectural interventions genuinely advance climate justice. Justice-oriented evaluation requires expanding performance frameworks beyond environmental efficiency to incorporate social distributional outcomes(El-Kholei & El-Shamy, 2026). Thermal comfort equity, for example, should be measured not only as an average indoor temperature range but as a distributed condition across different socio-economic housing typologies. Access to shaded public spaces, proximity to green infrastructure, exposure to flood-prone zones, and affordability of energy consumption must be integrated into performance analysis. These indicators reveal whether resilience strategies are equitably allocated or spatially concentrated in higher-income developments(Elrefaie, 2026). Moreover, affordability metrics should accompany sustainability certifications. A highly efficient building that remains financially inaccessible fails to advance climate justice. Thus, cost-of-occupancy, long-term maintenance burdens, and energy expenditure ratios must form part of the evaluation matrix. By embedding justice-sensitive indicators into architectural assessment tools, environmental performance becomes inseparable from social accountability.

This expanded evaluative lens transforms sustainability certification systems from purely technical benchmarks into instruments capable of measuring spatial equity and inclusive resilience.

4.2 Adaptive Governance and Intergenerational Responsibility

Long-term climate justice depends not only on design excellence but on governance continuity and adaptive institutional frameworks capable of responding to evolving environmental, economic, and social conditions. Buildings and urban systems are not static entities; they function within dynamic regulatory environments shaped by shifting climate projections, demographic transformations, technological innovation, and fluctuating economic capacities. Without structured mechanisms for monitoring, enforcement, and periodic reassessment, even well-designed climate-responsive projects risk becoming outdated, inaccessible, or misaligned with justice objectives (Hamada, Hong, & Raslan, 2026).

Adaptive governance, in this context, refers to a flexible, learning-oriented institutional approach that integrates environmental regulation, urban planning, and social policy. It requires regulatory systems that are not merely prescriptive but responsive—capable of incorporating new climate data, revising performance benchmarks, and addressing emerging vulnerabilities. Rather than relying on one-time compliance assessments, adaptive governance promotes iterative evaluation cycles, ensuring that buildings and urban developments remain environmentally efficient and socially equitable over time.

Institutional coordination is central to this model. Municipal authorities, planning agencies, housing ministries, environmental regulators, and academic institutions must operate within interconnected frameworks rather than isolated mandates. Collaborative data-sharing platforms can facilitate longitudinal monitoring systems that assess microclimatic conditions, energy affordability, indoor thermal comfort distribution, and infrastructure reliability. These indicators provide evidence-based insights into whether climate-responsive architecture is delivering tangible benefits to vulnerable populations or merely satisfying formal sustainability standards.

Moreover, governance mechanisms must address enforcement disparities. In many rapidly urbanizing contexts, building codes and environmental regulations exist but are unevenly implemented. Strengthening inspection systems, increasing transparency in development approvals, and integrating digital monitoring tools can enhance accountability. When justice objectives are embedded into regulatory processes—through mandatory minimum thermal standards, affordable housing sustainability requirements, or incentives for passive design adoption—architecture becomes structurally aligned with policy reform rather than dependent on voluntary innovation.

Intergenerational responsibility further deepens the ethical dimension of adaptive governance. Climate justice extends beyond present populations; it concerns the distribution of environmental risk and opportunity across time. Architectural decisions regarding material durability, structural resilience, and infrastructural integration influence not only current occupants but future generations who will inherit the built environment. Poorly designed buildings may require premature demolition, excessive retrofitting, or energy-intensive upgrades, transferring economic and environmental burdens forward(Alqhtani, Shaheen, & Aldali, 2026).

Designing with intergenerational foresight entails integrating long-term climate modeling into architectural planning, prioritizing durable and low-maintenance materials, and incorporating modular systems that facilitate adaptation. Retrofit-ready infrastructures—such as adaptable façades, flexible service cores, and expandable energy systems—allow buildings to evolve alongside technological advancements and changing climatic conditions. This forward-looking approach reduces lifecycle costs, minimizes resource waste, and safeguards future inhabitants from structural obsolescence.

Education and professional culture also play a pivotal role in sustaining intergenerational responsibility. Architectural curricula and continuing professional development programs must embed climate justice principles within design methodology, equipping practitioners with the analytical tools to assess long-term socio-environmental impact(Ayyat et al., 2026). Ethical frameworks within professional practice can reinforce the understanding that climate-responsive design is not a temporary trend but a permanent obligation.

By integrating adaptive governance structures with forward-looking design strategies, architecture transcends short-term mitigation and evolves into a durable justice infrastructure. Climate justice within the built environment is secured not solely through innovative materials or advanced technologies, but through continuous institutional engagement, regulatory accountability, and a sustained ethical commitment to future generations. In this integrated framework, the built environment becomes a living system—monitored, adjusted, and improved over time—to ensure that resilience remains inclusive and equitable across both present and future communities.

CONCLUSION

Climate change has fundamentally altered the intellectual and ethical terrain within which architecture operates. What was once framed primarily as a technical discipline concerned with structural integrity, spatial organization, and environmental performance has evolved into a field deeply entangled with questions of justice, governance, and long-term planetary responsibility. The built environment does not merely respond to climate change; it actively structures exposure to risk, mediates access to environmental resources, and shapes the lived realities of resilience and vulnerability. In this sense, architecture occupies a decisive position within contemporary debates on climate justice.

This chapter has argued that sustainable architecture, when critically reframed through a justice-oriented lens, transcends the conventional boundaries of environmental design. Sustainability, understood solely as energy efficiency or carbon reduction, risks remaining technocratic and socially neutral. Climate justice, however, introduces an ethical imperative that interrogates distribution: who benefits from resilient infrastructure, who bears environmental burdens, and who remains excluded from climate-adaptive development. By repositioning architecture within this normative framework, design becomes not only an environmental intervention but a redistributive mechanism embedded within socio-spatial systems.

The Egyptian context illustrates the urgency of this shift. Rapid urban expansion, climatic intensification, and socio-economic disparities intersect to produce uneven patterns of vulnerability. Informal settlements, energy-intensive construction practices, and insufficient regulatory enforcement reveal how climate exposure is spatially differentiated. Yet this same context also reveals opportunity. Vernacular climatic intelligence, passive design strategies, and context-sensitive planning offer pathways for reimagining sustainable architecture as an accessible and inclusive practice rather than a market-driven luxury.

Through examining climate-responsive design strategies, regulatory integration, and adaptive governance mechanisms, this chapter has demonstrated that the realization of climate justice requires multi-scalar coordination. Architectural innovation must align with policy reform, institutional accountability, and long-term monitoring systems. Environmental law and urban regulations establish normative frameworks, but it is through architectural materialization that these frameworks become tangible—manifested in thermal comfort, shaded streetscapes, durable housing, and reduced energy burdens.

Crucially, climate justice within the built environment must extend beyond immediate mitigation. It requires adaptive governance capable of responding to evolving climatic data, socio-economic transformations, and technological advancements. Buildings must be conceived as flexible systems, prepared for retrofitting and recalibration, rather than fixed objects locked into present assumptions. Intergenerational responsibility further deepens this commitment: today's architectural decisions will determine tomorrow's exposure to heat, flooding, resource scarcity, and infrastructural fragility. Designing with foresight thus becomes an ethical obligation that links environmental stewardship with temporal equity.

At its core, the integration of climate justice into architectural practice represents a redefinition of professional responsibility. Architects are not merely service providers responding to market demands; they are spatial actors participating in governance processes and shaping collective futures. By embedding justice-sensitive indicators, community engagement, affordability considerations, and regulatory alignment into design methodologies, architecture can evolve into a durable justice infrastructure—one capable of supporting equitable adaptation and sustainable development over time.

Ultimately, the pursuit of climate justice within the built environment demands more than isolated green projects or symbolic compliance with sustainability standards. It requires systemic transformation in how cities are planned, how buildings are regulated, and how environmental performance is evaluated. When environmental responsibility is inseparable from social inclusion, and when policy reform is translated into spatial equity, architecture becomes a transformative platform rather than a reactive response.

In this integrated vision, sustainable architecture is not simply a strategy for reducing emissions; it is a mechanism for reshaping urban futures toward resilience, fairness, and shared environmental security. Climate justice thus emerges not as an abstract aspiration but as a spatial practice materialized through design, sustained through governance, and carried forward through intergenerational ethical commitment. The built environment, when guided by this paradigm, holds the potential to become one of the most powerful instruments for advancing both environmental protection and social equity in an era of accelerating climatic uncertainty.

REFERENCES

Abdelhamid, M. M. (2026). Urban revitalization and its impact on urban tourism in the historic city center. An analysis of Elraml-station district, Alexandria, Egypt. *Journal of Cultural Heritage Management and Sustainable Development*, •••, 1–20. DOI: 10.1108/JCHMSD-06-2025-0181

Abraham, E. M., Ganyaglo, F. K., Andoh, J. A., Asamoah, E., & Koomson, S. (2025). The ethics of climate justice in emerging economies: exploring the intersection of development and environmental responsibility. *International Journal of Ethics and Systems*.

Al Ali, S., Al Shawabkeh, I., & Alqodsi, E. M. (2026). Facial Recognition Technology: Protecting Biometric Privacy in the Digital Age. *Journal of Legal Affairs and Dispute Resolution in Engineering and Construction*, *18*(2), 06526002. Advance online publication. DOI: 10.1061/JLADAH.LADR-1435

Al Shawabkeh, I. K. (2023). THE ROLE OF TAX DISPUTES RESOLUTION COMMITTEE IN SETTLING TAX DISPUTES IN THE UNITED ARAB EMIRATES. *Lawyer Quarterly*, *13*(1), 30-41. Retrieved from https://www.scopus.com/inward/record.uri?eid=2-s2.0-85150371431&partnerID=40&md5=01b627f61bde042ba0d007236feea9e4

Al-shawabkeh, I. K., Shouaib, M., Al Qudah, M., & Aldabbas, A. M. (2025). Tax declaration in UAE legislation: balancing the protection of taxpayers' rights and preserving public funds. *International Journal of Law and Management*. DOI: 10.1108/IJLMA-01-2025-0013

Alhuzail, N. A., & Majumdar, K. (2026). Bridging Traditions and Futures: Indigenous Women as Architects of Change. In *Indigenous Women's Resilience, Leadership and Social Change: Intersection of Knowledge, Tradition, and Justice* (pp. 335-347): Springer.

Alqhtani, L., Shaheen, B., & Aldali, K. (2026). Evaluating the Effectiveness of Camel Hair Filters in Wind Catchers for Air Quality Improvement and Natural Ventilation Comfort in Vernacular Saudi Architecture. *Buildings (Basel, Switzerland)*, *16*(4), 731. DOI: 10.3390/buildings16040731

Alqodsi, E. M. (2025). Extended producer responsibility in e-waste recycling: A secure path to a sustainable future for UAE. *Sustainable Development (Bradford)*, *33*(1), 309–318. DOI: 10.1002/sd.3107

Alqodsi, E. M., & Abdallah, A. K. (2024a). *Legal frameworks and educational strategies for sustainable development*.

Alqodsi, E. M., & Abdallah, A. K. (2024b). Preface. *Legal Frameworks and Educational Strategies for Sustainable Development*, xviii-xix. Retrieved from https://www.scopus.com/inward/record.uri?eid=2-s2.0-105007489920&partnerID=40&md5=6f9c26107dea4d2e5ab772594885b3af

Alshawabkeh, I. K. (2024). Using the tax to reduce environmental pollution in the United Arab Emirates: Possibility and challenges. In *Legal Frameworks and Educational Strategies for Sustainable Development* (pp. 1-12).

Amin, M., & Zaher, S. (2026). The critical nexus of water resources and food security underscoring climate justice in Egypt: Within-variable comparative analysis. *Review of Economics and Political Science*, *11*(1), 24–40. DOI: 10.1108/REPS-09-2024-0062

Araújo, J. (2026). Algorithms on fire: Leadership, power and climate collapse in the age of AI. *Leadership and Organization Development Journal*, *47*(2), 1–14. DOI: 10.1108/LODJ-06-2025-0542

Ayo-Odifiri, S. O., Ebekozien, A., & Aigbavboa, C. (2025). *Climate-Responsive Architectural Practices: Integrating Green Spaces into the Urban Environment to Achieve SDG 13.* Paper presented at the Global Forum for Sustainable Built Environment Conference.

Ayyat, M. S., Ismail, R. F., Abd El-Latif, K. M., Bassiony, S. S., Al-Sagheer, A. A., & Helal, A. A. (2026).. . *Egyptian Journal of Veterinary Science*.

Bampou, P. (2017). Green buildings for Egypt: A call for an integrated policy. *International Journal of Sustainable Energy*, *36*(10), 994–1009. DOI: 10.1080/14786451.2016.1159207

Bennett, J. (2024). Repair as Practice: Expanding Architectural Approaches to Climate Justice in Southern Africa. *Dimensions. Journal of Architectural Knowledge*(8).

Cole, R. J. (2020). Navigating climate change: Rethinking the role of buildings. *Sustainability (Basel)*, *12*(22), 9527. DOI: 10.3390/su12229527

Comelli, T., Samm-A, A., Poudel, D. P., Dabbeek, J., Nalau, J., Ensor, J., & Khan, M. M. (2026). The power of place-based Future Visioning: Centring climate justice in urban adaptation. *Cities (London, England)*, *170*, 106614. DOI: 10.1016/j.cities.2025.106614

Dahiyat, E. A. R. (2007). Intelligent agents and contracts: Is a conceptual rethink imperative? *Artificial Intelligence and Law*, *15*(4), 375–390. DOI: 10.1007/s10506-007-9047-z

El-Kholei, A. O., & El-Shamy, A. M. (2026). An interdisciplinary analysis of social and environmental justice in Cairo's New Communities: Implications for urban policy and governance. *Open House International*, •••, 1–27. DOI: 10.1108/OHI-12-2024-0405

Elbahrawy, E. A., & Gammaz, S. A. (2026). A practical framework for selecting pedestrian simulation tools in the Egyptian urban context. *Journal of Engineering and Applied Sciences (Asian Research Publishing Network)*, *73*(1), 16.

Elrefaie, H. B. (2026). Proposed Context-Specific, Integrated Policy Framework for Building Energy Conservation in Egypt. *International Design Journal*, *16*(1), 173–191. DOI: 10.21608/idj.2025.423303.1433

Eltinay, N. (2019). City-to-city exchange: Redefining "resilience" in the Arab region. *International Journal of Disaster Resilience in the Built Environment*, *10*(4), 222–238. DOI: 10.1108/IJDRBE-05-2019-0028

Esmail, O. K., & Afifi, S. (2026). Designing for Conviviality: A Context-Sensitive Framework for Evaluating Public Parks in Cairo, Egypt. *Space and Culture*, •••, 12063312251392376. DOI: 10.1177/12063312251392376

Fu, H., Hao, Y., Wu, Z., Xu, H., & Zuo, J. (2026). Human-centered collaborative design in green buildings: A comprehensive review of neurotechnology integration. *Renewable & Sustainable Energy Reviews*, *231*, 116772. DOI: 10.1016/j.rser.2026.116772

Gondi, D. S. (2026). Strategies for Implementing Sustainability Practices in Smart Hospitals. In *Responsible Innovation in Smart Healthcare: AI, IoT, and Ethical Sustainability Practices* (pp. 307-338): IGI Global Scientific Publishing.

Gonzalez, C. G. (2021). *Racial capitalism, climate justice, and climate displacement.* Paper presented at the Oñati socio-legal series, symposium on climate justice in the anthropocene. DOI: 10.35295/osls.iisl/0000-0000-0000-1137

Guo, W. (2026). Tracing policy change of the NEA in urban design through the advocacy coalition framework. *Cultural Trends*, *35*(1), 1–18. DOI: 10.1080/09548963.2024.2441782

Hafiz, E. (2025). Evaluating Egypt's Urban Emergency Response System: Toward Resilient Cities and Integrated Crisis Governance with special reference to south sinai. *JES. Journal of Engineering Sciences*, *53*(6), 810–835.

Hamada, A. A., Hong, S.-M., & Raslan, R. (2026). Distinguishing energy efficiency from substandard IEQ in Cairo's public administrative buildings: Audit-informed modelling and calibration of two cases. *Energy Reports*, *15*, 109066. DOI: 10.1016/j.egyr.2026.109066

Hamstead, Z. A. (2023). Critical heat studies: Deconstructing heat studies for climate justice. *Planning Theory & Practice*, *24*(2), 153–172. DOI: 10.1080/14649357.2023.2201604

Jadalhaq, I. M. (2019). Tort policy in a plural context: Pathways towards objective liability in UAE tort law. *Issues in Legal Scholarship*, *17*(1). Advance online publication. DOI: 10.1515/ils-2019-0001

Klinsky, S., & Mavrogianni, A. (2020). Climate justice and the built environment. *Buildings & Cities*, *1*(1), 412–428. DOI: 10.5334/bc.65

Li, Z., Zou, Y., Tian, M., & Ying, Y. (2022). Research on optimization of climate responsive indoor space design in residential buildings. *Buildings (Basel, Switzerland)*, *12*(1), 59. DOI: 10.3390/buildings12010059

Mahmoud, R. A. (2017). New Cairo's urban dynamics: Redefining gentrification as a main actor of social resilience. *Procedia Environmental Sciences*, *37*, 429–441. DOI: 10.1016/j.proenv.2017.03.013

Nag, A. (2026). Designing for Change: Sustainable and Culturally Responsive Housing With Shared Spaces. In *Sustainable Approaches in Spatial Design* (pp. 1-48): IGI Global Scientific Publishing.

Ofosu-Asare, Y. (2026). Architectures: Decolonial Design Practices in Africa. In *Encyclopedia of the anthropocene: Pluriversal perspectives* (pp. 1-23): Springer.

Othman, A. A. E., & Youssef, L. Y. W. (2021). A framework for implementing integrated project delivery in architecture design firms in Egypt. *Journal of engineering, design and technology, 19*(3), 721-757.

Pelsmakers, S., & Brown, J. B. (2025). Architectural pedagogies to address the climate emergency: Creaturely pedagogy, collaboration and communities of equity, empathy and interdisciplinarity. In *Architectural Thinking in a Climate Emergency* (pp. 103–187). Routledge. DOI: 10.4324/9781003293903

Soliman, A. M., & Soliman, Y. A. (2022). Exposing urban sustainability transitions: Urban expansion in Alexandria, Egypt. *International Journal of Urban Sustainable Development*, *14*(1), 33–55. DOI: 10.1080/19463138.2022.2056894

Subrahmanyam, S. (2026). Future of AI in Global Sustainability. In *AI and Sustainability for Global Security* (pp. 113–154). IGI Global Scientific Publishing. DOI: 10.4018/979-8-3373-1077-0.ch004

Tandon, A., & Nag, A. (2026). Designing Place-Based, Multisensory Gamified Systems for Inclusive Heritage and Collaborative Story Building in Architectural and Cultural Sites. In *Addressing Urban and Spatial Challenges Through Gamification* (pp. 127–160). IGI Global Scientific Publishing.

Taylor, J., Jokela, S., Laine, M., & Ferguson, L. (2025). Architects and climate justice. In *Architectural Thinking in a Climate Emergency* (pp. 19–41). Routledge. DOI: 10.4324/9781003293903-3

Vogt, M., Ritson, K., & Blanc, J. (2010). Sustainability and Climate Justice from a theological perspective. *RCC Perspectives*, (3), 30–46.

White, D. (2020). Just transitions/design for transitions: Preliminary notes on a design politics for a green new deal. *Capitalism, Nature, Socialism*, *31*(2), 20–39. DOI: 10.1080/10455752.2019.1583762

Chapter 15
Remote Sensing, AI, and Digital Forensics in Environmental Crime Investigations in India

Ankit Anand
http://orcid.org/0009-0003-5049-7967
SRM Institute of Science and Technology (KTR), India

Rajneesh Kumar
Pondicherry University, India

Gurminder Kaur
Pondicherry University, India

Shamima M. Parveen
Saveetha Institute of Medical and Technical Sciences, India

Hrishikesh Manu
Chanakya National Law University, India

ABSTRACT

In India, environmental crime is challenging climate justice, often impacting the marginalised communities and ecologically sensitive classes that depend on the environment for their lives and livelihoods. This chapter focuses on the increasingly common application of remote sensing, satellite imagery, artificial intelligence, and digital forensics in the detection, investigation, and prosecution of environmental crimes. By contextualising technology within a climate justice and rights-based

DOI: 10.4018/979-8-3373-6635-7.ch015

framework, this study examines the nature of data generated through technology and its compatibility with standards of evidence in India, specifically under the Bharatiya Sakshya Adhiniyam (BSA), 2023, and jurisprudence through the National Green Tribunal (NGT). Using doctrinal analysis, along with a review of Indian case studies and comparative insights from the EU and the US, the chapter identifies evidentiary uncertainties, algorithmic opacity, privacy risks, and institutional gaps.

INTRODUCTION

India is facing the largest and most complex environmental crisis it has ever encountered, combining ecological degradation, climate change, and systemic failures of governance, all of which are exacerbated by rapid technological advances. Environmental crimes that have accelerated climate vulnerability from illegal mining and deforestation to industrial pollution and river contamination have come into conflict with constitutional obligations toward life, dignity, and equality (*M.C. Mehta v. Union of India*, AIR 1988 Supreme Court 1037). These harms are not immeasurable or evenly distributed, but instead exacted on the most vulnerable among us, including some of the world's poorest people, such as forest dwellers, river line communities, agrarian families, and urban poor.

From this perspective, environmental crime in India should not only be viewed as an infraction of regulations but also as a serious climate justice problem that deserves to call for a fresh approach to the law and institutions. The pattern of environmental crime in India has been long-standing, ranging from illegal mining and uncontrolled deforestation to pollution, river despoliation, and continuous industrial violations of environmental norms. The mineral-rich regions of states like Odisha, Jharkhand and Chhattisgarh have been subjected to illegal mining, resulting in severe land degradation, loss of forests due to mining encroachment, depletion of groundwater, and massive displacement (Deb & Sarkar, 2017). Deforestation in ecologically sensitive areas, like the Western Ghats and the North-eastern states, has also hastened biodiversity reduction and disturbed the fragile climatic equilibrium (Barman et al., 2025). Even river systems like the Ganga and Yamuna are plagued by untreated industrial effluents, the discharge of municipal waste, and encroachment that compromise both ecological soundness and human health (Dey & Majumdar, 2024).

These environmental offences are a direct driver of climate vulnerability. Forest destruction reduces the ability to sequester carbon (Lal, 2005); illegal mining exacerbates land-use change (Prakash & Gupta, 1998), while water contamination weakens the resilience of adaptation for resource-based communities' livelihoods and survival (Robinson, 2016). Crucially, the costs of these harms are heavily loaded on marginalised and vulnerable communities that have little political voice and slim access to courts (Walker, 2009). It is the indigenous forest dwellers, traditional fishing and farming communities on river banks, and informal urban settlements in the close vicinity of special industrial clusters which bear the brunt of ecological degradation. This asymmetry situates environmental crime firmly within the field of climate justice, where issues of fairness and distributive equity, as well as intergenerational responsibility, form an important part of juridical reasoning.

Conventional methods for enforcing environmental compliance in India have not kept pace with the magnitude and complexity of modern-day environmental transgressions. As for regulatory structures, jurisdictionally broken government bodies, visits every few years of vehicles for self-reported compliance data, and some audits after the fact have been the historical method. Environmental governance is fragmented among various authorities, including central ministries, state pollution control boards, district administrations, and specialised tribunals, leading to competing jurisdictions, coordination failures, and compromised accountability (Gupta, 2014).

Traditional inspection-based enforcement is plagued by evidentiary and implementation challenges. Field checks are episodic, resource-intensive, and often focused on areas that are directly accessible, yet they leave large and remote regions unattended, hindering effective monitoring. Audited paper records and compliance reports are susceptible to tampering, underreporting, and selective omission by the entities they are intended to regulate. Softnesses are then exacerbated by regulatory slowness, bureaucratic bottlenecks, and, at times, outright political intervention, which leads to a weakening of enforcement actions against influential economic agents. Environmental abuses often go unnoticed due to lack of evidence; if a case does make it to court, the charges are often tossed out as shaky or irrelevant (Dias, 1994). Indian environmental adjudicatory bodies, particularly the National Green Tribunal and the Supreme Court, are increasingly confronted with technologically generated material such as satellite imagery, remote sensing outputs, and AI-assisted environmental risk classifications (Choudhary et al., 2025). Yet the judicial treatment of such material is neither automatic nor uncritical. Questions of admissibility, methodological reliability, evidentiary weight, expert authentication, and procedural safeguards determine whether technological outputs translate into legally actionable proof. This chapter therefore moves beyond technological optimism to examine how Indian courts reason about, evaluate, and conditionally accept geospatial and algorithmic evidence within constitutional and evidentiary frameworks.

In this context, technological solutions such as remote sensing, satellite surveillance, Artificial Intelligence (AI), and digital forensics are now being presented as game changers in environmental monitoring and enforcement. Through satellite imagery, monitoring of land-use change, patterns of deforestation, mining activity and fragmentation of protected areas becomes possible on a broader scale over time. Remote sensing methods help to monitor air and water pollution in large areas, ensuring the monitoring is continuous in time and space rather than discrete and periodic measurements (Hoff & Christopher, 2009).

These techniques are further augmented using AI-powered analytics to analyse large datasets and find patterns, outliers, and predict risk hotspots in the area (Babbar & Rani, 2025). The application of digital forensic procedures facilitates proving the authenticity, preservation, and presentation of data generated by technology in judicial contexts, facilitating new ways for proving evidence. Together, they offer the potential for more efficient, scalable and objective environmental crime detection and documentation.

However, the securitisation of environmental governance based on surveillance also triggers significant legal and ethical concerns. Continuous monitoring raises issues of privacy, ownership of the data, the opacity of algorithms, and the scope for selective or biased enforcement (Bahangulu & Owusu-Berko, 2025). In the context of AI, relying on probabilistic outputs forces us to revaluate the traditional evidentiary standards based on the concept of direct evidence and human confirmation. In the absence of sufficient procedural protections, technological monitoring can render environmental governance technocratic enterprise devoid of constitutional values and concerns for social justice. In this regard, the chapter also examines the role of Explainable Artificial Intelligence (XAI) as a doctrinal bridge between algorithmic governance and principles of natural justice. By enabling interpretability, auditability, and transparency in model decision-making, XAI strengthens compatibility with core procedural safeguards such as *audi alteram partem*, reasoned decision-making, and the right of affected parties to meaningfully challenge adverse technological findings.

The central question of this chapter is whether and how new technologies (remote sensing satellite imagery, AI, digital forensics) can be integrated with India's environmental enforcement and adjudicatory systems in a way that enhances climate justice by serving as guarantors of rights and procedural fairness. Although these technologies make important contributions to the detection and prosecution of environmental crime, Indian law currently lacks an integrated system that regulates their admissibility, verification, transparency and accountability.

This chapter therefore pursues five interconnected goals: first, to critically examine the legal status and evidentiary treatment of technological evidence in Indian environmental law; secondly, to consider how various institutions such as the NGT, Ministry of Environment, Forest and Climate Change, state pollution control boards cannot be overlooked when bringing technology into play; thirdly addressing ethical, privacy and equity concerns with an emphasis on vulnerable communities; fourthly drawing lessons from comparable experiences such as those from the European Union or United States experience and finally suggesting legal reforms so that the 'gains made by technology' can then be marshalled for principles like transparency, accountability and climate justice. The chapter further undertakes a structured analysis of landmark Indian cases involving satellite and geospatial evidence, assessing the judicial reasoning adopted in determining reliability and evidentiary value. It comparatively evaluates how the European Union integrates satellite monitoring within a structured regulatory and rights-based architecture, and how United States courts subject technical and algorithmic outputs to demonstrable methodological reliability standards before admissibility. By situating Indian developments within these comparative frameworks, the chapter seeks to articulate a calibrated model that privileges transparency, validation, and constitutional accountability over mere technological capability.

CONCEPTUAL AND THEORETICAL FRAMEWORK

An interdisciplinary conceptual framework brings together environmental criminal law, climate justice theory, evidentiary jurisprudence, and surveillance ethics. This framework has dual objectives: to locate environmental crime in India within a justice-oriented comprehension of climate governance, and to examine how new technologies, like remote sensing, satellite imagery, AI, and digital forensics, impact long-established legal conceptions of proof, accountability, and rights. In the absence of such a framework, tech adoption runs the risk of becoming ad hoc, normatively unmoored and possibly unjust.

In the Indian context, environmental crime must not be viewed solely as a contravention of statutes or administrative requirements. Illegal mining (Bhushan, 2008), deforestation (Bose, 2023), river contamination (Rajaram & Das, 2008), and industrial non-compliance (Chatterjee & Kanbur, 2015), create harms that are unequally distributed across society. The theory of climate justice provides an analytical lens with which to approach this asymmetry, as it centres the question of who pays for environmental impacts on one hand and who profits from them on the other (Audet, 2013).

From the viewpoint of distributive justice, environmental crime has an imbalance in its effects on indigenous peoples, forest dwellers, small farmers and riverine communities. Unregulated mining in the forests not only decimates local livelihoods but also pollutes water sources and exacerbates climate vulnerability, even as the profits flow into private pockets and the hand of middlemen with political clout (Kale, 2025). Untreated industrial effluents released into rivers such as the Ganga and Yamuna are also a source of health risks, environmental degradation, and financial burdens to communities downstream, who are not directly informed or involved in decision-making processes and enforcement actions (Srinivas et al., 2020). These harms are compounded by climate change: flooding, drought and ecological instability increase, turning local environmental crimes into system-wide climate injustices.

Procedural justice is equally central. Environmental governance in India has, for the most part, excluded affected communities from decision-making, monitoring and enforcement. These environmental impact assessments, forest clearances, and pollution consent mechanisms are often accused of being opaque and non-consultative. When environmental wrongs happen, the response is usually reactive, technocratic, and elite-driven, leaving little room for community voices. Climate justice theory insists that fairness is about more than just outcomes; it is also about inclusive, transparent processes, with clear avenues for accountability, which is a particularly relevant insistence as surveillance technologies are rolled out in ways that have never been explained publicly or consented to by the community (Schlosberg, 2012).

The situation of indigenous and forest communities adds a further twist to the justice analysis. Although constitutionally mandated and statutorily protected through legislation, like the Forest Rights Act, 2006, these communities are often equated with "encroachers" or offenders in surveillance technology-mediated enforcement regimes (Smadja, 2018). Satellite images and machine algorithms for land-use can pinpoint "encroachments" without acknowledging customary patterns of land use, seasonal habitation, or traditional ecological practices (Shrestha, 2015). Thus, the enforcement of environmental crime, if it is technologically driven but not justified by law and social reflection, can perpetuate itself over time rather than transcending its origins.

The increasing dependence on remote sensing, satellite imaging, AI analysis, and digital forensics represents a profound epistemic shift in the conduct of environmental investigations transitioning from human observation to data-driven inference. Yet, scientific or technical information is not automatically evidence that can be legally proven. The distinction underpins the evidentiary law of Canada and is central to understanding the use of technology in environmental trials (Wood, Tanner, & Richardson, 2010).

Data is produced by sensors, applications, and analysis algorithms that attempt to recognise patterns, anomalies, and likely events. Satellite photos record spectral data; AI systems make inferences about what is happening based on massive datasets; predictive models gauge future risks (Abutaleb, Abdelsalam, & Khaled, 2025). On the other hand, legal evidence must meet certain conditions, such as relevancy, veracity, authenticity, and admissibility, under a set of domestic laws, that could include the BSA, 2023. Courts do not just care about what the data says, but also how it was created, stored, analysed, and presented.

This disconnect becomes even more pronounced with predictive analytics. AI-generated outputs are, by nature, often premised on probabilistic reasoning, such as identifying hotspots, predicting the occurrence of illegal behaviour, or flagging areas at risk of non-compliance (Abdullahi, 2025). Predictions of this kind are valuable for regulatory planning and enforcement prioritisation, but they rest uncomfortably with legal norms that favour actual evidence over statistical probability. The Indian law of evidence has historically been sceptical about indirect or circumstantial evidence, particularly when it is used to establish liability.

Beyond all this, the opacity of many AI systems, the so-called "black box" problem, raises profound questions about the core principles of natural justice (Chaudhary, 2024). A recipient or affected community that cannot comprehend how an algorithm reached a conclusion cannot effectively contest it. It also begs the question of whether AI output can be used as expert evidence, enhanced material or third-party information. Accordingly, the chapter does not consider technology as a surrogate for legal reasoning, but rather as a proposition that poses new knowledge, whose epistemic value needs to be transparently mediated through doctrinal safeguards. In this sense, we can also understand the conceptual distinction between technology as a detection tool and technology as evidence, warning us against collapsing these two without rigorous legal scrutiny.

The use of satellite imagery, AI, and digital forensics in environmental governance also raises broader ethical issues about surveillance. Environmental surveillance is increasingly related to the permanent or intermittent observation of soil, water, and human activities, with aggressions often made without informing or seeking permission from the population (Young et al., 2022). The surveillance is for the greater good of protecting the environment and combating climate change, but it raises complex questions about power, accountability, and rights.

One moral concern revolves around consent and information asymmetry. Forest-dwellers or riverine communities are seldom aware of the scale and purpose of technological scrutiny (Deb, 2013). Surveillance may occur without procedural protections or notification, and there may be no practical recourse, both of which erode democratic values. It is especially troubling when surveillance data are used at a later date to initiate enforcement actions or legal proceedings.

There are also technological-solutions, which suggest that technological advances can address long-standing governance problems. Over-dependency on AI and satellite data can also hide other structural weaknesses, such as rampant corruption, regulatory capture, or anaemic institutional capacity (Köbis, Starke, & Rahwan, 2022). Technology may shed more light on violations, but visibility does not automatically translate into accountability and justice. In the absence of ethical governance frameworks, tools may be developed that serve to empower the state without a counterbalance.

Environmental surveillance, third, represents questions of selective attention. Marginalised areas and communities may be disproportionately policed, while polluting activities by powerful agents go under-enforced. This selective way of looking at things can reinforce inequality in the name of scientific objectivity.

LEGAL FRAMEWORK GOVERNING ENVIRONMENTAL PROTECTION AND DIGITAL EVIDENCE IN INDIA

The reception of remote sensing, satellite imagery, AI, and digital forensics, as well as their effectiveness in the investigations of environmental crime in India, is warring under the existing Indian legal framework (Kumar, 2025). In the case of India, environmental governance is determined through a complex statutory framework that encompasses regulatory, civil, and criminal aspects. At the same time, issues related to proof are compiled under evidence law dating back to the 19th century, which has not incorporated algorithmic or geospatial technologies into its framework. The NGT, being located at the interface between these, therefore introduces another layer of complication with its mildly adaptable procedure mandate. This part examines how these legal frameworks facilitate, limit, and, in some instances, mediate the technological evidence-making process regarding the environment.

Environmental Statutory Regime

India's environmental legal regime is fundamentally regulatory, involving licensing, monitoring, and enforcement of unauthorised use, rather than proactive criminal investigation. The Environment (Protection) Act, 1986 (EPA), as the mother legislation, lays down institutional mechanisms to coordinate Central and State policies and the enforcement of such policies at the state level to achieve the objectives contained in these Acts (Mushtaq, Bandh, & Shafi, 2020). The EPA is authorised to conduct inspections, sampling, and data gathering, thereby establishing a statutory foundation for the use of technology-based devices, such as satellite surveillance and remote sensing. However, as discussed in the following section, it does not specifically consider AI-generated analytics or satellite imagery as evidence; therefore, its legal status has mostly been left to judicial interpretation.

The Forest (Conservation) Act 1980 (FCA), in particular, is crucial for the regulation of deforestation and the diversion of forest land for non-forest purposes (Saxena, 2024). Monitoring for compliance under the FCA is now increasingly using satellite imagery to measure forest cover changes, identify encroachments, and verify adherence to clearance conditions. Although this technical monitoring has actually increased the detection capacity, regarding the administrative approach, it is FCA based. Enforcement tends to be after-the-fact regularisation or punishment, rather than criminal prosecution, leaving open the question of how satellite observations become legally sustainable evidence of a violation.

The Air (Prevention and Control of Pollution) Act, 1981 and the Water (Prevention and Control of Pollution) Act, 1974, established pollution control boards, as well as systems for consent, monitoring, and enforcement (Prasad, 2006). The last few years have witnessed the introduction of continuous emission and effluent measurement systems (CEMS/OCEMS) in these regulatory dispositions, combined with remote sensing or AI-based pollution modelling. However, the enforcement of such Acts is again predominantly based on sampling reports, inspections and laboratory analysis. However, satellite-detected pollution plumes or AI-predicted hotspots are more often a prompt for investigation than standalone evidence, illustrating the wariness of some legal systems regarding machine-mediated proof.

This tension is common to all of these statutes. Despite the fact that technology has become more integrated into regulatory practice, the recognition of evidence in digital form as technological rather than explicit evidence has not been reflected in statutory language. It is cause for concern when environmental infractions escalate into contentious actions, especially when a defendant seeks to contest criminal liability, penalties, or shutdowns.

Bharatiya Sakshya Adhiniyam (BSA), 2023 and Digital Evidence

The BSA, 2023, addresses the relevance and admissibility of evidence and proof. Although adjusted to the standards of electronic records, it is conceptually based on a predigital world. Section 2 defines the term "evidence" to include both oral evidence and documentary evidence, and treats digital documents as a type of document through judicial construction and legislative amendment. Principles have this structure, although in the real world, it is difficult to apply them to satellite images and digital files.

Section 39, the provision on expert evidence, is particularly applicable to technological and environmental evidence. Interpretation of satellite imagery, AI analytics, and geospatial models all require an expert explanation of the methodology, accuracy, and limitations. Such evidence may be admissible as expert opinion by the courts; however, its probative value would depend on the credibility of the expert, a clear methodology, and the opportunity for cross-examination. The AI-generated outputs further add a nuance to this decision problem, since the "expert" is, in fact is a concisely irreversible algorithm whose mental processes are as much hidden from its creators as they are from its users.

The admissibility of electronic records is prescribed under sections 62 and 63, which need to be certified to prove their authenticity and veracity. The certification under Section 63 specifically requires disclosure, inter alia, of the method of production and its reliability, those in whom confidence was placed, the nature of the process, and whether it did not involve any human intervention. Although this system is suitable for traditional electronic records, it fails to accurately reflect the complexity of data from satellites or AI. For whom is the "person in control" of a satellite system, how to maintain custody of the data generated continuously, and whether proprietary algorithms can meet disclosure requirements without divulging trade secrets are among the questions that must be resolved.

Those restrictions are exacerbated when we have AI-generated evidence. Predictive results, probabilistic models and pattern recognition belong to neither binary reasoning in terms of admissible v/s inadmissible evidence. The Indian courts have historically required a demonstrable nexus between the evidence and the fact in issue. AI forecasts, on the other hand, tend to refer more to probabilities rather than certainties. Accordingly, such outcomes may risk being considered circumstantial or confirmatory at best, rather than conclusive evidence.

For satellite imagery, there are relatively few conceptual obstacles to this kind of usage, but the resolution, timeliness, and interpretative subjectivity become even more problematic(Burke et al., 2021). In the absence of standardised procedures for authentication and interpretation, satellite images may receive cautious treatment by courts that question their probative value (Macrory & Purdy, 2001).

Standardisation of Environmental Digital Forensics: International Protocols and Evidentiary Reliability

While the admissibility of electronic and digital records under the Bharatiya Sakshya Adhiniyam, 2023 provides a statutory foundation for the use of technological evidence, evidentiary reliability ultimately depends upon adherence to recognised forensic standards governing identification, acquisition, preservation, and analysis. In this regard, international digital forensic protocols offer important guidance for environmental investigations involving satellite imagery, drone footage, geospatial datasets, continuous emission monitoring system (CEMS) logs, and AI-generated analytical outputs (Muhammed Siddik et al., 2024).

ISO/IEC 27037 establishes guidelines for the identification, collection, acquisition, and preservation of digital evidence, emphasizing integrity, traceability, and documentation. Its chain-of-custody principles require detailed recording of how digital material is obtained, transferred, stored, and accessed, thereby minimizing risks of tampering or evidentiary contamination (Ajijola et al., 2014). Complementing this framework, ISO/IEC 27041 provides standards for ensuring the assurance and validation of digital forensic methods, while ISO/IEC 27042 sets out structured requirements for the analysis and interpretation of digital evidence. ISO/IEC 27043 further addresses incident investigation principles, offering systematic approaches to digital evidence management in complex regulatory contexts. In the United States, the National Institute of Standards and Technology (NIST) digital forensic guidelines similarly emphasize methodological reliability, reproducibility, and documentation safeguards(Veber & Klíma, 2014).

For environmental enforcement in India, compliance with such internationally recognised standards is particularly significant. Satellite datasets must preserve original metadata; drone recordings require secure storage with hash-based integrity verification; CEMS and AI-generated environmental analytics must maintain audit trails capable of independent technical scrutiny. Without adherence to validated forensic acquisition protocols, digital environmental evidence risks challenge on grounds of authenticity, reliability, or procedural irregularity.

Accordingly, the institutionalisation of certified environmental digital forensic laboratories, standard operating procedures aligned with ISO/IEC norms, and trained forensic personnel become essential to strengthening judicial confidence. Alignment with globally recognised forensic standards not only enhances evidentiary credibility but also ensures that technological innovation operates within a legally defensible and procedurally robust framework.

NGT and Evidentiary Flexibility

The NGT Act, 2010, outlines a unique procedural framework for the adjudication of environmental disputes in India. The Tribunal 19 contains a non-obstante clause to the effect that it 'shall not be bound by the procedure laid down in the Code of Civil Procedure, 1908 or by the rules of evidence contained in BSA, 2023' and would have regard to principles of natural justice. This flexibility has facilitated the NGT's invocation of scientific and technical subjects, such as satellite imagery, expert committee reports, more easily (Shrotria, 2015).

In reality, the NGT has consistently employed satellite imagery, geospatial analysis, and scientific inquiry to identify environmental damage, tracking compliance with its instructions, and issuing remedial orders (Satapathy et al., 2008). It is a pragmatic recognition of the vastness and complexity of environmental degradation in India. However, flexible evidence is not the same as no evidence at all. The decisions of the Tribunal remain legally binding and carry serious legal repercussions that include monetary sanctions, project stoppages, and directions impacting lives and livelihoods.

Thus, the extent to which the NGT is flexible then begs important questions. Although the Tribunal may receive material of a technological nature without being bound by strict rules of evidence, it must nonetheless observe substantial justice, procedural fairness, and give reasons for decisions. Over-dependence on expert committees or satellite imagery, without providing them and the victims an opportunity to challenge them, would erode 'natural justice'. Further, the NGT's liberal approach neither endorses nor guarantees ready acceptance by superior courts when issues go up for appellate scrutiny (Rao et al., 2023).

The boundaries of that flexibility are revealed when technological claims are disputed. Arguments over how to interpret satellite images, the accuracy of data, or algorithmic bias reveal that there are no common standards. The NGT's reactive, petition-based structure leaves it even less capable of developing a body of consistent evidentiary doctrine, resulting instead in fact-specific reasoning without a systemic scope.

REMOTE SENSING AND SATELLITE IMAGERY IN ENVIRONMENTAL ENFORCEMENT

Remote sensing and satellite imagery have become the cornerstone of India's environment enforcement architecture, providing a scale, continuity, and objectivity that traditional inspection-based mechanisms cannot match (Rahman, 2007). Whether it is measuring forest cover and river pollution, or even ascertaining illegal mining through satellites, such technologies are being used more extensively than one can imagine by regulators and tribunals. However, their scientific nature, legal status, and evidential significance pose intricate issues at the heart of environmental adjudication and climate justice.

Remote sensing is the process of deriving information about the Earth's surface without direct physical contact, and it is done typically through sensors aboard satellite from a distance (Lein, 2009). These sensors provide information in different spectral bands, making it possible to detect land cover/land use status, vegetation removal, water quality indicators, and surface changes associated with mining and industrial activities. In India, governmental agencies such as Indian Space Research Organisation (ISRO) and the National Remote Sensing Centre (NRSC), as well as those private satellite operators, underpin environmental monitoring efforts.

One of the most important scientific parameters is space resolution, i.e., how small an object can be detected. High-resolution imagery can identify small-scale intrusions or mine pits, while low-resolution data is more suitable for macro-scale trend analysis. However, higher resolution tends to be at the expense of a lower temporal frequency and increased cost. Temporal resolution impacts the ability to observe transient or rapidly changing environmental violations. Between satellite passes, illegal mining or effluent discharge may occur, causing under-detection or a lack of evidence.

Remote sensing data is also limited by the atmosphere and environment. Cloud, haze and seasonal changes can mask land characteristics, especially in forested, monsoon-affected areas(Chen et al., 2022). Synthetic Aperture Radar (SAR)-type technologies alleviate some of these problems, but at the cost of interpretation that requires specialist knowledge. Misclassifications are exacerbated where spectral signatures overlap, such as between natural soil disturbance and mining.

Equally important are interpretive constraints. Satellite imagery does not "speak for itself;" it must be actively interpreted by humans or algorithms. There are many different ways to interpret such a dataset, and two analysts may come up with contradictory answers, according to insiders, particularly in the case of contested questions of forest rights or traditional land use. These limitations of science illustrate why satellite imagery should not be shifted from a monitoring measure to the basis for legal liability.

Legal Status of Satellite Imagery

In India, satellite imagery is considered an electronic record that falls under the category of 'Documentary Evidence' in evidence law and is therefore governed by the rules of relevance, authenticity, and admissibility. Authentication is a foundational requirement. Courts and tribunals must ensure that the image is authentic, unaltered, and correctly depicts the place and time it claims to be. It usually involves verifying the metadata, declaring the satellite along with its sensor characteristics, and describing the image processing method.

Expert evidence is crucial in turning satellite data into legally significant findings. Geospatial professionals must be prepared to discuss the resolution based on the available information, uncertainty, and interpretation biases. Without that expert testimony, satellite imagery is at risk of being dismissed as inconclusive or speculative. However, grounding in experts brings its own issues, such as competing readings and questions about institutional independence, which in fact recur where experts are linked to regulatory agencies (Ginzky, 2001).

The chain of custody of files in satellite imagery is an underdeveloped aspects of Indian environmental litigation to date. Satellite evidence differs from traditional physical evidence, which is frequently encountered and processed at multiple institutional levels. Creating a verifiable and complete chain, from collecting numbers to presenting them in court, is challenging when private vendors or various separate agencies are involved. Unfortunately, the lack of standardised protocols increases the likelihood of a challenge pertaining to trial evidence. Nonetheless, courts have also been wary of accepting satellite imagery as definitive evidence. This conservative attitude reflects scientific uncertainty and doctrinal reticence, especially in cases involving criminal punishment or high-stakes civil sanctions (Rychlak & Taylor, 2022).

The application of satellite imaging in illegal mining cases in Odisha and Jharkhand is indicative of both the potential and limitations of satellite imagery . Remote sensing has been used by regulatory agencies to monitor unauthorised excavation pits and encroachment beyond the permissible limit. Carrington, in many cases, has led to the imposition of fines and remedial notices. However, controversies have emerged over timing; people accused of violations claim that the imagery captured lawful activity or ongoing disturbances. Such cases emphasise the need to place satellite data into context with on-the-ground validation and documentary evidence.

In the Western Ghats, satellite imagery has helped uncover deforestation, encroachment and violations of eco-sensitive zone regulations. With the high-definition images, policymakers have been able to map forest loss and provide guidance on restoration as well. Forest-dwelling communities, for their part, have questioned these claims, asserting that the satellite imagery does not accurately capture traditional land uses and seasonal occupation. The dangers of satellite-enforced marginalisation are clear; such conflicts highlight the risk that satellite-enforced systems can side-line customary rights without integrated participatory verification.

Satellite maps are also being used to monitor river pollution, with close monitoring of the health of the Ganga and Yamuna. Pollutant plumes have been identified, drainage outlets mapped, and changes in water quality parameters have been recorded using remote sensing. These uses have provided the basis for judicial orders and policy proposals, such as orders to implement improved effluent treatment. However, the circumstantial nature of satellite-based pollution proxies must be confirmed by sampling and laboratory analysis, and they cannot stand alone as direct evidence.

Indian courts and the NGT have largely adopted satellite imagery in a pragmatic manner, acknowledging its usefulness while being cautious of its limitations. Satellite maps and photographs are generally considered to be supportive or supplementary evidence to findings based on visits, expert reports, or registers (Norris et al., 2016). Seldom does anyone treat them as dispositive on their own.

An underlying judicial concern is that of trustworthiness. The courts have emphasised the importance of transparent documentation regarding the image's origin, processing, and expert interpretation. Where such details are unavailable, satellite information has been given little value. A further difficulty arises from spatial and temporal distortion; images are commonly shown without a scale or explanation of the display delay.

This judicial scepticism is exacerbated in adversarial contexts, particularly where satellite imagery is opposed by parties affected by the proceedings. Conflicting expert testimony, allegations of evidence suppression and lagging claim suggests a lack of consistent evidentiary procedures. These are further complicated in an environment of rapid technological development that constantly outpaces the evolution of legal principles.

AI IN ENVIRONMENTAL CRIME DETECTION AND PREDICTION

AI has been used with growing regularity by the enforcement agencies in India as a response to the scale, complexity, and temporal urgency of environmental crimes. While satellite technology primarily records descriptive notes on the environment, artificial-intelligence-based systems work analytically and predictively. They crunch large data sets, often incorporating remote sensing, sensor networks, and legacy records of compliance, to identify patterns, flag threats, or predict areas at risk (Yetilmezsoy et al., 2011). While the ability to do so represents a tremendous step in regulatory capacity, it also poses profound legal and epistemic challenges, especially with regard to evidence, accountability, and climate justice.

The use of AI in environmental enforcement for India is primarily seen through three related domains: detection, prediction, and prioritisation. Machine learning models are employed to monitor land-use changes, identify illegal mining pits, forecast deforestation rates, and track industrial emissions. In controlling pollution in rivers, AI models utilise satellite imagery data, sensor readings, and records of historical pollution to forecast levels of contamination and pinpoint likely sources of discharge (Singh & Singh, 2025).

One strong application of AI is in land-use and land-cover change analysis, where satellite images are classified over a time period to identify non-permitted uses. These systems can detect small changes that go unnoticed by humans, such as gradual mining expansion or slow forest loss (Gu & Zeng, 2024). Similarly, pollution control authorities are increasingly using AI instruments to scrutinise data from continuous emissions monitoring and detect patterns of non-conformance.

Another important application is in the area of predictive enforcement. AI systems produce risk scores or "hotspot" maps that indicate where certain types of environmental breaches are likely to occur. They enable regulators to better target scarce inspection resources and reduce the need for reactive enforcement, shifting towards an anticipatory approach to regulation. In principle, this change improved regulatory efficiency and environmental protection (Berk, 2021).

However, these programs also expose structural deficiencies. AI models are extremely sensitive to the quality, representativeness, and completeness of the underlying data. In areas where monitoring is sparse or enforcement in the past has been lax, datasets may be both sparse and biased, resulting in naively biased predictions (Van Der Ven et al., 2024). It is especially worrying in environmentally sensitive and socially marginalised areas, where an incompleteness of information could be indicative of poor governance rather than environmental compliance.

The application of AI in environmental enforcement raises some basic questions concerning the admissibility of evidence under Indian law. BSA, 2023, is based on proven facts, direct evidence, or circumstantial evidence, and its probative value. AI predictions, unlike human ones, are by and large probabilistic; they represent a degree of confidence rather than an absolute truth. This epistemic divergence is problematic when AI-inferred findings are to be accepted in courtrooms.

One of the fundamentals requirements of admissibility is reliability. There must be evidence that courts can test and challenge. At the same time, AI systems, particularly those using complex machine learning models generate decisions that can be statistically robust but legally incomprehensible. The lack of a specific causal explanation that connects input data to the output decision makes it difficult for courts to assess general reliability.

Furthermore, AI results can be based on undisclosed or difficult-to-interpret training data. If data sources, error rates, and validation processes are not transparent, courts will have a low level of confidence in AI-generated content. It is particularly so in matters with potential adverse implications, such as sanctions, project termination, and criminal exposure.

Relevance to the facts in issue is also a consideration under Indian evidentiary law. AI predictions can serve as strong evidence that a crime is likely to have occurred, but they do not prove that anyone broke any laws at this time or place. For this reason, AI results are more easily described as investigative leads rather than evidence, until they are confirmed independently by another type of evidence, such as inspections, sampling, or documentary records.

One of the reasons judges might be slow to accept AI evidence is the issue of its algorithmic black box. A lot of AI systems operate as "black boxes"; even for their developers, the internal logic is often obscure or entirely opaque (Brevini, 2020). This opaqueness violates fundamental tenets of procedural fairness and natural justice, including the right of those accused to know and test the evidence against them.

Indian courts have traditionally been sceptical of scientific and technical evidence that cannot be subjected to any real scrutiny. This caveat is exacerbated in the case of AI because of proprietary algorithms, trade secrets, and claims to intellectual property that may discourage disclosure. In the absence of access to source code, training data, and decision rules, courts struggle to evaluate whether an AI system is biased, inaccurate, or contextually inappropriate.

Judicial hesitation is also influenced by concerns about the delegation of decision-making authority. When enforcement is driven by predictions from AI, there is a danger that the outputs of an algorithm replace human reasoning. It raises constitutional issues of arbitrariness and lack of application of mind particularly when such decisions affect fundamental rights or livelihoods.

The legal vacuum surrounding AI governance creates even more uncertainty. Unlike practitioners in jurisdictions where early articulations of algorithmic accountability frameworks are emerging, Indian law lacks benchmarks for the validation, audit, or contestation of AI systems deployed to enforce environmental norms. It leaves the position open for courts to tiptoe around these issues and more in favour of judicial finality.

As a conceptual matter, it is important to differentiate AI as a tool of investigation from AI as evidence. In the current traditional-legal-institutional context, AI in India primarily operates in this latter capacity.

Predictions made by AI are critical for pattern recognition, prioritisation of inspections and policymaking. They improve regulation and can be essential in the early identification of harm to the environment. However, when posited in adjudication, these predictions encounter formidable doctrinal obstacles. Given the probabilistic aspects of AI outputs (in addition to concerns about transparency and explainability), such products also have limited capacity to satisfy reasonable standards of proof.

With that said, AI predictions can become admissible when presented in a supportive or contextual manner. For instance, an AI-detected industrial pollution hotspot, when complemented by sampling and inspection data or reports, may reinforce the evidential matrix. Because of this kind of case, AI is not a substitute for traditional evidence, but rather an overlay.

A panoply of legal reforms would be required for AI predictions to be considered evidence in their own right. These include codified benchmarks for algorithmic trustworthiness, mandated reporting of validation metrics, and procedural protections that allow stakeholders to dispute AI outputs. Without such reforms, the deployment of AI evidence will introduce imbalances into both due process and climate justice, especially in forward-deployed communities that are already over-policed and overexposed to enforcement.

DIGITAL FORENSICS AND VERIFICATION OF ENVIRONMENTAL DATA

With the use of satellite imagery, AI analytics, drones, sensors and digital records on the upswing to enforce environmental laws in India (along with other laws), digital forensics has become a critical factor in turning technology's detection into legally acceptable proof. It offers methodological protections to ensure that electronic environmental data is accurate, reliable, and legally defensible. Without a strong basis for forensic validation, the data can be argued to be speculative or procedurally inadequate. Thus, digital evidence in environmental crime will be a function of the integrity of metadata, adherence to chain-of-custody, and the institution's ability to serve as a custodian (Suggs & Yarbrough, 2004).

Data from digital forensic science is based on metadata, which captures important information, including the time, location, sensor measures (configuration), processing, and storage (Casali et al., 2022). For environmental issues, metadata has become fundamental for proving evidence because it provides proof of the authenticity and integrity of digital data. However, environmental data undergo through several processing steps before it is available for use, including acquisition, analysis, and presentation. Both hash values and secure access controls serve as forensic tools to ensure data integrity, but they are used erratically by Indian regulatory agencies, that rely mainly on internal or external third-party reports without a complete set of forensic evidence. Consequently, courts may undervalue technically accurate data that is not adequately explained and tested with expert testimony.

The reproduction and transferability of digital data also complicate the preservation of an unbroken chain of custody. Satellite imagery and AI results can be shared among agencies, reside in multiple locations or come from private suppliers, thus leaving documentation and responsibility blind spots open. India has not developed standard practices for handling environmental digital evidence, and these practices have been ad hoc, more a matter of administrative convenience than forensically sound. Concerns about opacity and disputability arise when third-party contractors develop analytic guidelines without providing access to the data or methodologies used for rendering. As litigation reveals, such gaps can erode evidentiary strength and make a decision vulnerable to appellate challenge.

Broader institutional capacity weaknesses add to these challenges. Because many of our enforcement agencies lack digital forensic experts, they must rely heavily on outside professionals and have only a cursory understanding of the processes used to acquire technical evidence. Courts likewise struggle to develop a mechanism for understanding complicated digital claims in the absence of expertise, and thus, methods of evaluation are not uniform. The lack of specialised environmental digital forensic labs is another example of marginalising environmental crime within the overall criminal justice system. As a result, we are awash in environmental data gleaned from technological systems, yet without forensic oversight and institutional capability, our legal reliance on these data is compromised. It is therefore imperative that the capacity for digital forensics be built in order to support effective environmental enforcement and the pursuit of climate justice (Roy et al., 2025).

INSTITUTIONAL MECHANISMS AND GOVERNANCE GAPS

The use of tools such as remote sensing, AI analytics and digital forensics in the fight against environmental crime is highly dependent on the institutional setting within which they operate. Environmental governance in India is managed by various tribunals, ministries, and regulatory authorities that have overlapping responsibilities. Although this scattered system was intended for holistic supervision, it has resulted in coordination gaps and lapses in responsibility.

National Green Tribunal (NGT)

The NGT is at the forefront of India's environmental adjudicatory system (Rengarajan et al., 2018). Conceived to facilitate concentrated, quick, and expert administration of environmental justice, the NGT has been a vigilant pioneer in injecting scientific and technological elements into judicial adjudication. Being tasked by statute to give effect to principles such as sustainable development, the precautionary principle, and the polluter pays principle, it has allowed it to take a more flexible and purposive view of evidence, including technologically produced data.

A major institutional innovation of the Tribunal has been its willingness to consider non-traditional evidentiary sources, including satellite imagery, expert committee reports, and scientific assessments. This relaxation in the rules of evidence has enabled the NGT to address massive and intricate environmental violations that cannot be detected by ordinary inspections. The Tribunal has often relied on techno-scientific monitoring in cases of illegal mining, deforestation, and river pollution, through which patterns of damage and regulatory violations could be established (Roy et al., 2025).

However, that flexibility goes both ways, leaving the institution open to accusations of exploitation. The role of expert committees and external technical reports in the work of the NGT raises concerns about standardisation, transparency, and accountability. The Guidelines are not a set of common rules for the collection, verification, and presentation of technical information to the Tribunal. As such, the quality and dependability of evidence might differ quite considerably from case to case. Although the NGT has tried to strike a balance between speed and science, the absence of clear evidence markers allows way for inconsistent application as well as possible appeals.

Additionally, the Tribunal's ability to analyse complicated technology evidence is limited by disparities in resources and technical expertise among its members. The NGT has members with expertise and experience; however, the new technology-intensive world of AI and high-resolution remote sensing requires continual training. In the absence of formal institutional structures for technical training and guidance, it may be challenging to maintain the Tribunal's aspirational identity as a science-informed adjudicator.

Ministry of Environment, Forest and Climate Change (MoEFCC) and State Pollution Control Boards

The MoEFCC, along with the State Pollution Control Boards, serves as the executive arm for environmental governance in India. These organisations are concerned with policy development, regulation, monitoring, and enforcement. Over the last few years, the MoEFCC and SPCBs have, to a large extent, resorted to technological solutions as a complimentary means to the conventional regulatory approach.

At the central level, while progress has been promising, it has not progressed significantly beyond initiatives such as satellite-based forest cover monitoring, e-based environmental clearance mechanisms, or even electronic detailing of grievances and complaints. These are small steps towards acknowledging that technology can make us more transparent, responsive, and efficient. Yet, the translation of policy purpose into effective implementation is patchy. New technologies are also introduced without sufficient integration to existing legal and administrative structures, with the consequence that implementation is fragmented.

State Pollution Control Boards (SPCBs) are in a worse plight. A significant proportion of SPCBs operate under extreme resource constraints, including inadequate skilled personnel and obsolete infrastructure. Some boards have implemented continuous emission monitor readers and digital reporting, but still lack the capacity to understand or act upon the data received. This inequality results in a fragmented policy environment, with states having different levels of compliance with and enforcement of their environmental laws (Chungyalpa, 2020).

Another governance vacuum is created by the mismatch between data production and the enforcement. If technological oversight identifies violations, enforcement may be slow or diluted by bureaucratic lethargy, political influence, and legal complexities. Without established guidelines that connect digital evidence to enforcement outcomes, the deterrent impact of technology monitoring is compromised.

Cross-Agency and Jurisdictional Failures

Environmental crime in India is rife with disregard for administrative and jurisdictional constraints (Roy, 2020). A single illegal mining activity may cross multiple districts or states. River pollution commonly involves upstream and downstream authorities, whereas deforestation sometimes cuts across forest, revenue, and local governance. Remedying such harms requires collective efforts of the agency and government level. Cross-agency coordination remains the weakest link in India's environmental governance mechanism. Information produced by one agency is not easily accessible to others, and there is no common place to share or integrate technological evidence. Such a fragmented approach results in duplication of effort, enforcement gaps, and missed opportunities for comprehensive action. Jurisdictional fragmentation also complicates accountability. When more than one agency has concurrent or ambiguous jurisdiction, responsibility for enforcement may be diffused. This diffusion is problematic, especially for advanced technologies, where issues of data possession, control, and interpretation may arise. Without established lines of hierarchical authority, technological evidence runs the risk of being viewed as a source of information rather than a trigger for legal action.

These governance failures have wider implications for climate justice. By and large, it is the already marginalised communities who bear the brunt of environmental violence, and they are also the ones most hit by institutional failures. The drag is felt in the persistence of environmental violations and problems, despite social and ecological vulnerabilities, because few are held accountable for sharing responsibility when agencies repeatedly fail to coordinate or take robust action based on technological evidence.

JUDICIAL TREATMENT OF SATELLITE AND GEOSPATIAL EVIDENCE IN INDIAN ENVIRONMENTAL ADJUDICATION

Indian environmental courts, particularly the National Green Tribunal and the Supreme Court of India, have progressively engaged with satellite imagery and geospatial technologies in resolving ecological disputes. However, judicial acceptance has not been uncritical; instead, courts have developed a cautious but evolving evidentiary approach grounded in verification, institutional credibility, and corroborative validation.

A significant illustration is found in *Samaj Parivartana Samudaya v. State of Karnataka*, AIR 2013 Supreme Court 3217, which concerned extensive illegal iron ore mining in forest areas. Satellite imagery and geo-referenced boundary mapping were used to identify encroachments beyond licensed areas. The Supreme Court did not treat the imagery as automatically determinative; rather, it relied upon reports prepared by the Central Empowered Committee and forest authorities who had cross-verified satellite findings with ground inspections. The Court accorded substantial evidentiary weight to the technological material because it was institutionally authenticated, methodologically explained, and corroborated through field validation. The decision demonstrates that technological evidence becomes legally persuasive when embedded within structured expert evaluation.

Similarly, in the continuing mandamus proceedings of *T.N. Godavarman Thirumulpad v. Union of India*, AIR 2005 Supreme Court 4256, satellite-based forest cover assessments prepared by the Forest Survey of India played a critical role in identifying forest land and monitoring diversion. While the Court acknowledged the scientific credibility of remote sensing data, it emphasised that digital classification must align with statutory definitions under forest law and must not replace legal interpretation. Satellite imagery was thus treated as a powerful regulatory and monitoring instrument, but not as self-proving evidence detached from statutory context and on-ground verification.

In pollution control jurisprudence, including *Paryavaran Suraksha Samiti v. Union of India*, (2017) 5 SCC 326, courts have relied upon geospatial mapping and monitoring data to assess compliance with effluent treatment obligations. Remote sensing and GIS-based identification of discharge points informed judicial directions, yet the Court insisted upon compliance reports, regulatory certification, and accountability mechanisms before issuing binding orders. The evidentiary value of technological outputs was therefore reinforced through procedural safeguards and documentary substantiation.

The National Green Tribunal has also relied extensively on time-series satellite imagery in cases involving illegal sand mining, riverbed encroachments, and coastal regulation violations. In several matters, geo-spatial overlays comparing pre- and post-activity landscapes were used to establish environmental degradation. Nevertheless, the Tribunal has clarified that satellite images submitted by private parties cannot be accepted at face value without authentication from recognised governmental remote sensing agencies or technical experts. Where metadata was incomplete, time-stamping unclear, or chain of custody unestablished, the Tribunal has either reduced the evidentiary weight of such material or required further verification. This demonstrates a jurisprudential balance between embracing technological capability and maintaining evidentiary integrity.

Collectively, Indian adjudicatory practice reveals that satellite and geospatial evidence is neither rejected nor treated as infallible. Its persuasive force depends upon methodological transparency, expert authentication, institutional certification, and corroboration through ground-truthing. Courts have thus positioned technological material as strong corroborative evidence capable of shaping environmental outcomes, while simultaneously insisting upon procedural safeguards consistent with evidentiary fairness and rule-of-law principles.

RIGHTS, PRIVACY, AND EQUITY CONCERNS

The integration of remote sensing, AI, and digital forensics in Indian environmental governance raises serious questions about rights, privacy, and equity. While these technologies present the potential to enhance enforcement and environmental protection, they also create risks of surveillance abuse, unfair treatment, and procedural injustice, especially among disadvantaged communities (Jha, Markandey, Gupta, & Mishra, 2025).

The right to life and personal liberty, under Article 21 of the Indian Constitution, has been broadly interpreted by the courts to include the rights to privacy, bodily integrity, and a dignified life. Environmental surveillance using such technologies can require the collection of large datasets about individuals and communities. Satellite images can monitor land use at the household/plot level, and AI-assisted analysis can identify patterns of action/resource use or movement in forested/riverine areas.

This surveillance, although environmental in nature, raises issues of informational privacy. Metadata harvesting, geotagged photos, and AI-based surveillance may collect a person's data without their consent, which can be seen as a violation of Article 21. The Supreme Court in *Justice K. S. Puttaswamy (Retd.) and Anr. v. Union of India and Anr.*, AIR 2017 Supreme Court 4161, recognised that privacy is an inalienable part of the right to life, and state action that violates it must meet the parameters of legality, necessity, and proportionality. However, India does not have a comprehensive regulatory regime that specifically deals with the privacy aspects of environmental surveillance technology. Moreover, the danger is not limited to personal privacy. The process of data gathering in environmental science often involves natural resource-based communities, the collective behaviour and patterns of those who are available for scrutiny by the state. When unaccompanied by appropriate safeguards, this surveillance can degenerate into coercive monitoring, disproportionately impacting communities that depend on forest products, riverine resources, or common lands.

Indigenous communities, forest dwellers and other marginal rural inhabitants are particularly vulnerable to distributive and procedural injustices in technologically mediated environmental governance. Satellites and AI can be used to predictively police these areas for "encroachments" or illegal resource extraction. Although aimed at preserving ecosystems, these tools might render traditional and subsistence livelihoods illegal.

Algorithmic bias exacerbates these vulnerabilities. AI models trained on incomplete or biased data may systematically overlook harms in the environment caused by actors with power, like industrial polluters and disproportionately identify routine activities that local communities undertake as violations (Tariq, 2025). For instance, subsistence activities based on forests, such as the collection of firewood, hunting or limited shifting cultivation, could be erroneously classified as illegal deforestation. This structural bias risks reinforcing historic inequalities, limiting access to resources, and subjecting marginalised groups to disproportionate legal consequences.

Procedural participation also raises issues of equity. Those who bear the brunt of enforcement led by surveillance are generally excluded from consultation or avenues for recourse. Enforcement action notices can be ineffective; a notice of enforcement action may not be effectively communicated, and technical reports might be unavailable because of language or technical difficulty. As a result, the very populations the law is designed to protect may be further victimised as power disparities are incorporated into technology uptake.

In addition to privacy and equity concerns, the use of AI and satellite surveillance in environmental law presents profound due process problems (Kumar et al., 2025). Procedural protections, such as notice, the right to contest evidence, and the right to be heard, are fundamental aspects of both domestic law and international human rights standards. However, much of the mechanical data that is becoming available challenges these truisms.

To begin with, there are limitations to the transparency of AI algorithms and the interpretation of satellite data. Agencies might use in-house software, black boxes, or third-party technical reports, leaving affected individuals to find it hard to verify any evidence on which they are being judged. Without access to the base methodologies or raw data, contesting enforcement actions becomes virtually impossible. Second, remedies are not improved in terms of justice and time value. When environmental offences are identified by AI or satellite imagery, the enforcement process can move quickly, leaving little time for communities to act or appeal. This is a danger to summary penalisation and diminishes both material justice and procedural justice. Third, the lack of explicit legislative or regulatory guidance for technology-mediated evidence supports various techniques. Courts or tribunals could also apply different standards when assessing AI-predicted outcomes, such as remote sensing product outputs, resulting in unequal treatment of the same matter. This legal ambiguity further undermines the capacity of marginalised communities to effectively protect their rights.

Finally, the ethical precept of participatory governance in environmental law, exemplified not only by instruments like the Rio Declaration but also by Indian constitutional practice, mandates meaningful participation for communities affected by environmentally related decisions. If left unaddressed, technology-mediated enforcement runs the risk of overlooking these communities and limiting their ability to influence environmental policies and protections that impact their well-being.

TECHNOLOGICAL EVIDENCE AND ENVIRONMENTAL LIABILITY: ADJUDICATORY IMPACT AND GOVERNANCE IMPLICATIONS

The integration of remote sensing, satellite imagery, artificial intelligence, and digital forensic reconstruction into environmental adjudication in India represents more than a technological upgrade; it marks a structural reconfiguration of how environmental liability is established, contested, and enforced. While technological tools significantly enhance detection capacity, their true legal significance lies in how they reshape evidentiary thresholds, burden allocation, and remedial outcomes. Indian environmental jurisprudence, grounded in the doctrines of strict and absolute liability and animated by the precautionary and polluter pays principles, provides a fertile normative framework within which technological evidence operates (Dhaka, 2025). High-resolution satellite imagery and geospatial time-series analysis, for instance, convert ecological degradation into quantifiable and temporally traceable data, enabling tribunals and courts to move from suspicion-based reasoning to documentary verification. In civil and regulatory proceedings, such evidence often satisfies the standard of preponderance of probabilities, strengthening the basis for environmental compensation and interim injunctive relief. However, in criminal prosecutions where proof beyond reasonable doubt is required, algorithmic outputs cannot independently establish culpability; they must be corroborated, authenticated, and subjected to expert scrutiny. Thus, technology alters evidentiary architecture but does not displace foundational procedural safeguards.

Judicial and tribunal practice illustrates this recalibration. In proceedings concerning illegal mining and deforestation, satellite imagery has been relied upon to establish spatial expansion beyond approved boundaries, quantify environmental harm, and justify application of the polluter pays principle. Similarly, real-time emission monitoring systems supported by AI analytics have been treated as prima facie indicators of regulatory non-compliance, prompting directions for compensation, closure, or technical audits. These case-based experiences reveal that technological evidence most strongly influences interim and regulatory outcomes such as suspension orders, environmental compensation, and monitoring directions rather than criminal conviction rates. The evidentiary clarity offered by visual and data-driven documentation reduces judicial hesitation in granting preventive relief, thereby operationalizing the precautionary principle in anticipatory form. Yet, these gains are mediated by institutional capacity: enforcement agencies frequently lack the technical expertise to interpret complex datasets, calibration disputes undermine reliability, and appellate scrutiny often demands conventional corroboration. The empirical pattern therefore reflects cautious judicial receptivity rather than technological determinism.

At the same time, the governance implications of algorithmic and satellite-based enforcement require deeper constitutional engagement. The deployment of remote sensing technologies by regulatory authorities implicates privacy, proportionality, and legality requirements. Any surveillance-oriented environmental monitoring must be anchored in statutory authorization and satisfy necessity and proportionality standards, particularly where drone or AI-based tools capture data relating to private land use. Moreover, reliance on proprietary or opaque algorithms raises administrative law concerns regarding transparency, non-arbitrariness, and reasoned decision-making. If environmental liability or regulatory prioritization is influenced by algorithmic classification systems whose logic remains undisclosed, affected parties may be deprived of meaningful opportunity to challenge adverse findings, thereby undermining principles of natural justice. The absence of standardized certification protocols for AI-based environmental analytics further complicates evidentiary integrity, raising questions about chain of custody, validation methodology, and judicial comprehension of technical material.

Equality considerations also demand scrutiny. Data-driven enforcement strategies may inadvertently reproduce structural disparities by disproportionately targeting regions characterized by informal extraction economies or marginalized forest-dependent communities. Satellite classification errors or algorithmic bias in land-use identification can lead to mischaracterization of customary land practices as illegal encroachments. Without participatory verification mechanisms and contextual sensitivity, technological governance risks privileging visual data over lived socio-legal realities. In this sense, climate justice must be assessed not merely in terms of improved detection of environmental harm but also in terms of distributive fairness and procedural inclusivity. Technology can enable accountability, but it can also intensify asymmetries if institutional safeguards remain underdeveloped.

Furthermore, it is essential to temper technological optimism with institutional realism. Detection does not automatically translate into prosecution; quantification of harm does not guarantee recovery of compensation; predictive modelling does not substitute for regulatory will. State Pollution Control Boards and local enforcement bodies frequently face resource constraints, inadequate digital literacy, and limited forensic infrastructure. Courts, while increasingly receptive to technological material, may lack structured training in algorithmic evaluation, leading to either over-reliance or excessive skepticism. In this context, technological tools should be understood as evidentiary enhancers rather than transformative solutions. Their capacity to advance climate justice depends upon parallel reforms in procedural codification, expert accreditation, transparency mandates, and judicial training.

Ultimately, the incorporation of remote sensing, AI, and digital forensics into environmental crime investigations reshapes the evidentiary ecosystem of environmental law in India. It strengthens the factual foundations upon which precautionary and compensatory doctrines operate, expands the state's capacity for anticipatory governance, and enhances the visibility of ecological harm. Yet its legitimacy and effectiveness are contingent upon constitutional compliance, algorithmic accountability, institutional competence, and equitable implementation. The future of technology-enabled environmental adjudication therefore lies not in uncritical adoption but in calibrated integration where scientific sophistication is matched by legal rigor, procedural fairness, and governance transparency.

COMPARATIVE PERSPECTIVES: EU AND US

European Union

The European Union has established a highly sophisticated system for monitoring environmental issues, which is integrated with efforts to deploy new technologies and legal mechanisms. The EU's flagship Earth observation initiative, the Copernicus Programme, offers high-resolution satellite images of different environmental parameters such as land use, water quality, deforestation, and pollution patterns. Copernicus data are increasingly used by EU member states, both for regulatory compliance and in the assessment of policies and public services (Jutz & Milagro-Perez, 2020).

The Copernicus Programme operates in coordination with the European Space Agency (ESA), embedding satellite monitoring within a structured regulatory and administrative architecture rather than treating it as isolated evidentiary material. Environmental compliance assessments, agricultural subsidy verification, deforestation monitoring, and pollution tracking are institutionally integrated with geospatial validation frameworks, ensuring traceability and standardized metadata documentation. In practice, satellite-derived evidence in the EU is not merely observational but forms part of systematic regulatory compliance systems supported by technical certification and quality assurance protocols (Thépaut et al., 2018).

EU courts approach admissibility of satellite imagery and AI-assisted analytics through principles of administrative legality, proportionality, and procedural fairness. Emphasis is placed on demonstrable data integrity, reproducibility of results, and transparency in algorithmic processing before evidentiary weight is assigned. The recently adopted EU AI Act reinforces this approach through a risk-based regulatory model that classifies certain environmental monitoring systems as high-risk, thereby imposing obligations of documentation, explainability, human oversight, and bias mitigation (Butt, 2024). Complementing this framework, the General Data Protection Regulation (GDPR) regulates surveillance-oriented data collection, particularly where environmental monitoring intersects with identifiable individuals or private land use (Goddard, 2017). Together, these frameworks illustrate that the EU integrates satellite monitoring within a structured regulatory architecture grounded in transparency and accountability.

European Union Member States apply very strict standards of admissibility and reliability to satellite and AI-produced evidence. Regulatory frameworks must mandate the certification of data sources, audit trails for AI algorithms, and the incorporation of metadata to verify results (Seizov & Wulf, 2020). The courts and lawmakers, however, consider satellite imagery as corroborative evidence, often mandating expert authentication and proof of the chain of custody. The EU focuses on algorithmic responsibility and transparency through devices such as the AI Act, which imposes obligations on high-risk AI systems, including environmental monitoring devices (Laux, Wachter, & Mittelstadt, 2024). Agencies need to show that they have deployed explainable, risk-assessed, and bias-minimised algorithms. It would prevent predictive analytics from unfairly impacting disadvantaged groups, and such outcomes could be challenged in court or otherwise invalidated.

Public participation is a basic principle of environmental governance in the EU (du Plessis, 2008). Environmental data can be made available to communities during enforcement or planning. This level of openness also align with procedural and distributive justice, ensuring that technologically mediated monitoring will not result in de facto marginalisation among affected populations.

United States

The US government has a rich tradition of incorporating technology and enforcement in the environmental space, notably through the EPA and state compliance regimes (Gunningham, 2009). For clean Air Act and Clean Water Act compliance, hazardous waste management, and land-use violations, remote sensing, AI analytics, and geospatial modelling are widely applied to monitor adherence.

Under the Federal Rules of Evidence, particularly Rule 702, expert testimony based on scientific or technical knowledge must demonstrate reliability, relevance, and methodological soundness (Oh, 1997). The Daubert standard further requires courts to evaluate whether the underlying methodology is testable, peer-reviewed, subject to known error rates, and generally accepted within the scientific community (Kurtz & Pintarelli, 2024). Consequently, satellite imagery, AI models, and digital forensic outputs are admissible only when supported by demonstrable scientific validation and expert authentication.

The Daubert standard guides US courts. Under this rule, scientific evidence must be both reliable and relevant to the matter at hand, and it must have undergone peer review. If they can fulfil those criteria, satellite images, AI models, and digital forensics are legitimate. Probabilistic outcomes are frequently acceptable as leads in an investigation, but they must be confirmed through field verification or expert testimony in order to constitute legally binding evidence (Walker & Monahan, 1996).

In environmental enforcement, the Environmental Protection Agency (EPA) and state agencies routinely rely on remote sensing, geospatial modelling, and digital monitoring in proceedings involving hazardous waste, land-use violations, and water pollution. However, US courts insist on preservation of metadata, documented chain-of-custody procedures, and compliance with National Institute of Standards and Technology (NIST) digital forensic guidelines before according evidentiary weight. Emerging judicial caution toward AI-generated outputs further underscores the requirement of methodological transparency and explainability. Courts increasingly scrutinize algorithmic systems to ensure that conclusions are auditable and not products of opaque "black-box" processes (Palmiotto, 2020).

The US places a premium on methodological transparency, as agencies document the sources of information, the algorithms used to process the data, and the assumptions made by models. AI-drawn predictive models are auditable, and stakeholders can contest the method or result in administrative hearings.

EPA initiatives promote citizen science and community monitoring to supplement technological enforcement efforts. Sharing geospatial data with the public fosters accountability and communities that are able to participate in environmental decision-making.

Lessons for India

The comparative analysis demonstrates that the EU integrates satellite monitoring within a structured regulatory architecture supported by legislative oversight and algorithmic accountability, while US courts require demonstrable methodological reliability before admitting technologically generated outputs. In both jurisdictions, transparency, validation, and procedural safeguards are prioritized over mere technological capability. These models indicate that effective reform in India requires codified evidentiary standards, institutional validation mechanisms, and judicial scrutiny frameworks to ensure that satellite and AI-generated evidence withstands constitutional and procedural review.

The EU and US experiences have lessons directly applicable to India. In both jurisdictions, more rigorous evidentiary tests have been developed for using satellite imagery and outputs generated by AI, such as certification rather than metadata conformity, and expert validation of AI output, indicating a comparable requirement in India to develop admissibility norms for technological evidence. They also stress the need for algorithmic transparency and accountability wherein AI systems must be explainable and auditable in order to prevent bias, guarantee fairness and enable meaningful contestations by affected parties, an area India is yet to institutionalise. A system of governance rooted in rights also defines these systems, including mechanisms based on participation for the protection of community rights through consultations, consent principles and access to information, especially for marginalised communities. On the institutional front, environmental agencies in the EU and the US function within a clearly defined mandate that minimises fragmentation, which means India could utilise dedicated technological evidence units in institutions like the NGT and SPCBs. Their models of cross-boundary environmental data sharing (specifically in the EU) have been useful too, for our efforts to consider helping to manage shared rivers and regional pollution in India. Finally, the long-term commitment to capacity building, including ongoing training of judges, regulators and technical staff, signals the importance of ensuring institutional actors are able to meaningfully interpret and utilise satellite and AI-derived environmental data.

EVIDENTIARY TENSIONS AND LEGAL GAPS IN INDIA

As India becomes increasingly more reliant on remote sensing, artificial intelligence, and digital forensic tools for environmental crime investigations, several legal and evidentiary obstacles have emerged. While these technologies significantly improve the ability to monitor and detect, their admissibility, reliability, and use are not adequately addressed within the current system. Moreover, these tensions are not only technical; they intersect with a range of much broader issues, including those associated with climate justice, institutional competence, fairness in procedures, and human rights, which make their direct application in environmental enforcement a complex and contested terrain.

One key problem is the verification, authenticity and credibility of machine-manufactured evidence. While regulatory authorities and judicial tribunals, like the National Green Tribunal, often use satellite images, AI products, and digital forensic reports as evidence, there is a lack of consistency in their legal status. Satellite data is frequently considered corroborative evidence, rather than primary sources; however, debates often challenge issues such as spatial resolution, temporal accuracy, and interpretive significance in regions where degradation has been gradual or covert. The lack of established chain-of-custody practices further erodes evidentiary value, particularly when multiple agencies have a role in collecting and analysing the data. The same issues can be seen in AI-based evidence, where the likelihood of illegal mining or deforestation is calculated and confronted by the uncertainty and opacity of the model. Courts grapple with reconciling these predictive outputs with a focus on factual proof in the BSA, 2023, and face a challenge because proprietary algorithms are difficult to explain, let alone validate by an expert. Digital forensic evidence is also plagued by uneven practices in metadata certification and data management, which are exacerbated by the absence of properly trained personnel to authenticate intricate geospatial and algorithmic datasets.

Of course, artificial intelligence, in particular, presents us with our old friend, prediction-proof dichotomy. While AI tools are useful for flagging likely hotspots and patterns of environmental violations, they cannot, on their own, prove that an offence has taken place. The Indian system demands direct, goal-oriented evidence of advocacy as a necessary evil, thus creating dissonance between preventive enforcement programs and tests of judicial admissibility. This problem is exacerbated by the opacity of many AI systems, which act as "black boxes" and give little explanation for their reasoning. Consequently, it is understandable that the judiciary has been unwilling to base decisions on such evidence in the absence of interpretable models and overqualified expert interpretation. Moreover, bias risks are significant, as AI systems trained on incomplete or biased datasets can lead to adverse equity impacts including tribal populations and forest dwellers potentially being targeted or marginalised. In the absence of legislative acceptance and procedural direction, AI predictions, therefore, should function not as replacements for legally admissible evidence but as investigative pointers.

These evidentiary complications are further confounded by data sovereignty and transboundary environmental governance concerns. A significant amount of environmental damage in India, especially that impacting crucial river systems like the Ganga, Kosi, and Brahmaputra, has cross-border aspects that necessitate the use of foreign satellite data, which we bring to the collaboration table. The reliance on such data also poses issues of authenticity, chain of custody, and legal admissibility, particularly when that imagery or analysis is restricted by national security controls or proprietary restrictions imposed by foreign-sourced entities. India does not have an overarching law that covers the cross-border sharing of information, verification, and dispute resolution in environmental enforcement. Compartmentalised jurisdiction, along the lines of the Ministry of Environment, Forest, and Climate Change, State Pollution Control Boards, and the National Green Tribunal, tends to create additional hurdles when cross-border evidence is contested.

In sum, such tensions reveal that even advanced technology masks but cannot reverse the structural deficiencies that fuel the cauldron of environmental governance in India. Problems related to authenticity and integrity undermine the legal value of digital evidence. Additionally, AI is a predictive system; hence, it has to be involved with primitive existing laws for both proof standards and due process. Cross-border and data sovereignty issues need a mutual legal, institutional, and diplomatic response. It is crucial to work on bridging these gaps if technological interventions are to play a meaningful role in credible environmental enforcement and in the pursuit of climate justice more generally.

EXPLAINABLE AI AS A BRIDGE BETWEEN ALGORITHMIC GOVERNANCE AND NATURAL JUSTICE

The concerns associated with "black-box" artificial intelligence systems have prompted increasing scholarly and regulatory attention toward Explainable AI (XAI) as a technical-legal bridge between algorithmic governance and procedural fairness. Explainable AI refers to model architectures and post hoc interpretability techniques that render algorithmic decision-making processes transparent, intelligible, and reviewable. Through mechanisms such as feature-importance mapping, decision pathway tracing, sensitivity analysis, and model documentation, XAI frameworks enable stakeholders to understand how specific inputs lead to particular outputs (Vitale et al., 2025).

In the context of environmental enforcement, where AI systems may classify land-use violations, predict pollution risks, or identify non-compliant industrial units, explainability assumes constitutional significance. Transparency regarding data sources, weighting parameters, and error margins enhances auditability and allows independent technical verification. Such interpretability strengthens compatibility with judicial review, as courts can meaningfully assess whether the methodology underlying algorithmic conclusions is rational, proportionate, and non-arbitrary.

The principles of natural justice particularly *audi alteram partem* (the right to be heard)require that affected parties be informed of the basis of adverse decisions and be given a fair opportunity to contest them. Where environmental liability or regulatory penalties rely on opaque AI classifications, the inability to interrogate the logic of the system may undermine reasoned decision-making and procedural fairness. Explainable AI frameworks help satisfy these requirements by enabling affected parties to challenge adverse findings, question underlying assumptions, and seek expert rebuttal where necessary.

Without explainability, AI-generated environmental risk assessments risk violating core principles of procedural fairness, especially if predictive analytics directly influence enforcement orders, compensation calculations, or criminal investigation triggers. Embedding explainability standards within environmental governance thus becomes essential not merely as a technical improvement, but as a constitutional safeguard ensuring transparency, accountability, and meaningful participation in technologically mediated regulation.

POLICY AND LEGAL REFORM RECOMMENDATIONS

Although remote sensing, AI, and digital forensics offer transformational potential for the detection of environmental crime in India, the traction of these in practice is constrained by evidentiary, institutional, and ethical challenges. In order to ensure that new technologies serve climate justice, rather than creating legal and social rifts, we need a wide range of reforms at the both policy and legal levels. These recommendations cover five key areas: admissibility, accountability for algorithms, institutional development, rights-based safeguards, and cross-border environmental data governance.

Admissibility Standards

One of the prerequisites for technology to play an effective role in environmental litigation is establishing criteria or standards for admissibility. Indian courts and tribunals, including the NGT, also make such ad hoc findings on satellite images, AI outputs, and digital forensics. To fill this void in the law, the BSA, 2023, should be amended to include clear guidelines for authentication and a reliable chain of custody documentation. Satellite imagery, drone photographs, and AI-produced reports should include standardised metadata on the certification of the data source, validation of the source, and temporal accuracy. In addition, predictions made by the AI model should be treated as leads in an investigation, which are dependent on verification through experts and validation of the algorithms, as well as corroboration with physical or direct evidence. It will establish sound legal standards that courts and regulators can apply to technology-based evidence, which means less dispute over whether something is genuine or not.

Algorithmic Accountability

Detecting pollution hotspots, predicting deforestation, or illegal mining are now increasingly delegated to AI applications in the field of environmental monitoring. However, machine-learning models can be opaque, and concerns surrounding fairness, reliability and bias emerge. What India needs is a law on algorithmic accountability that enforces AI systems used in enforcement to be auditable, explainable and transparent. It would consist of obligations for documentation of training data, performance metrics of the models, and validation reports that could enable courts and regulators to evaluate the accuracy (or reliability) of the bill. The compliance of public-interest algorithms can be independently audited by government-appointed or accredited technical experts, and the algorithms concerned should be made available for examination by civil society and academics. Algorithmic responsibility also requires procedures to identify and correct bias, especially when underprivileged or indigenous communities are more affected by predictive surveillance. Environmental governance can further promote citizen trust and mitigate tech-enabled injustices by integrating ethical AI principles, such as fairness, transparency, and inclusivity.

Institutional Strengthening

The technological fixes are only as strong as the ability of institutions to process, interpret, and act on data. The NGT, MoEFCC, and SPCBs are still in a nascent phase, with the adoption of AI tools based on satellite data, because of a lack of funds, trained personnel, and uncoordinated efforts. Building institutional capacity involves creating specialised technology-evidence units within these agencies, staffed with geospatial analysis, AI and digital forensics experts. The training of judges, investigators, and regulators can be as important as the technical evidence when it comes to interpreting it, understanding algorithmic constraints, or identifying procedural safeguards. In addition, interdepartmental coordination mechanisms should be institutionalised to facilitate smooth data sharing, verification processes, and enforcement actions between the central, state, and local levels. These steps will improve technology-based environmental enforcement by making it more effective, efficient, and credible.

Rights-Based Safeguards

Technology can be helpful in monitoring and enforcing preventive measures. However, it also yields potential downsides, as privacy, autonomy, and a sense of engagement are also affected by it, especially amongst marginalised populations. Protections premised on rights must prevent such surveillance, which would disproportionately affect indigenous peoples, forest-based communities or rural populations. There would be a legal requirement for prior notice, consent, and participatory processes before AI or satellite monitoring can be deployed in sensitive areas. Finally, technology-affected communities need mechanisms to make and review claims about the data used for predictive AI, dispute the outcomes of AI predictions, and be involved in the decision-making. Social impact assessments of environmental AI tools provide a means to anticipate and address unintended harms, ensuring that technological development remains consistent with principles of equity, procedural justice, and climate justice.

Cross-Border Environmental Data Governance

Many environmental issues, including pollution of transboundary rivers and regional emissions, cross national borders and require cooperation with other countries. India needs to establish laws and mechanisms to integrate foreign satellite and AI data for domestic enforcement, without compromising data sovereignty and the principle of national security. Such mechanisms could involve reciprocal arrangements with neighbouring states, standard protocols for data verification, and institutional protocols for joint monitoring and enforcement. Combining all these elements, the creation of a National Environmental Data Authority would simplify the coordination of other existing data collection activities controlled by domestic agencies, along with several transboundary inputs and international satellites. Such a regime would enhance cross-border accountability, legal admissibility, and prospects for climate justice prospects, while also establishing India's leadership in regional environmental governance and collective action on climate change.

CONCLUSION

The convergence of environmental degradation, societal vulnerability to climate change, and rapid technological advancements presents India with the chance for long-term opportunities, but also poses the task of effective environmental governance. Technologies such as remote sensing and satellite imagery, artificial intelligence, and digital forensics can revolutionise the detection, monitoring, and prosecution of environmental crime. These technologies enable the surveillance of illegal mining, deforestation, and industrial pollution of large river systems at a scale that has been historically difficult to achieve through conventional enforcement systems, due to limited geographical coverage, jurisdictional inconsistencies, and evidentiary limitations. Beyond detection, such technological evidence increasingly influences regulatory liability, interim relief, and environmental compensation by strengthening application of the precautionary and polluter pays principles.

Nevertheless, technology alone is insufficient to deliver climate justice. Its impact is shaped by the legislative, institutional, and moral context that governs its application. Under the Indian legal system, which holds unambiguous competence over environmental disputes, it is evident that Indian courts and tribunals, including the NGT, are still confronted with challenges in ascertaining to what extent technological evidence products can be considered authentic, reliable, and admissible by law. The predictive outcomes of AI systems regularly fail to meet the burden of proof mandated under the BSA, 2023, and deficiencies in chain-of-custody procedures and forensic verification further erode judicial confidence. While tribunals show flexibility in civil and compensation matters, criminal prosecutions continue to require strict corroboration and expert validation of digital evidence. The lack of a clear legal framework for AI and information from the digital environment also increases the chances that law enforcement will be applied in an uneven way and weaken the public trust.

Ethical considerations when deploying technology are just as important. In particular, surveillance technologies may have negative implications for the privacy and rights of indigenous peoples, forest dwellers and other marginalised communities who bear the brunt of environmental harms. Algorithmic unfairness, explanation opacity, and uneven data coverage account for the effects of existing social biases, unless transparency, accountability, and procedural fairness are actively implemented in enforcement. In addition, reliance on opaque algorithmic systems without clear statutory standards raises concerns of proportionality, fairness, and meaningful opportunity to challenge digital findings.

Comparative experiences from the EU and the US underscore that there is no alternative to connecting technology uptake with strong evidentiary standards, algorithmic accountability, and rights-based safeguards. For India, these hybrid legal-technical models can pave the way for overcoming the divide between invention and judicial acceptability, while making sure that technology offers sustenance to environmental justice. Indian case experiences in illegal mining surveillance and emission monitoring indicate that technology strengthens detection and regulatory intervention, though long-term enforcement still depends on institutional capacity.

The institutions need to be empowered and coordinated in this effort. Focused technological evidence units in the Ministry of Environment, SPCBs and NGT, coupled with training of judges and regulators, will facilitate better understanding and maximise the use of digital-based data. When designed alongside participatory safeguards, grievance mechanisms and cross-border data-sharing agreements, such reforms can keep technology within a regime of accountability. Without such institutional strengthening, technological detection may not translate into effective prosecution or sustained compliance.

Comparative experiences from the EU and the US underscore that there is no alternative to connecting technology uptake with strong evidentiary standards, algorithmic accountability, and rights-based safeguards. For India, these hybrid legal-technical models can pave the way for overcoming the divide between invention and judicial acceptability, while making sure that technology offers sustenance to environmental justice. Indian case experiences in illegal mining surveillance and emission monitoring indicate that technology strengthens detection and regulatory intervention, though long-term enforcement still depends on institutional capacity.

Finally, remote sensing, AI, and digital forensics can be powerful tools for climate justice in India only if embedded within legally solid, institutionally strong, and rights-respecting governance architecture. Technology, used responsibly, can help enable equitable enforcement that protects vulnerable communities and ensures sustainable environmental stewardship for the future. The objective is therefore calibrated integration, where technological innovation operates within clear doctrinal, constitutional, and procedural safeguards rather than replacing them.

REFERENCES

Abdullahi, A. (2025). Artificial intelligence (AI)-driven sanctions: Transforming global governance mechanisms in response to non-compliance. *Irish International Journal of Law. Political Sciences and Administration*, *9*(3), 149–169.

Abutaleb, K. A., Abdelsalam, A. A., & Khaled, M. A. (2025). Unleashing environmental intelligence through AI, image processing, and big data: Paving the path to a sustainable future. In *Modelling and advanced earth observation technologies for coastal zone management* (pp. 315–354). Springer Nature Switzerland., DOI: 10.1007/978-3-031-78768-3_12

Ajijola, A., Zavarsky, P., & Ruhl, R. (2014). A review and comparative evaluation of forensics guidelines of NIST SP 800-101 Rev. 1: 2014 and ISO/IEC 27037: 2012. In *Proceedings of the World Congress on Internet Security (WorldCIS-2014)* (pp. 66–73). IEEE. DOI: 10.1109/WorldCIS.2014.7028169

Audet, R. (2013). Climate justice and bargaining coalitions: A discourse analysis. *International Environmental Agreement: Politics, Law and Economics*, *13*(3), 369–386. DOI: 10.1007/s10784-012-9195-9

Babbar, H., & Rani, S. (2025). AI-driven integration of remote sensing and geospatial data for enhanced environmental monitoring and decision-making. *International Journal of Remote Sensing*, •••, 1–17. DOI: 10.1080/01431161.2025.2514820

Bahangulu, J. K., & Owusu-Berko, L. (2025). Algorithmic bias, data ethics, and governance: Ensuring fairness, transparency, and compliance in AI-powered business analytics applications. *World Journal of Advanced Research and Reviews*, *25*(2), 1746–1763. DOI: 10.30574/wjarr.2025.25.2.0571

Barman, D., ... (2025). *Forest degradation: Its impact and conservation efforts in India*. In G. Shukla, A. K. Manohar, A. Raj Kizha, P. Panwar, & S. Chakravarty (Eds.), *Forest degradation and management* (Forestry Sciences, Vol. 87). Springer. https://doi.org/DOI: 10.1007/978-3-031-84055-5_21

Berk, R. A. (2021). Artificial intelligence, predictive policing, and risk assessment for law enforcement. *Annual Review of Criminology*, *4*(1), 209–237. DOI: 10.1146/annurev-criminol-051520-012342

Bhushan, C. (2008). Rich lands, poor people: The socio-environmental challenges of mining in India. *Indian Economic Review*, *2008*, 44–53.

Bose, P. (2023). Equitable land-use policy? Indigenous peoples' resistance to mining-induced deforestation. *Land Use Policy*, *129*, 106648. DOI: 10.1016/j.landusepol.2023.106648

Brevini, B. (2020). Black boxes, not green: Mythologizing artificial intelligence and omitting the environment. *Big Data & Society*, *7*(2), 2053951720935141. DOI: 10.1177/2053951720935141

Burke, M., Driscoll, A., Lobell, D. B., & Ermon, S. (2021). Using satellite imagery to understand and promote sustainable development. *Science*, *371*(6535), eabe8628. Advance online publication. DOI: 10.1126/science.abe8628 PMID: 33737462

Butt, J. (2024). Analytical study of the world's first EU Artificial Intelligence (AI) Act. *International Journal of Research Publication and Reviews*, *5*(3), 7343–7364. DOI: 10.55248/gengpi.5.0324.0914

Casali, M. B., Travaini, G. V., Di Francesco, C. V., & Genovese, U. R. (2022). Environment, environmental crimes, environmental forensic medicine, environmental risk management and environmental criminology. *Health Care*, *10*(2), 263. PMID: 35206877

Chatterjee, U., & Kanbur, R. (2015). Non-compliance with India's Factories Act: Magnitude and patterns. *International Labour Review*, *154*(3), 393–412. DOI: 10.1111/j.1564-913X.2015.00027.x

Chaudhary, G. (2024). Unveiling the black box: Bringing algorithmic transparency to AI. *Masaryk University Journal of Law and Technology*, *18*(1), 93–122. DOI: 10.5817/MUJLT2024-1-4

. Chen, J., Chen, S., Fu, R., Li, D., Jiang, H., Wang, C., Hicks, B. J., & others. (2022). *Remote sensing big data for water environment monitoring: Current status, challenges, and future prospects. Earth's Future, 10*(2), e2021EF002289. https://doi.org/DOI: 10.1029/2021EF002289

Choudhary, A., Sharma, S., Singh, R., & Choudhary, A. (2025). AI in environmental sustainability: A review of applications and challenges. *Journal of Multidisciplinary Research Advancements*, *3*(2), 68–75. DOI: 10.3126/jomra.v3i2.90513

Chungyalpa, W. (2020). Examination and analysis of the Central Pollution Control Board and the State Pollution Control Board: Indian administrative arm for environmental protection. *Indian Journal of Sustainable Development*, *6*(1), 11.

Deb, D. (2013). The value of forest: An ecological economic examination of forest people's perspective. In *Challenges and opportunities for the world's forests in the 21st century* (pp. 123–159). Springer Netherlands., DOI: 10.1007/978-94-007-7076-8_7

Deb, M., & Sarkar, S. C. (2017). Issues of sustainable development in the mines and minerals sector in India. In *Minerals and allied natural resources and their sustainable development: Principles, perspectives with emphasis on the Indian scenario* (pp. 519–551). Springer., DOI: 10.1007/978-981-10-4564-6_11

Dey, S., & Majumdar, A. (2024). Current status of pollution in major rivers and tributaries of India and protection–restoration strategies. In Kanhaiya, S., Singh, S., Dixit, A., & Singh, A. K. (Eds.), *Rivers of India*. Springer., DOI: 10.1007/978-3-031-49163-4_4

Dhaka, S. (2025). Sustainable development and the role of the Indian judiciary in promoting it with special reference to the precautionary principle and the polluter pays principle. *International Journal of Technology. Health and Sustainability, 1*(1), 8–12.

Dias, A. (1994). Judicial activism in the development and enforcement of environmental law: Some comparative insights from the Indian experience. *Journal of Environmental Law, 6*(2), 243–262. DOI: 10.1093/jel/6.2.243

. du Plessis, A. (2008). Public participation, good environmental governance and fulfilment of environmental rights. *Potchefstroom Electronic Law Journal/Potchefstroomse Elektroniese Regsblad, 11*(2), 1–34.

Ginzky, H. (2001). Satellite images as evidence in legal proceedings relating to the environment. *Droit et Ville, 51*(1), 37–68. DOI: 10.3406/drevi.2001.1765

Goddard, M. (2017). The EU general data protection regulation (GDPR): European regulation that has a global impact. *International Journal of Market Research, 59*(6), 703–705. DOI: 10.2501/IJMR-2017-050

Gu, Z., & Zeng, M. (2024). The use of artificial intelligence and satellite remote sensing in land cover change detection: Review and perspectives. *Sustainability (Basel), 16*(1), 274. DOI: 10.3390/su16010274

Gunningham, N. (2009). Environmental law, regulation and governance: Shifting architectures. *Journal of Environmental Law, 21*(2), 179–212. DOI: 10.1093/jel/eqp011

Gupta, S. (2014). *Environmental policy and governance in a federal framework: Perspectives from India*. In *Environmental policies in Asia: Perspectives from seven Asian countries* (pp. 15–42). https://doi.org/DOI: 10.1142/9789814590488_0002

Hoff, R. M., & Christopher, S. A. (2009). Remote sensing of particulate pollution from space: Have we reached the promised land? *Journal of the Air & Waste Management Association, 59*(6), 645–675. DOI: 10.3155/1047-3289.59.6.645 PMID: 19603734

Jha, M. K., Markandey, D. K., Gupta, R., & Mishra, P. (2025). The role of technology in environmental governance. In *Intersecting environmental governance with technological advancements* (pp. 287–322). IGI Global Scientific Publishing.

Jutz, S., & Milagro-Perez, M. P. (2020). Copernicus: The European Earth observation programme. *Revista de Teledetección*, (56), V–XI.

Kale, S. S. (2025). Governing subterranean wealth in India: The shifting boundaries around mining's sectoral political network. *Studies in Comparative International Development*, *60*(2), 1–26. DOI: 10.1007/s12116-025-09460-8

Köbis, N., Starke, C., & Rahwan, I. (2022). The promise and perils of using artificial intelligence to fight corruption. *Nature Machine Intelligence*, *4*(5), 418–424. DOI: 10.1038/s42256-022-00489-1

Kumar, N. (2025). The role of artificial intelligence in criminal investigations in India. *The Journal of Legal Studies*, *35*(49), 98–111. DOI: 10.2478/jles-2025-0006

Kumar, N., Singh, D., Sharma, P., Chaurasia, P. K., & Singh, S. (2025). Ethical and privacy challenges in AI-driven water management. In *The handbook of AI for clean water* (pp. 327–340). CRC Press. DOI: 10.1201/9781003600718-28

Kurtz, J. E., & Pintarelli, E. M. (2024). The Daubert standards for admissibility of evidence based on the Personality Assessment Inventory. *Psychological Injury and Law*, *17*(2), 105–116. DOI: 10.1007/s12207-024-09508-5

Lal, R. (2005). Forest soils and carbon sequestration. *Forest Ecology and Management*, *220*(1–3), 242–258. DOI: 10.1016/j.foreco.2005.08.015

Laux, J., Wachter, S., & Mittelstadt, B. (2024). Three pathways for standardisation and ethical disclosure by default under the European Union Artificial Intelligence Act. *Computer Law & Security Review*, *53*, 105957. DOI: 10.1016/j.clsr.2024.105957 PMID: 41322429

Lein, J. K. (2009). Implementing remote sensing strategies to support environmental compliance assessment: A neural network application. *Environmental Science & Policy*, *12*(7), 948–958. DOI: 10.1016/j.envsci.2009.08.001

Macrory, R., & Purdy, R. (2001). The use of satellite images as evidence in environmental actions in Great Britain. *Droit et Ville*, *51*(1), 69–88. DOI: 10.3406/drevi.2001.1766

M.C. Mehta v. Union of India (Oleum Gas Leak Case), (1987) 1 SCC 395.

Muhammed Siddik, A., Varghese, G., & Megson, D. (2024). Need for comprehensive guidelines for environmental forensic investigations. *Detritus*, *27*(27), I–III. DOI: 10.31025/2611-4135/2024.19388

Mushtaq, B., Bandh, S. A., & Shafi, S. (2020). Environmental acts and legislation. In *Environmental management: Environmental issues, awareness and abatement* (pp. 149–184). Springer Singapore., DOI: 10.1007/978-981-15-3813-1_4

Norris, J. R., Allen, R. J., Evan, A. T., Zelinka, M. D., O'Dell, C. W., & Klein, S. A. (2016). Evidence for climate change in the satellite cloud record. *Nature*, *536*(7614), 72–75. DOI: 10.1038/nature18273 PMID: 27398619

Oh, P. B. (1997). Assessing admissibility of nonscientific expert evidence under federal evidence rule 702. *Defense Counsel Journal*, *64*, 556.

Palmiotto, F. (2020). The black box on trial: The impact of algorithmic opacity on fair trial rights in criminal proceedings. In *Algorithmic governance and governance of algorithms: Legal and ethical challenges* (pp. 49–70). Springer International Publishing.

Prakash, A., & Gupta, R. P. (1998). Land-use mapping and change detection in a coal mining area: A case study in the Jharia coalfield, India. *International Journal of Remote Sensing*, *19*(3), 391–410. DOI: 10.1080/014311698216053

Prasad, P. M. (2006). Environment protection: Role of regulatory system in India. *Economic and Political Weekly*, *41*(13), 1278–1288.

Rahman, A. (2007). Application of remote sensing and GIS technique for urban environmental management and sustainable development of Delhi, India. In *Applied remote sensing for urban planning, governance and sustainability* (pp. 165–197). Springer., DOI: 10.1007/978-3-540-68009-3_8

Rajaram, T., & Das, A. (2008). Water pollution by industrial effluents in India: Discharge scenarios and case for participatory ecosystem-specific local regulation. *Futures*, *40*(1), 56–69. DOI: 10.1016/j.futures.2007.06.002

Rao, N. V., Bhaskaran, V., & Nagendra, H. (2023). Can green tribunals help to resist neo-liberalism in environmental governance? The case of India. *Land Use Policy*, *131*, 106739. DOI: 10.1016/j.landusepol.2023.106739

Rengarajan, S., Palaniyappan, D., Ramachandran, P., & Ramachandran, R. (2018). National Green Tribunal of India—An observation from environmental judgements. *Environmental Science and Pollution Research International*, *25*(12), 11313–11318. DOI: 10.1007/s11356-018-1763-2 PMID: 29572740

Robinson, E. J. (2016). Resource-dependent livelihoods and the natural resource base. *Annual Review of Resource Economics*, *8*(1), 281–301. DOI: 10.1146/annurev-resource-100815-095521

Roy, K., Fatima, S., Chaudhary, S. R., Sonal, S., Mishra, A., & Chanda, S. (2025). Forensic environmental science and legal admissibility: Bridging the gap between data and justice. *International Journal of Environmental Sciences*, *11*(12s), 546–563. DOI: 10.64252/6pq8m965

Roy, S. (2020). Environmental crimes—An analysis of legal provisions with reference to India. *NUJS Journal of Regulatory Studies*, *5*, 30.

Rychlak, R. J., & Taylor, S. P. (2022). Satellite data and imagery as evidence. *Journal of Space Law*, *46*, 351.

Satapathy, D. R., Katpatal, Y. B., & Wate, S. R. (2008). Application of geospatial technologies for environmental impact assessment: An Indian scenario. *International Journal of Remote Sensing*, *29*(2), 355–386. DOI: 10.1080/01431160701269002

Saxena, K. B. (2024). The Forest (Conservation) Amendment Act 2023: A threat to forest lands and forest dwellers. *Social Change*, *54*(1), 108–117. DOI: 10.1177/00490857231222101

Schlosberg, D. (2012). Climate justice and capabilities: A framework for adaptation policy. *Ethics & International Affairs*, *26*(4), 445–461. DOI: 10.1017/S0892679412000615

Seizov, O., & Wulf, A. J. (2020). Artificial intelligence and transparency: A blueprint for improving the regulation of AI applications in the EU. *European Business Law Review*, *31*(4), 611–640. DOI: 10.54648/EULR2020024

Shrestha, S. (2015). *Assessment of Bagmati River encroachment through application of GIS and remote sensing* (Master's thesis, Pokhara University). Pokhara University Repository.

Shrotria, S. (2015). Environmental justice: Is the National Green Tribunal of India effective? *Environmental Law Review*, *17*(3), 169–188. DOI: 10.1177/1461452915595548

Singh, S., & Singh, M. (2025). Role of artificial intelligence for environment protection: An analysis. *Artificial Intelligence and Law*, •••, 165–175.

Smadja, J. (2018). A chronicle of law implementation in environmental conflicts: The case of Kaziranga National Park in Assam (North-East India). *South Asia Multidisciplinary Academic Journal*, *17*. Advance online publication. DOI: 10.4000/samaj.4422

Srinivas, R., Singh, A. P., & Shankar, D. (2020). *Understanding the threats and challenges concerning the Ganges River basin for effective policy recommendations towards sustainable development.* *Environment, Development and Sustainability, 22*(4), 3655–3690. DOI: 10.1007/s10668-019-00361-0

Suggs, J. A., & Yarbrough, K. R. (2004). *Environmental crime*. In *Proceedings of the 14th International Forensic Science Symposium, Interpol* (p. 360). Lyon, France.

Tariq, M. U. (2025). The challenge of artificial intelligence in environmental sustainability: Opportunities, risks, and policy implications. In *Nexus of environmental quality and technology innovation* (pp. 267–292). IGI Global Scientific Publishing. DOI: 10.4018/979-8-3373-1822-6.ch012

Thépaut, J. N., Dee, D., Engelen, R., & Pinty, B. (2018, July). The Copernicus programme and its climate change service. In *IGARSS 2018—2018 IEEE International Geoscience and Remote Sensing Symposium* (pp. 1591–1593). IEEE. DOI: 10.1109/IGARSS.2018.8518067

Van Der Ven, H., Corry, D., Elnur, R., Provost, V. J., Syukron, M., & Tappauf, N. (2024). Does artificial intelligence bias perceptions of environmental challenges? *Environmental Research Letters, 20*(1), 014009. DOI: 10.1088/1748-9326/ad95a2

Veber, J., & Klíma, T. (2014). Influence of standards ISO 27000 family on digital evidence analysis. In *Proceedings of the 22nd Interdisciplinary Information Management Talks* (pp. 103–114).

Vitale, S., Kulhari, D., & Caneparo, P. (2025). AI integration in legal decision-making: Innovations and challenges. *Qubahan Techno Journal, 4*(4), 29–43. DOI: 10.48161/qtj.v4n4a77

Walker, G. (2009). Environmental justice and normative thinking. *Antipode, 41*(1), 203–205. DOI: 10.1111/j.1467-8330.2008.00663.x

Walker, L., & Monahan, J. (1996). Daubert and the reference manual: An essay on the future of science in law. *Virginia Law Review, 82*(5), 837–857. DOI: 10.2307/1073712

Wood, S., Tanner, G., & Richardson, B. J. (2010). What ever happened to Canadian environmental law. *Ecology Law Quarterly, 37*, 981–1010.

Yetilmezsoy, K., Ozkaya, B., & Cakmakci, M. (2011). Artificial intelligence-based prediction models for environmental engineering. *Neural Network World, 21*(3), 193–218. Advance online publication. DOI: 10.14311/NNW.2011.21.012

Young, N., Roche, D. G., Lennox, R. J., Bennett, J. R., & Cooke, S. J. (2022). Ethical ecosurveillance: Mitigating the potential impacts on humans of widespread environmental monitoring. *People and Nature*, *4*(4), 830–840. DOI: 10.1002/pan3.10327

Chapter 16
Transboundary Climate Litigation and Constitutional Responsibility

Amr Ezzat Elhaw
http://orcid.org/0009-0000-1254-2323
Sharjah Police Science Academy, UAE

Tarek Abo El Wafa
http://orcid.org/0000-0002-3923-5187
United Arab Emirates University, UAE

ABSTRACT

This chapter examines how international climate commitments are incorporated into domestic law through constitutional environmental rights and how climate justice reshapes state accountability. It addresses the gap between non-binding agreements, such as the Paris Agreement, and the enforcement of constitutional provisions in rights-based climate litigation. Drawing on a comparative analysis of cases from Germany, Colombia, South Africa, and the Netherlands, it shows that courts increasingly treat environmental degradation as a violation of fundamental rights, including life, health, and intergenerational equity.The chapter argues that constitutional environmental rights are becoming enforceable legal norms and that judges rely on international principles to require stronger climate action. It also examines judicial and normative borrowing, highlighting the emergence of diagonal accountability in climate litigation. Finally, it calls for legal reforms to embed climate justice in national systems through constitutional recognition, procedural safeguards, and institutional mechanisms.

DOI: 10.4018/979-8-3373-6635-7.ch016

INTRODUCTION

Climate change presents an unprecedented legal and normative challenge that transcends national boundaries, thereby requiring a reorganization of both international and constitutional legal frameworks. The global system of responsibilities for climate change is described in the "Paris Agreement" and other frameworks, but its normative force is often insufficient to establish enforceable national requirements (Peel and Lin, 2019). At the same time, it is becoming an obligation for domestic courts to safeguard basic rights, including the rights to life, health, and a healthy environment, against environmental degradation. This emerging field of transboundary climate litigation raises an important legal question: what can countries do to reconcile their international responsibilities with their constitutional responsibilities as the risks of climate rise?

This chapter examines a disjunction between international climate commitments and constitutionally guaranteed rights-based protections in the doctrine and institutional structure. This contradiction is particularly clear when the courts are called upon to determine whether a state's failure to mitigate climate change violates fundamental rights. The problem is exacerbated by the lack of efficient legal tools to implement environmental obligations across borders and by the absence of a binding international legal system capable of balancing constitutional norms with international legal principles (Savaresi and Auz, 2019; Buser, 2023).

This chapter addresses the question: How may legal systems implement climate justice through judicial, legislative, and international mechanisms while complying with transnational duties and constitutional principles?

The chapter employes the three-phase methodological structure. The first step in a descriptive study would be to examine key legal documents, including international climate agreements (e.g., the Paris Agreement, the UNFCCC, and relevant human rights treaties), to define the formal legal basis for environmental protection. Secondly, a comparative study examines major judicial and legislative developments in the jurisdictions of "Germany, Colombia, South Africa, and the Netherlands" and reveals patterns and gaps in the judicial recognition of climate-related constitutional rights (Setzer and Vanhala, 2019; Preston, 2018). An analytical method is used to evaluate case law and doctrinal research to identify how constitutional adjudication and international environmental regulation relate to one another (Meguro, 2020; De Bellis, 2024).

The chapter investigates how constitutional and international law interrelate in establishing governmental responsibility for climate action. Second, it aims to assess the possibility of legislative creativity and policy change to bridge domestic environmental frameworks with international climate justice concepts. The chapter examines methodologies used in judicial litigation over climate, particularly when courts in a country apply not only national constitutional provisions but also international environmental standards. Finally, it proposes normative models for aligning environmental law with the principles of intergenerational equality, distributive justice, and sustainable development, which are now regarded as critical to international jurisprudence (Peel and Osofsky, 2018; Mayer, 2021).

Furthermore, the chapter underscores the rise of diagonal accountability, wherein individuals can pursue legal proceedings not only against their own governments but also against global injustice, so promoting climate justice as an international legal standard.

(De Bellis, 2024; Main-Klingst & Ott, 2023).

PART I -THEORETICAL FRAMEWORK

1.1. The Normative Foundations of Climate Justice

Climate justice as a concept has emerged to become a paradigm of change in international and constitutional law. It is a significant break from customary approaches to environmental protection, which have generally focused on ecosystem conservation without considering broader social concerns. Climate justice redefines the climate crisis as a legal and ethical crisis intrinsically linked to problems of human rights, intergenerational fairness, and global distributive justice. The idea is that disadvantaged groups are disproportionately affected by climate change, especially those least responsible for its effects, and that existing legal frameworks need to evolve to address these structural inequalities. (Peel, J., & Osofsky, H. M. (2018)

Climate Justice has been expressed in terms of CBDR, formulated in the United Nations Framework Convention on Climate Change (UNFCCC, 1992) under "Article 3". This theory acknowledges that the responsibility of global warming belongs to all states; however, the states that emitted in the past have a disproportionate obligation. The CBDR principle provides a conceptual-normative statement: that climate obligations are not only environmental but also highly ethical and juridical, due to the long-term inequalities between the Global North and the Global South.

These developments have been jurisprudentially replicated by the introduction of climate justice through an international court ruling that has made environmental conservation a condition for the actualization of basic rights. One of the milestones in that regard was the advisory opinion of the "Inter-American Court of Human Rights" in 2017, which found that the right to a healthy environment was independent and enforceable, and recognized that states could be liable for transboundary environmental harm (IACtHR, 2017). Critical thinking led to an important change:

the environment is no longer seen as an object of regulation, but as a legal interest that helps fulfill life, dignity, and other fundamental human rights.

Domestic courts, enabled by the inclusion of environmental rights in countries' constitutions, have been empowered to reconsider state obligations considering climate justice. Some constitutions, such as the Colombian Constitution “Article 79” and the South African Constitution (Section 24), recognize the right to an ecologically balanced environment as a fundamental right. In addition, courts in certain jurisdictions have gone beyond declaratory interpretations to further impose an obligation on states to take certain climate action. The “German Federal Constitutional Court” concluded in the case of “Neubauer et al. v. Germany” (2021) that the state's climate laws burdened future generations excessively by delaying reductions in emissions, thereby violating the rights of future generations guaranteed by the Constitution. This ruling was an administrative recognition of intergenerational fairness as an objective of policy and a constitutional necessity (Buser, 2023).

A similar form of climate justice as intertemporal responsibility was published in “Future Generations v. Ministry of Environment” (2018), in which the Colombian Supreme Court accepted the standing of young plaintiffs to promote the rights of the environment on behalf of the present and future generations. The Court instructed the government to adopt a national action plan on the conservation of the Amazon rainforest, and this was a direct relationship between climate degradation and the violation of fundamental constitutional rights.

The issue of climate justice and fundamental human rights has been solidly proven in several jurisdictions. The biggest decision so far is the one referred to as “Urgenda Foundation v”. In the case of the “Dutch Supreme Court”, which held that a lack of sufficient measures to reduce emissions imposed by the government violated the right to life and the right to private and family life in “Articles 2 and 8” of the ECHR, specifically, this right was inadequacy, the State of the Netherlands” (2019). The Court held that the state had a positive duty to protect its citizens against foreseeable environmental risks, including those caused by climate change. This decision was a landmark court case: a court in the United States issued legally binding climate obligations grounded in human rights law (Mayer, 2021).

These transformations mark the beginning of a major redefinition of environmental harm as not only a technological or regulatory failure, but also a violation of justice, fairness, and constitutional legitimacy. It has become the common law by which courts are reconsidering the role of the government, establishing environmental stewardship as an enforceable right, and incorporating cross-border and intergenerational responsibilities into the law.

1.2. International Climate Obligations and Treaty Frameworks

The international legal framework on climate change is essentially grounded in two main tools: the UNFCCC signed in 1992, and the "Paris Agreement", signed in 2015 under its auspices. All these documents outline countries' responsibilities in the context of greenhouse gas (GHG) mitigation, adaptation, and financial support. The substantive law of these duties and their enforceability under international law remains the subject of a current doctrinaire debate.

The UNFCCC establishes a more generalized form of cooperation by operating under the principles of equity, common but differentiated responsibilities (CBDR), and respective capabilities. Under "Article 4", which defines the overall role of all Parties, includes the necessity to develop and periodically update national lists of greenhouse gas emissions, and to implement measures aimed at reducing climate change. These needs are essentially procedural and facilitative rather than prescriptive. They are a form of soft law, in which obedience is expected but not legally binding in a court of law (Rajamani, 2006). The Convention created a distinction between industrialized and non-industrialized (developing) nations, a distinction that later influenced the burden-sharing system of the Paris Agreement.

The Paris Agreement marked the beginning of a new era, adopting a more universal approach to climate governance with a more domestic definition. Under article 4(1), all the Parties shall prepare, communicate, and keep successive nationally determined contributions (NDCs) that reflect their highest ambition. The language of choice of obligation, namely the word shall, suggests an obligatory procedural obligation to file and update NDCs. The content and promissory nature of the NDCs are informal and open to the discretion of the individual states (Voigt and Ferreira, 2016). This paradigm combines legal paradigms: binding procedures and democratically defined substantive components. In addition, the introduction of an improved transparency framework under Article 13 of the Agreement obligates countries to file biannual reports and undergo expert reviews, formalizing accountability not through adjudication but through monitoring. With regards to adaptation, in the "Paris Agreement "Article 7", it is referred to as a global goal and "All Parties have a duty to engage in adaptation planning and delivery.

However, the adaptation policies remain largely aspirational, with exhortative terms like 'should' and 'encouraged to,' and lack the binding nature that shapes procedural mitigation. As a result, the aspect of adaptation remains a field of normative incompetence, in terms of enforceability and justiciability (Bodansky, 2016). The concept of climate finance, especially in its "Article 9" form, is an essential part of the Paris framework. The developed countries are legally required to provide financial support to support Parties in developing countries in mitigation and adaptation activities. Though such an obligation is formally formulated in terms required, the quantitative elements of it, such as the promise to raise 100 billion a year, are not necessarily binding and are to be negotiated in politics. Reporting on the finances provided is mandatory, but the level and allocation of the finances are at the organization's discretion (Pickering et al., 2015).

The legality of such agreements, as per international law, is therefore determined by the legality of the agreements. Whereas both the UNFCCC and the Paris Agreement are formally treaties under the Vienna Convention on the Law of Treaties (1969), they are only semi-binding because they lack robust dispute-resolution mechanisms, and the issue of climate obligations is a political one. The Paris Agreement is particularly deficient in a court or a system of penalties. "Article 15" will provide a non-punitive, facilitative commission to improve compliance.

Therefore, procedural obligations (e.g., filing of NDCs, reporting, transparency) are legally binding, but content obligations (e.g., emission reduction targets, financial amounts, adaptation measures) are politically binding but not legally binding in the traditional sense. However, the national courts have increasingly interpreted these international commitments to affect constitutional and administrative duties, particularly where states have strong environmental rights laws. In "Neubauer v. The article, titled "Germany" (2021), refers to the "German Federal Constitutional Court, which concluded that the state's climate policy had not done enough to protect future generations, given that it was based on the state's commitments under the Paris Agreement." It shows that international climate standards can be enforced domestically when integrated into constitutional or legislative frameworks (Voigt, 2021).

In turn, although neither the Paris Agreement nor the UNFCCC establishes clearly enforceable international standards, they give rise to normative expectations and procedural obligations that are becoming increasingly relevant in national legal frameworks. Their influence is not based on coercion, but on their ability to systematize legal reasoning, influence administrative judgment, and support the adjudication of rights-based claims at the national level.

1.3. Environmental Rights and State Duties in the Constitution

The evolution of environmental rights in comparative constitutional systems marked the beginning of a substantial shift in the legal status of ecological conservation. Previously, environmental protection was considered a matter of administrative policy, but it has gradually been recognized as a constitutional right, subject to review on both the merits and the procedures. Over the past four decades, enough constitutions, particularly in the Global South and post-conflict democracies, have incorporated the right to a healthy environment in their normative frameworks.

This trend shows a growing consensus that environmental degradation is a direct threat to basic human rights, including the right to life, the right to health, the right to water, and the right to dignity. As a result, constitutional environmental rights have become an important legal avenue for people and communities to seek judicial redress against governments' actions and inaction regarding climate justice. Comparative constitutional analysis identifies two dominant models: the rights-based model, in which the environment is treated as an intrinsic human right, and the duty-based model, in which the state is primarily responsible for environmental management. The courts in both models have been able to interpret these rules to set binding obligations on the public agencies. For example, the South African Constitution (1996), Section 24, states that everyone has the right to a non-hazardous environment that does not endanger his or her health or well-being, and that the state must take reasonable measures, both legislative and other, to realize this right. It is a regular use of this article by the so-called courts of South Africa, and especially the Constitutional Court, to review environmental licensing cases, aligning it with the rest of the socio-economic rights guaranteed by the Constitution (Ferreira, 2019).

One of the most developed frameworks on constitutional environmentalism in Colombia has been created by the so-called Constitutional Court. "Article 79" of the Constitution of 1991 holds that every citizen is entitled to a healthy environment and makes the protection of ecological resources a legal responsibility of the state. Tutela actions have strongly supported this right and made it easier to obtain judicial intervention in environmental matters. In a landmark case, Future Generations v. Ministry of Environment case (2018), the Court ruled that the Colombian Amazon is a legal person and that the inactivity of the government in fighting deforestation violated the constitutional rights of the current and successive generations (Corte Suprema de Justicia (Sala de Casación Civil), STC4360-2018 (Apr. 5, 2018).

The Court also concluded that intergenerational equity is a valid constitutional principle that imposes pressing obligations on the state (Cepeda Espinosa, 2020). The German Basic Law (Grundgesetz), although traditionally conservative in its environmental language, has seen significant growth in its interpretation. The 1994 amendment to "Article 20a" provides that the state shall protect the natural sources of life and fauna by law, a duty the state must enforce through the executive and judicial branches, in accordance with the law and fairness. This paragraph, which was once viewed as a programmatic phrase, has acquired normative meaning today, especially in environmental and climate-related lawsuits. In Neubauer et al. v. Germany (2021), the Federal Constitutional Court concluded that the Federal Climate Protection Act interfered with the rights to life and personal liberty according to the provisions of "Articles 2(1)" and 2(2) because it did not sufficiently protect the future generation against the long-term effects of climate change.

The Court established that, in the context of proportionality, "Article 20a" imposes a constitutional responsibility on the state to take proactive climate action (Morgenthaler, 2022). The trends represented by these developments reflect a wider pattern: constitutional environmental rights have ceased to be understood as mere aspirations and are increasingly recognized as real legal rights. The role of the state has shifted from discretionary policy to compulsory constitutional obligations, especially in relation to climate policy. Courts, in cases involving the establishment of such rights, have the power to review legislative and administrative practices for compliance with environmental norms, thereby strengthening the accountability of official institutions and solidifying the concept of ecological justice.

PART II ANALYSIS AND DISCUSSION

2.1. The Judicialisation of Climate Governance: From Political Discretion to Legal Accountability

The growing involvement of the courts in climate regulation marks an essential shift from the traditional domain of political discretion to that of constitutional and international law accountability. Such a practice is commonly known as the judicialization of climate governance and is indicative of an expanding trend in which plaintiffs challenge the adequacy of governmental action in response to climate change by asserting constitutional and human rights claims. Judicial intervention has been eminent where legislative or administrative authorities have performed poorly in implementing climate objectives in line with national commitments or international environmental regulations.

In "Urgenda Foundation v. State of the Netherlands, the Dutch Supreme Court held that the failure of the government in enacting appropriate mitigation actions breached its affirmative duties under "articles 2 and 8" of the ECHR. What is meaningful about this case, besides its outcome, is its legal reasoning: the Court relied on scientific evidence (IPCC reports), international law (UNFCCC and Paris Agreement), and human rights principles, which shows the polyphonic basis of modern climate litigation (Spier, 2020; Salmi et al., 2025).

The Urgenda judgment brought about major changes in policy: the Dutch government thus adjusted to the tighter emissions-reduction targets and hastened investments in renewable energy. The case is an example of how judicial decisions can drive policy change, making climate change not only a political but also a legal imperative. Such a system of judicial responsibility has increasingly gained significance in the Global South. In "Leghari v". The "High Court of Lahore", Federation of Pakistan," international law applied the concepts of constitutional rights to life and dignity to compel the federal government to implement its National Climate Change Policy.

The Court's finding was groundbreaking in both form and content: not only did it recognize the rights violations, but it also created a judicially reviewed Climate

Change Commission, which marked the beginning of structural rather than declaratory orders (Haider, 2016). Similar tendency can be observed in Latin America, in Future Generations v. The "Colombian Supreme Court" in its publication titled "Ministry of Environment" (2018) recognized the rights of future generations, as well as the Amazon rainforest, requiring the government to adopt a final deforestation reduction strategy- this step is the example of the expansion of constitutional climate requirements in the Global South.

Even though these judicial interventions have the potential to be transformative, there are no institutional restrictions on them. Courts in many jurisdictions operate under institutional capacity constraints, political opposition, and enforcement challenges. The adherence of the executive to the rigid obligations imposed by the judgment could depend on political goodwill within the executive department and on the existence of favorable administrative systems. Without such avenues, even high-profile judicial pronouncements can be purely symbolic and fail to lead to systemic transformation. In the future, international judicial institutions can improve climate regulation laws. Current advisory proceedings before the ICJ and ITLOS seek to clarify whether a state has greater or lesser responsibility for climate change and environmental harm. Should such cases establish that the inaction on climate violates existing legal norms, they can serve as interpretive principles for national courts, thereby enhancing the validity and enforceability of domestic climate judgments.

2.2. Fundamental Rights as a Framework for Climate Justice

A salient characteristic of modern climate litigation is its growing dependence on fundamental rights frameworks to delineate and implement climate responsibilities. This rights-based approach reconceptualizes environmental degradation as a direct violation of rights safeguarded by constitutional and international human rights law, including the rights to life, health, dignity, and intergenerational equity. It allows plaintiffs to transition from theoretical environmental policy to tangible legal responsibility, therefore encouraging more rigorous court examination and improving access to effective remedies (Peel & Osofsky, 2018; Meguro, 2020; Afify et al., 2023).

By anchoring climate assertions in fundamental rights, courts are enabled to interfere in governmental actions even without explicit legislative directives or enforceable international institutions. The right to a healthy environment is progressively regarded not merely as an aspirational objective, but as a "justiciable right" whose infringement invokes constitutional repercussions. This advancement is especially important in systems where environmental rights are officially incorporated into national constitutions or when courts interpret general rights provisions considering environmental threats (Boyle, 2019; Buser, 2023).

A prominent illustration of this rights-based shift is the seminal case "Neubauer et al. v. Germany" (2021). The German Federal Constitutional Court determined that the state's climate policy was unconstitutional as it imposed significant mitigation responsibilities on future generations, infringing upon their rights to life and personal development as stipulated in "Article 2(1)" of the Basic Law. Upon reviewing "Article 20a," which obligates the state to safeguard the natural foundations of life, the Court determined that the government's climate legislation did not adhere to the constitutional principles of proportionality and intergenerational equality (Morgenthaler, 2022). The ruling was informed by both domestic constitutional principles and international commitments under the Paris Agreement, demonstrating how transnational standards can influence constitutional reasoning and constrain the state's discretion (Voigt, 2021).

In "Future Generations v. Ministry of Environment" (2018), the Colombian Supreme Court rendered a landmark decision that broadened the scope of rights-based climate litigation. The Court acknowledged the Amazon rainforest as a legal entity possessing its own rights and determined that the government's inaction in mitigating deforestation infringed against the essential rights of both current and future generations as stipulated in "Article 79" of the "Colombian Constitution" (Cepeda Espinosa, 2020). The Court's rationale integrated environmental conservation, indigenous rights, and ecological ethics into a comprehensive framework, directing the government to formulate and execute a national action plan to cease deforestation. This case illustrates how Latin American courts are promoting a variant of transformational constitutionalism that integrates environmental justice within the overarching concepts of dignity, sustainability, and intergenerational responsibility (Preston, 2018; Boyle & Peart, 2020).

Such cases reveal an increasing judicial consensus that environmental damage, especially resulting from climate inaction, is not merely a failure of political duty but also a transgression of fundamental rights. Judicial bodies are progressively seeing climate-related litigation to uphold fundamental constitutional principles and to guarantee that environmental deterioration does not compromise the prerequisites for the exercise of other rights. The legal rationale behind these rulings underscores that the rights to life and health cannot be effectively safeguarded without an accompanying obligation on states to avert environmental degradation (Peel & Lin, 2019; Mayer, 2021).

The rights-based approach to climate justice signifies a normative and doctrinal advancement in public law. It empowers courts to transcend mere symbolic pronouncements and to mandate substantive obligations on states—necessitating progressive, science-informed, and fair climate policy. Although these decisions do not replace political processes, they serve as an essential corrective mechanism that aligns climate governance with constitutional and international human rights standards. By doing so, courts are transforming the legal landscape by acknowledging that environmental protection is fundamental to human rights. (Lee, S. (2025).

2.3. Judicial Dialogue and Transnational Borrowing in Climate Jurisprudence

One of the most remarkable features of climate litigation in the last decade has been the emergence of transnational judicial dialogue, whereby courts reference foreign and international decisions to shape their own reasoning and outcomes. This process has led to the diffusion of shared principles, such as intergenerational equity, the precautionary principle, the polluter pays principle, and the recognition of climate harms as human rights violations. (Iyengar, S. (2023).

The "Inter-American Court of Human Rights' Advisory Opinion" OC-23/17 stands as a cornerstone in this transnational jurisprudence. The Court not only affirmed the autonomous justiciability of the right to a healthy environment under the American Convention but also expanded its extraterritorial application to include transboundary environmental harm. This opinion now serves as a persuasive authority across Latin America and has been cited by national courts in "Ecuador, Costa Rica, and Colombia" when interpreting constitutional environmental duties (Boyle, 2019).

In Europe, ongoing litigation before the "European Court of Human Rights", such as "*KlimaSeniorinnen v. Switzerland* and *Carême v. France*"—further exemplifies this trend. These cases argue that state inaction on climate change constitutes a violation of "Articles 2 and 8" of the HCHR, invoking the precedent of "*Urgenda*". If the Court affirms these arguments, it will likely consolidate a European-wide human rights approach to climate governance, with potentially binding implications for all Council of Europe member states (Lavrysen, 2023).

2.4. Normative Pathways for Reform: Embedding Climate Justice in Legal Systems

The expanding corpus of climate jurisprudence has demonstrated legal deficiencies in current governance frameworks while offering normative opportunities for altering the underpinnings of environmental and constitutional law. As courts increasingly intervene to enforce climate obligations, legal systems face growing pressure to adjust their frameworks to provide clearer accountability standards, improve institutional coherence, and broaden access to environmental justice (Preston, 2018; Buser, 2023; El Wafa & El Maslouhi, 2024).

A fundamental reform is the constitutional recognition of environmental rights. Incorporating the right to a healthy environment or a stable climate into national constitutions transforms environmental protection from optional policy to a binding legal obligation. Provisions in the constitutions of Colombia "Art. 79", South Africa (Sec. 24), and clearly elucidated in Germany via "Article 20a", have empowered courts to examine governmental inactivity and compel climate-related actions (Morgenthaler, 2022; Cepeda Espinosa, 2020). Constitutionalizing environmental rights converts them into enforceable norms, establishing a legal basis for judicial scrutiny and enabling courts to serve as active protectors of intergenerational equality and substantial environmental safeguarding (Boyle & Peart, 2020; Voigt, 2021).

Nonetheless, constitutional recognition is inadequate without climate-specific procedural assurances that facilitate access to justice. In this context, national legal frameworks ought to conform to international agreements like the "Aarhus Convention," which establishes the rights to access environmental information, engage in public participation in environmental decision-making, and obtain judicial or administrative remedies (Richardson & Razzaque, 2019). These procedural rights function as legal entitlements and democratic safeguards, enabling civil society, environmental advocates, and marginalized people to hold nations accountable for climate damage. Integrating Aarhus-type procedural rules enhances deliberative legitimacy and fosters institutional transparency, which are vital for democratic climate governance. (Richardson, B. J., & Razzaque, J. (2019)

Furthermore, there is a persuasive argument for the creation of autonomous climate accountability institutions. Courts, as essential participants in promoting climate justice, function within the limitations of justiciability, case-specific mandates, and institutional capacity. Governments should establish specialized entities, such as climate ombudspersons, commissions, or independent advisory committees, to augment judicial action, emulating the "Pakistani Climate Change Commission" or the "UK Climate Change Committee". (Vieira, M. A. (2022)

These institutions can oversee the execution of judicial rulings, evaluate policy alignment with national and international commitments, and offer expert insights to inform both litigation and policymaking (Haider, 2016; Main-Klingst & Ott, 2023). Their autonomy is crucial to guarantee that scientific facts, human rights issues, and intergenerational interests remain pivotal in governmental responses.

Subsequent reform must focus on legal standing and transboundary access to justice, particularly considering the global implications of climate effects. Legal institutions must adapt to enable individuals and groups, especially those from vulnerable or transnational areas, to contest environmental damage regardless of strict territorial connections. The advisory decision OC-23/17 issued by the Inter-American Court of Human Rights has established a significant precedent by recognizing extraterritorial environmental obligations, which domestic courts may utilize to enhance access to remedies (Boyle, 2019; De Bellis, 2024).

The incorporation of climate justice into legal frameworks necessitates a normative reconfiguration of public law that prioritizes sustainability, intergenerational accountability, and distributive equity as constitutional principles. Legal education, judicial training, and cross-jurisdictional interaction must facilitate this transition by providing legal institutions with the resources to interpret environmental standards through a rights-based and climate-conscious perspective. (Sulyok, K. 2024).

In conclusion, climate litigation is not a substitute for policy; rather, it serves as a crucial corrective mechanism that compels governments to adhere to their constitutional and international duties. To increase its efficacy, reform must transition from reactive litigation to the systemic integration of climate justice inside legal frameworks. This encompasses the constitutionalization of environmental rights, the institutionalization of procedural safeguards, and the establishment of systems for sustained oversight and policy coherence. Such reforms enable legal systems to guarantee that the judiciary functions not in isolation, but as an integral component of a coordinated institutional framework capable of addressing the magnitude and immediacy of the climate problem (Peel & Lin, 2019; Mayer, 2021).

CONCLUSION

The intersection of transboundary climate litigation and constitutional accountability represents a significant shift in the international legal approach to the climate problem. Climate commitments are now evolving beyond mere policy choice or diplomatic discussion, becoming enforceable legal duties rooted in fundamental rights, constitutional law, and international norms. This chapter illustrates how domestic courts, via rights-based adjudication and normative borrowing, are closing the structural gap between non-binding international climate obligations and the necessity for justiciable and accountable state action.

The comparative examination of cases in Germany, Colombia, the Netherlands, and South Africa demonstrates that courts are not solely reading environmental rules. A new normative order is being established wherein the environment is recognized as a legal entity, and the safeguarding of climatic systems is deemed essential for the fulfillment of human rights. This legal transformation affirms that climate justice is not merely ethically preferable but legally essential, especially when considered in terms of intergenerational equity, distributive justice, and the safeguarding of marginalized populations (Mayer 2021; Boyle 2019).

Furthermore, the chapter has shown how transnational judicial communication and diagonal accountability are reshaping the parameters of legal duty in environmental governance. By doing so, courts are initiating a global dialogue in which constitutional rights are employed to substantively enforce international climate commitments, thus reestablishing equilibrium between global harm and local remedies (De Bellis 2024; Main-Klingst and Ott 2023).

However, for this judicial momentum to effect systemic change, legal institutions must undergo structural renovation. This encompasses the constitutionalization of environmental rights, the incorporation of procedural safeguards like access to information and public engagement, and the establishment of autonomous institutions tasked with monitoring compliance and facilitating interbranch coordination (Preston 2018; Richardson and Razzaque 2019). Judicial bodies alone cannot bear the burden of climate governance. Their efficacy relies on the existence of enabling legal frameworks and a political will that adheres to judicial directives.

This chapter asserts that climate litigation transcends a mere reactive legal approach. It is a pivotal tool for international legal advancement. The court is becoming a pivotal entity in the developing realm of climate accountability by harmonizing local constitutional protections with international environmental obligations and framing climate damage as violations of fundamental rights. The way forward requires sustained dedication to normative innovation, institutional cooperation, and legal creativity to ensure that the commitments of climate justice are rendered both feasible and enforceable.

REFERENCES

Afify, A. M., Tannar, H., Zakria, A. A., Hashish, A., & El-Wafa, T. A. (2023). Unveiling the Right to Health in Egypt: Exploring the Transformations and Challenges in Egyptian Constitutional Law and Policy. *Academic Journal of Interdisciplinary Studies, 12*(6), 95–112. DOI: 10.36941/ajis-2023-0156

Bodansky, D. (2016). The legal character of the Paris Agreement. *Review of European, Comparative & International Environmental Law, 25*(2), 142–150. DOI: 10.1111/reel.12154

Boyle, A. (2019). Litigating climate change under human rights law. *The International and Comparative Law Quarterly, 68*(3), 759–786. DOI: 10.1017/S0020589319000224

Boyle, A., & Peart, R. (2020). Climate change and the right to a healthy environment. In *The Human Right to a Healthy Environment* (pp. 210–238). Oxford University Press.

Buser, A. (2023). National climate litigation and the international rule of law. *Leiden Journal of International Law, 36*(3), 593–615. Advance online publication. DOI: 10.1017/S0922156522000772

Cepeda Espinosa, M. J. (2020). The Colombian Constitutional Court and environmental rights: The Amazon case. *The Harvard Environmental Law Review, 44*, 47–86. https://harvardelr.com/2020/03/30/the-amazon-case

Corte Suprema de Justicia (Sala de Casación Civil). (2018, April 5). *STC4360-2018*. https://cortesuprema.gov.co/corte/wp-content/uploads/2018/04/STC4360-2018.pdf

De Bellis, M. (2024). Transnational climate litigation: Emergence and limits of a diagonal protection of fundamental rights. *Italian Journal of Public Law, 16*(1), 173–199. https://www.academia.edu/118624460

El-Wafa, T. A., & El Maslouhi, A. (2024). PARLIAMENTARY QUESTIONS IN THE PRACTICE OF UNITED ARAB EMIRATES (UAE)'S FEDERAL NATIONAL COUNCIL (FNC). *Malaysian Journal of Syariah and Law, 12*(1), 228–244. DOI: 10.33102/mjsl.vol12no1.498

Ferreira, G. (2019). The South African Constitution and climate change: An environmental law perspective. *Potchefstroom Electronic Law Journal, 22*, 1–28. DOI: 10.17159/1727-3781/2019/v22i0a4806

Haider, S. (2016). Leghari v. Federation of Pakistan: A judicial expansion of climate rights in the Global South. *Transnational Environmental Law, 5*(1), 229–236. https://doi.org/10.1017/S2047102515000342

Iyengar, S. (2023). Human rights and climate wrongs: Mapping the landscape of rights-based climate litigation. *Review of European, Comparative & International Environmental Law*, *32*(2), 299–309. Advance online publication. DOI: 10.1111/reel.12498

Lavrysen, L. (2023). Climate change and the European Court of Human Rights: The pending cases. *Human Rights Law Review*, *23*(1), 1–26. DOI: 10.1093/hrlr/ngac028

Lee, S. (2025). Human Rights and Climate Litigation: Trends, Challenges, and Implications. *The Korean Journal of International and Comparative Law*. https://doi.org/DOI: 10.1163/22134484-12341219

Main-Klingst, L., & Ott, H. E. (2023). Climate litigation, extraterritoriality of human rights and the German constitution. *Chinese Journal of Environmental Law*, *7*(2), 273–300. DOI: 10.1163/24686042-12340110

Mayer, B. (2021). Climate change mitigation as an obligation under human rights treaties? *The American Journal of International Law*, *115*(3), 409–446. DOI: 10.1017/ajil.2021.9

Meguro, M. (2020). Litigating climate change through international law: Obligations strategy and rights strategy. *Leiden Journal of International Law*, *33*(4), 771–792. DOI: 10.1017/S0922156520000473

Morgenthaler, A. (2022). Climate change and fundamental rights: The German Federal Constitutional Court's Neubauer decision. *European Constitutional Law Review*, *18*(2), 279–296. DOI: 10.1017/S157401962200014X

Peel, J., & Lin, J. (2019). Transnational climate litigation: The contribution of the Global South. *The American Journal of International Law*, *113*(4), 679–726. DOI: 10.1017/ajil.2019.48

Peel, J., & Osofsky, H. M. (2018). A rights turn in climate change litigation. *Transnational Environmental Law*, *7*(1), 37–67. DOI: 10.1017/S2047102517000292

Pickering, J., Jotzo, F., & Wood, P. J. (2015). Sharing the global climate finance effort fairly with limited coordination. *Global Environmental Politics*, *15*(4), 39–62. DOI: 10.1162/GLEP_a_00325

Preston, B. J. (2018). The evolving role of environmental rights in climate change litigation. *Chinese Journal of Environmental Law*, *2*(2), 131–160. DOI: 10.1163/24686042-12340030

Rajamani, L. (2006). *Differential treatment in international environmental law*. Oxford University Press. DOI: 10.1093/acprof:oso/9780199280704.001.0001

Richardson, B. J., & Razzaque, J. (2019). Public participation in environmental decision-making: Strategies for climate justice. *Transnational Environmental Law*, *8*(1), 123–149. DOI: 10.1017/S2047102518000263

Salmi, A., Bhattacharya, B., Bhattacharya, S., & Abo El Wafa, T. (2025). *The Role of Statutory Law in Regulating Artificial Intelligence: Balancing Innovation and Responsibility*. Access to Justice in Eastern Europe.

Savaresi, A., & Auz, J. (2019). Climate change litigation and human rights: Pushing the boundaries. *Climate Law*, *9*(3), 244–262. DOI: 10.1163/18786561-00903006

Setzer, J., & Vanhala, L. C. (2019). Climate change litigation: A review of research on courts and litigants in climate governance. *Wiley Interdisciplinary Reviews: Climate Change*, *10*(3), e580. DOI: 10.1002/wcc.580

Spier, J. (2020). The Urgenda judgment: Its consequences for the UK and beyond. *Environmental Law Review*, *22*(1), 3–10. DOI: 10.1177/1461452920901985

Sulyok, K. (2024). *Transforming the Rule of Law in Environmental and Climate Litigation: Prohibiting the Arbitrary Treatment of Future Generations*. Transnational Environmental Law., DOI: 10.1017/S2047102524000116

Vieira, M. A. (2022). *Litigating for Climate Justice–Chasing a Chimera?* https://doi.org/DOI: 10.1007/978-981-19-1987-9_10

Voigt, C. (2021). Intergenerational equity in climate litigation: The German Federal Constitutional Court's decision on the Climate Change Act. *Review of European, Comparative & International Environmental Law*, *30*(3), 361–366. DOI: 10.1111/reel.12393

Voigt, C., & Ferreira, F. (2016). The Paris Agreement: An international law perspective. *Climate Law*, *6*(1–2), 1–16. DOI: 10.1163/18786561-00601002

Chapter 17
Tropical Forest Forever Facility:
Climate Justice Through Legally Robust, Results-Based Forest Payments

Silvio Andrae
http://orcid.org/0000-0002-8586-7812
Independent Researcher, Germany

ABSTRACT

In light of ongoing global deforestation and the structural asymmetry between global climate benefits and local conservation costs, this chapter examines the Tropical Forest Forever Facility (TFFF) as an emerging transnational governance regime. It analyzes the legal, institutional, and judicial feasibility of the TFFF by integrating legal doctrine analysis, comparative policy research, and case-based governance assessments. The chapter shows that the long-term effectiveness of the TFFF depends less on the amount of financial resources mobilized than on the coherence of its legal architecture, the credibility of its monitoring systems, the institutionalization of indigenous peoples' participation rights, and the design of risk-sharing and conditionality mechanisms. It also develops reform pathways to strengthen legal enforceability, financial resilience, and distributive justice, thereby positioning the TFFF as a potential blueprint for sustainable, rights-based climate finance.

DOI: 10.4018/979-8-3373-6635-7.ch017

INTRODUCTION

Tropical forests are among the key stabilizing factors in the global climate system. They act as carbon sinks, biodiversity reservoirs, regulators of hydrological cycles, and the basis of life for millions of people. Nevertheless, despite decades of international commitments, deforestation and degradation are progressing at an alarming rate in many regions. Long-term analyses show that deforestation in the Amazon is already responsible for most of the regional decline in precipitation and a significant proportion of warming, regardless of global emission trends (Franco et al., 2025). Existing instruments of international climate and environmental policy have so far failed to effectively reconcile global climate goals with the political, economic, and social realities of forested countries.

At the heart of this failure lies a structural asymmetry: while the ecological benefits of forest protection accrue globally, the economic opportunity costs, fiscal burdens, and governance risks are borne predominantly by countries in the Global South. They are expected to preserve ecosystems of planetary importance while also facing development pressures, budgetary constraints, and institutional deficits. This constellation represents not only a financing problem but also a fundamental fairness issue in international climate policy. This asymmetry reflects fragmented global climate governance, characterized by overlapping regimes, competing institutions, and unclear responsibilities (Keohane & Victor, 2011; Ostrom, 2010; Young, 2017).

Against this backdrop, the Tropical Forest Forever Facility (TFFF), initiated by Brazil and positioned as a flagship initiative for COP30 in Belém, marks one of the most ambitious attempts to date to reorganize global forest financing (see Concept Note 3.0). By mobilizing extensive public and private funds and distributing investment returns as performance-based payments for forest protection, the TFFF aims to establish a permanent, predictable, and scalable financing mechanism. Unlike project-based approaches or CO_2 compensation models, it is primarily based on an area-based logic and rewards the conservation of standing forests.

This development is part of a broader shift in global forest governance, which recent analyses describe as a transition to a ‘third generation’ of regulatory arrangements. While early approaches relied primarily on intergovernmental soft law and later instruments on voluntary standards and market-based certification, the current generation of governance combines binding demand regulation with global redistribution mechanisms. In this context, the Tropical Forest Forever Facility is also classified as a central element of a new hybrid governance regime (Muradian & May, 2026).

The TFFF's institutional aspirations are correspondingly high. It aims to combine climate finance, biodiversity conservation, and human rights within an integrated governance architecture. At the same time, it systematically links forest protection to global capital markets and transforms ecological protection services into a long-term financing regime. This hybrid structure fundamentally distinguishes the TFFF from earlier instruments such as REDD+ or classic programs for payments for ecosystem services. At the same time, it makes the initiative particularly vulnerable to financial volatility, political conflicts, and institutional erosion.

Despite increasing political attention, scientific analysis of the TFFF has been fragmentary to date. The discourse so far has focused primarily on fund volumes, expected returns, and potential donor contributions. Much less attention has been paid to the legal foundations, governance structures, and equity implications that will determine the institution's long-term viability. Without such an institutional perspective, there is a risk that the TFFF will be evaluated primarily based on short-term mobilization successes rather than its long-term governance performance.

This chapter addresses this research gap. It analyzes the TFFF as an emerging transnational governance regime. It argues that its success depends less on the amount of funds mobilized than on the quality of its institutional design. Legal architecture, fiduciary structures, monitoring systems, and participation mechanisms are not complementary elements, but rather constitutive prerequisites for effective climate finance. More specifically, it asks how the interaction among legal design, capital market integration, and conditional payment mechanisms shapes the TFFF's institutional resilience and justice performance.

Based on interdisciplinary approaches from law, institutional economics, and climate and forest governance, the chapter develops three central arguments. First, it shows that the financial architecture of the TFFF creates inherent tensions between market rationality and distributive justice, which can only be resolved through binding legal safeguards. Second, it highlights that monitoring, reporting, and verification systems function as epistemic infrastructures that shape power relations and distributional effects. Third, it argues that lasting climate justice can only be achieved if political commitments are translated into enforceable institutional commitments.

Methodologically, the chapter combines legal analysis with comparative policy research and institutional design. It draws on official concept papers, civil society assessments, and emerging implementation frameworks. In addition, case vignettes from Brazil, Indonesia, and selected African forest countries are used to illustrate the interplay between global architectural decisions and national governance contexts. This chapter positions the TFFF within the broader transformations in climate finance, including the transition from project-based funding to capital market integration and rule-based redistribution. In doing so, it contributes to current debates on the legalization and stabilization of transnational environmental governance.

The article is structured as follows: Section 2 reconstructs the financial and institutional architecture of the TFFF. Section 3 analyzes its legal foundations and fiduciary structures. Section 4 examines the anchoring of various dimensions of climate justice in the institution's design. Section 5 deals with monitoring and integrity risks. Section 6 analyzes financial and political vulnerabilities. Section 7 subjects the TFFF to a legal stress test. Section 8 builds on this to develop context-sensitive reform paths. Section 9 concludes with a discussion of strategic policy implications.

By situating the TFFF within broader debates on climate justice and transnational governance, this chapter aims to go beyond instrumental assessments of forest financing. It understands the facility as a key institutional experiment in anchoring global environmental responsibility. Whether this experiment succeeds will depend less on political symbolism than on the robustness of its legal and governance foundations.

THE ARCHITECTURE OF THE TFFF

The Tropical Forest Forever Facility is designed as a long-term, capital market-based financing instrument that aims to secure global forest conservation services through an institutionally stabilized payment architecture. Earlier policy studies had already developed an almost identical model of area-based, sanction-bound, and capital market-based forest financing, including endowment logic, MRV procedures, and asymmetric penalty mechanisms, which makes the TFFF appear to be an institutional further development of an established governance approach.

Unlike project-based funding programs or annually renegotiated budgetary resources, the TFFF aims to embed forest protection in a permanent, investment-based structure. Its architecture combines elements of public climate finance, private capital mobilization, and results-based management into a hybrid governance regime (Moura Costa, 2024).

Capital Structure and Risk Layering

At the heart of this design is the capital structure. The model currently under discussion aims to establish an investment pool of approximately US$125 billion, consisting of a public sponsor tranche and mobilized private capital. The public tranche has an explicitly catalytic function: it serves to absorb risk, increase creditworthiness, and leverage additional private funds. Private investors are involved through market-based return expectations, while returns from long-term capital investment—and not the fund's substance—are to be used to finance annual forest payments.

This endowment logic differs fundamentally from traditional instruments of international climate finance. While programs in the context of REDD+ primarily rely on project-based grants or short-term budgetary funds, the TFFF pursues a long-term investment horizon to generate stable, predictable distributions. Based on optimistic return assumptions, these could reach an annual volume of up to four billion US dollars. This potentially decouples forest protection from volatile budget cycles and institutionalizes it.

At the same time, empirical studies on results-based climate finance show that incentive problems, uncertainties regarding additionality, and limited administrative capacities in recipient countries pose recurring implementation hurdles (Angelsen et al., 2018; Corbera & Schroeder, 2011; Gupta et al., 2012; Streck, 2012; Andrae, 2025). These experiences illustrate that financial volumes alone are not sufficient for effectiveness; they must be embedded in robust institutional arrangements.

Risk layering is central to this context. The more public funds are used in a subordinate position—for example, through guarantees, loss-absorption mechanisms, or hybrid capital—the greater the security for private investors. At the same time, however, the risk of potential fluctuations in returns shifts to the sponsor countries. This structural decision has considerable normative implications, as it determines whether recipient countries receive reliable cash flows or are effectively residual creditors whose claims are considered only after capital market-based obligations have been serviced.

Specifically, the plan is to structure the public tranche as a long-term, subordinated capital instrument with an explicit loss absorption function. These funds are to be provided in the form of junior loans with maturities of up to forty years and are primarily intended to protect private investors. Institutionally, the investment function is separated from the operational management of the TFFF and anchored in an independent Tropical Forest Investment Fund (TFIF) administered by the World Bank. Although this functional fragmentation increases the model's capital market viability, it also exacerbates coordination, liability, and governance issues between the investment and payment regimes (Barber et al., 2025).

The design of the capital structure is thus directly linked to issues of climate justice and the political credibility of the TFFF. If market risks are transferred unilaterally to recipient countries, this undermines the compensatory claim of the instrument. A balanced architecture, therefore, requires mechanisms to secure base returns and smooth out economic fluctuations.

In contrast to earlier forest finance mechanisms such as REDD+ and the Green Climate Fund, which depend largely on periodic public budget allocations and project-based disbursements, the TFFF institutionalizes forest protection through a permanent capital endowment combined with rule-based distribution logic. Whereas previous instruments were characterized by funding volatility, fragmented implementation, and limited enforceability, the TFFF seeks to stabilize incentives structurally through financial layering and juridified conditionality.

Payment Logic

The TFFF's payment logic is based on this financial foundation. At its core is area-based results-based financing, which rewards the conservation of forest area as such rather than avoided emissions. The benchmark currently under discussion, around four US dollars per hectare per year, aims to provide a continuous, low-threshold, and comprehensive incentive for forest conservation (Sutherland et al., 2026). In conjunction with the targeted fund volume, this model marks a significant leap in scale compared to previous instruments. While REDD+ mechanisms have mobilized less than $500 million annually on average since 2008, the TFFF aims to increase this volume by tenfold.

Conceptually, area-based compensation represents a conscious departure from a primarily carbon-centric perspective. Forests are not treated solely as reservoirs of quantifiable emission equivalents, but rather as multifunctional ecosystems that provide ecological, social, and climatic services. The area-based approach thus avoids classic problems of additionality, permanence, and baseline manipulation, which have caused significant credibility deficits in voluntary carbon markets.

This shift from carbon-offset logic to area-based conservation payments represents a conceptual reorientation. Instead of monetizing marginal emission reductions, the TFFF treats standing forests as ongoing providers of global public goods. In doing so, it reduces baseline manipulation risks but simultaneously abstracts from heterogeneous opportunity costs across regions.

At the same time, this model brings new challenges. In regions with high opportunity costs, a flat-rate per-hectare payment may be structurally inadequate. In addition, it reflects ecological differences among primary forests, degraded areas, and regenerating stands only to a limited extent. This creates a conflict among the objectives of administrative simplicity, global scalability, and ecological differentiation.

Sanction Mechanism

Another key control element of the TFFF is its asymmetric sanction mechanism. For each hectare of deforested land, a payment reduction of 100 times the annual compensation is envisaged. This non-linear penalty logic is intended to reflect the long-term loss of ecosystem services in economic terms and to overcompensate for short-term gains from land conversion. While continuous forest conservation yields moderate, predictable returns, deforestation results in abrupt, disproportionate financial losses.

In institutional terms, the sanction mechanism transforms forest governance from a soft incentive regime into a rule-based accountability system. Its effectiveness, however, depends on credible enforcement and symmetrical application across participating states. Without robust institutional safeguards, selective or politically contingent sanctioning risks undermining the regime's legitimacy.

Effective sanctioning further presupposes the integrity of monitoring systems, sufficient political enforcement capacity, and a coherent domestic transmission of incentives and penalties to subnational and private actors. Absent these conditions, the penalty logic risks imposing primarily fiscal burdens on public budgets without substantially altering underlying land-use decisions.

The asymmetric 100:1 sanction logic has already been justified in earlier policy drafts with reference to the long ecological regeneration times of tropical forests and was explicitly designed as an instrument for internalizing irreversible environmental losses (Moura Costa, 2024). This ecological rationale strengthens the mechanism's normative legitimacy but does not substitute for institutional enforceability.

The effectiveness of the payment and sanction architecture ultimately depends on the reliability of the monitoring, reporting, and verification system. A multi-level procedural framework combining satellite-based remote sensing, national reporting, and independent scientific validation is envisaged, supplemented by formal review and appeal mechanisms. As Section 5 demonstrates, the integrity of this system represents one of the central structural vulnerabilities of the TFFF.

Governance and Institutional Embedding

The financial and operational architecture of the Tropical Forest Forever Facility is inextricably linked to its institutional governance structure. To date, the organizational embedding of the TFFF has not been conclusively clarified. Both independent forms of institutionalization and links to existing multilateral development banks are being discussed. Regardless of the formal location, however, key governance functions must be guaranteed, in particular the fiduciary management of funds, independent data verification, transparent decision-making processes, balanced participation of donor and recipient countries, and the institutional representation of indigenous and local communities.

The structure of voting rights and decision-making structures is particularly sensitive in this context. Disproportionate influence by financially strong sponsor countries carries the risk that the TFFF will degenerate into a geopolitical control instrument. Conversely, an overly fragmented governance architecture can significantly impair the fund's ability to assert itself and take operational action. The institutional challenge is therefore to limit power asymmetries without losing the ability to act collectively.

Institutionally, the TFFF thus moves between the logic of a financial fund and the development of an independent governance regime. Its control structures must be compatible with financial markets, secured by the rule of law, and designed in accordance with human rights. Transparency mechanisms also play a central role in this context. The planned annual publication of deforestation rates, payment decisions, and sanctions is intended to generate targeted reputational pressure and strengthen national political accountability.

The governance structure thus directly influences the functionality of all other architectural components. It determines whether payments are made reliably, sanctions are enforced credibly, participation rights are effectively institutionalized, and long-term political stability is achieved. Capital structure, payment logic, sanction mechanisms, and governance form an integrated system in which weaknesses in any of the components can undermine the overall integrity.

The TFFF, therefore, faces less of a primarily financial issue than an institutional architectural one. Only if design, law, and justice are coherently conceived and organizationally anchored can they develop into a permanently stable instrument of global forest governance.

Taken together, capital structure, payment logic, sanction mechanisms, and governance arrangements form an integrated institutional architecture. Weaknesses in fiduciary design, monitoring integrity, or sanction enforcement may cascade across the entire regime. The TFFF, therefore, represents not merely a financing instrument but an architectural experiment in institutionalizing long-term ecological responsibility. Its effectiveness depends on whether these elements are coherently coordinated and institutionally secured. The TFFF should therefore be understood less as a single financial instrument and more as a comprehensive governance regime for the long-term stabilization of global forest protection services.

THE TFFF AS A TRANSNATIONAL LEGAL CONSTRUCT

The Tropical Forest Forever Facility is not merely a financial innovation project, but a normative institution in the making. Its viability depends not only on the economic plausibility of its return model, but also on the legal quality of its construction. If the TFFF is to have a lasting, reliable, and equitable impact, it must be designed as a transnational legal system with clear legal personality, fiduciary asset commitment, defined liability distribution, and enforceable conditionality. Otherwise, it will remain a politically ambitious but legally fragile promise.

The following analysis views the TFFF as a hybrid legal construct at the intersection of international law, national budgetary law, capital market law, trust law, and human rights protection standards. Its stability depends on whether these levels are coherently intertwined.

Legal Personality and Institutional Anchoring

The first and fundamental question concerns the legal nature of the TFFF. Without its own legal personality, it would have neither legal capacity nor standing to sue. It could not conclude contracts, hold assets, or assert or defend rights in court. A mere coordinating platform or political agreement without its own legal personality would structurally undermine the instrument's functionality.

The decision on institutional embedding, therefore, determines liability regimes, immunity issues, supervisory mechanisms, and the scope of judicial control. Three basic models are conceivable:

First, an entity established under international law by a multilateral treaty, with its own international legal personality. This model gives the TFFF high political legitimacy and symbolic equality among member states. At the same time, however, it gives rise to the classic problems of international organizations: limited justiciability, immunity issues, and potential deficits in individual enforcement.

Secondly, a foundation or trust structure under national law. Such a model regularly ensures strong asset commitment and clearly defined fiduciary duties. However, it is subject to the jurisdiction of a single legal area and thus to possible extraterritorial influences.

Thirdly, the establishment of a legally separate special vehicle within a multilateral development bank. This solution offers institutional experience, administrative infrastructure, and market confidence, but carries the risk of structural power asymmetries in favor of dominant shareholders.

From a legal perspective, it is not the symbolic form that is decisive, but the functional design: the TFFF must act as an independent legal entity, strictly separate its assets from the budgets of the sponsor states, and have clearly defined bodies with normatively defined powers. Without this institutional clarity, there is a risk of liability diffusion, ambiguities of competence, and political instrumentalization.

Trust Structure and Asset Binding

At its core, the Tropical Forest Forever Facility operates as a permanent capital vehicle whose investment returns finance results-based forest conservation payments. This design presupposes an insolvency-remote, purpose-bound, and institutionally insulated asset structure that protects financial resources from short-term political appropriation and budgetary volatility.

From a legal-institutional perspective, the TFFF can be interpreted as a transnational trust and asset-management arrangement whose central function is the juridical separation of fund assets from sponsor-state finances and their long-term dedication to forest protection. Comparable fiduciary models in corporate and charitable governance demonstrate that such separation is a key precondition for institutional credibility and intertemporal commitment (Langbein, 1995; Hansmann & Kraakman, 2000; DeMott, 2019; Sitkoff & Dukeminier, 2021).

The fund's assets must therefore remain strictly segregated from public budgets, exclusively allocated to performance-based forest payments, and shielded from third-party claims. Without such safeguards, the TFFF risks degenerating into a politically adjustable special fund rather than a durable governance institution.

Fiduciary duties imposed on governing bodies constitute the normative core of this trust arrangement. These include duties of care in investment decisions, loyalty toward beneficiaries, impartiality between recipient states, and prudence regarding long-term capital preservation. Together, these obligations constrain discretionary power and transform political commitments into legally structured responsibilities.

A particularly sensitive issue concerns the legal status of beneficiaries. If forest countries are treated merely as recipients of discretionary transfers, payments remain politically contingent. If, by contrast, disbursements are strictly tied to verifiable performance criteria, recipient states acquire conditional legal entitlements resembling enforceable benefit rights. The latter model strengthens institutional reliability and reduces vulnerability to political bargaining.

Importantly, the trust structure also has distributive implications. By stabilizing payment expectations and limiting discretionary interference, fiduciary asset binding contributes to procedural and distributive fairness. It creates predictable institutional pathways through which forest protection efforts can translate into durable financial recognition. In this sense, trust law does not merely serve technical asset management but functions as a juridical infrastructure of climate justice.

Liability Allocation and Risk Distribution

The Tropical Forest Forever Facility operates at the intersection of public climate finance and private capital markets. This hybrid positioning generates complex liability and risk-allocation structures that are decisive for the regime's institutional stability and normative legitimacy. The legal architecture of the TFFF must therefore establish clear responsibility frameworks that define how financial, political, and operational risks are distributed among investors, sponsor states, recipient countries, and governing bodies.

From the perspective of private investors, legal certainty regarding the ranking of capital tranches, the scope of public guarantees, regulatory classification, and disclosure obligations is essential. When sponsors provide subordinated guarantees or first-loss capital, investor risk is significantly reduced, while fiscal exposure is transferred to public balance sheets. Such arrangements may qualify as contingent liabilities under domestic budgetary law and therefore require enhanced parliamentary oversight and transparency.

At the same time, liability relations vis-à-vis recipient states raise fundamental distributive concerns. If disbursements are tied strictly to market performance, forest countries become residual risk bearers whose income streams fluctuate with global financial volatility. Absent legally secured minimum payment thresholds or stabilization mechanisms, this structure privileges capital protection over ecological performance and undermines the redistributive rationale of the TFFF.

Beyond financial exposure, governance-related risks play a critical role. Erroneous monitoring data, methodological inconsistencies, or politically selective sanctions may result in substantial economic losses and reputational harm. Without clearly defined accountability mechanisms, affected states and communities lack effective legal remedies. Robust institutional design, therefore, requires independent review bodies, formal objection procedures, reason-giving obligations, and binding dispute-resolution mechanisms, including international arbitration where appropriate.

Importantly, liability allocation also has a temporal dimension. Investment losses may materialize immediately, while ecological degradation and governance failures often unfold over longer periods. If short-term financial risks are systematically externalized to public actors and recipient states, while long-term environmental risks remain weakly internalized, the regime reproduces structural asymmetries characteristic of earlier climate finance instruments.

A coherent liability framework must therefore balance investor protection, fiscal responsibility, and beneficiary security. Only if financial risk-sharing, procedural safeguards, and corrective mechanisms are institutionally aligned can the TFFF function as a credible vehicle for long-term, rights-based forest governance.

Legal Bindingness of Payment and Sanction Mechanisms

The economic logic of the TFFF is built on clearly defined performance triggers: compliance with agreed deforestation thresholds generates payment entitlements, while threshold violations trigger sanctions, reductions, or suspensions. The legal character of these mechanisms is decisive for the regime's credibility. The central question is not merely whether payments are conditional, but whether such conditionality operates as a rule-bound legal consequence or as politically adjustable discretion.

A legally robust model requires that performance indicators, threshold values, and calculation methodologies be precisely defined ex ante. Ambiguous benchmarks or discretionary reinterpretations risk transforming conditionality into a flexible political instrument rather than a predictable governance tool. Rule-based design enhances legal certainty and stabilizes expectations among both recipient states and investors.

At the same time, legal bindingness does not necessarily exclude all forms of discretion. Limited interpretative margins may be required to address force majeure events, data anomalies, or extraordinary ecological circumstances. However, such discretion must be procedurally constrained through transparency obligations, reason-giving requirements, and formal documentation standards. Without procedural discipline, discretion may erode institutional trust.

Equally important is the availability of effective review mechanisms. Recipient states must have access to impartial bodies capable of assessing contested monitoring results or sanction decisions. Whether through specialized review panels, arbitration clauses, or hybrid dispute-settlement mechanisms, the possibility of independent scrutiny constitutes a core element of juridification.

Selective or inconsistent enforcement represents a particular systemic risk. If sanction decisions vary according to geopolitical alliances, fiscal interests, or diplomatic considerations, the regime's normative authority deteriorates. The legitimizing force of conditional finance derives precisely from the equal and general application of predefined rules.

In this sense, the bindingness of payments and sanctions is not a technical detail but a structural foundation of the TFFF. Only where performance-based finance is anchored in enforceable, transparent, and reviewable legal commitments can it transcend the volatility and fragmentation that have characterized earlier forest governance instruments.

Clawback Mechanisms and Repayment Logic

In the debate over securing long-term protective effects, there is growing discussion of suspending payments in the event of serious or repeated deforestation, and of converting them into repayable, interest-free loans. Such a clawback mechanism can strengthen permanence assurance, as it reduces short-term misguided incentives and institutionally anchors long-term responsibility. At the same time, this instrument raises significant questions under international, budgetary, and contract law.

In particular, it must be clarified whether the reclassification of grants as repayment obligations creates new government debt, how such claims are to be recorded in the balance sheet, what repayment terms and deferral options apply, and to what extent insolvency or debt restructuring risks are affected. Without clear legal regulations, hybrid grant-loan structures threaten to create considerable legal uncertainty and further exacerbate the existing debt problems of many forest countries. A viable clawback regime, therefore, requires precise contractual definitions, transparent accounting rules, and close coordination with international debt and debt relief mechanisms.

Additionality, Fiscal Integrity, and Human Rights Commitments

A central legal precondition of the TFFF architecture is the fiscal additionality of sponsor-state contributions. Payments to the facility must not substitute existing climate finance obligations or development assistance commitments but must constitute verifiable net additions to public spending on global environmental protection. Without binding additionality rules, the TFFF risks becoming an accounting device for relabeling pre-existing budget lines rather than a genuine redistributive mechanism.

From a public finance perspective, this requirement necessitates clear budgetary classification, harmonized reporting standards, and robust safeguards against double-counting. Contributions must be transparently aligned with Official Development Assistance criteria and the New Collective Quantified Goal on Climate Finance. The indiscriminate classification of sovereign bond investments or fund shares as climate finance would undermine the integrity of international reporting systems and weaken mutual accountability among donor states.

Fiscal integrity is therefore inseparable from the credibility of the global climate finance architecture. If major contributors systematically re-label financial instruments without expanding real fiscal effort, collective trust deteriorates, and burden-sharing arrangements erode. The TFFF consequently depends on autonomous accounting and disclosure rules that ensure international comparability and institutional legitimacy.

Beyond fiscal discipline, the TFFF is embedded within binding international human rights frameworks. Forest protection policies directly affect land tenure, livelihood security, cultural identity, and political participation of indigenous peoples and local communities. As a result, payment and governance structures must comply with established human rights standards, including free, prior, and informed consent, non-discrimination, and effective participation rights.

The proposed minimum allocation of 20% of funds to indigenous peoples and local communities (IPLCs) raises further questions about their status as third-party beneficiaries. If legally codified, such a quota may transform IPLCs from political stakeholders into holders of conditional legal entitlements. This would require clarification of enforceable claims, direct access modalities, the implementation of FPIC standards, and independent monitoring at national and international levels.

Without robust legal embedding, the twenty percent allocation risks remaining symbolic. Comparative evidence shows that secure land tenure and effective participation rights constitute key prerequisites for durable forest protection (Larson et al., 2013; Rights and Resources Initiative, 2020). Human rights safeguards are therefore not peripheral add-ons but structural conditions for the TFFF's long-term effectiveness and legitimacy.

CLIMATE JUSTICE IN THE DESIGN OF THE TFFF

The legal and institutional architecture of the Tropical Forest Forever Facility constitutes the structural precondition for its political and ecological effectiveness. However, this architecture is not normatively neutral. Its legitimacy unfolds only in relation to questions of climate justice. The TFFF seeks to address a central structural contradiction of global climate governance: while tropical forests generate global public goods in the form of carbon sequestration, biodiversity conservation, and climate regulation, the opportunity costs, governance burdens, and social conflicts associated with their protection are concentrated in forest-rich countries of the Global South (Andrae, 2026; Salcedo-La Viña et al., 2025).

Recent governance research emphasizes that territorial conservation rules and demand-side regulations are unlikely to curb deforestation sustainably unless they are complemented by international redistributive mechanisms that strengthen enforcement capacities and fiscal resilience in producer countries (Muradian & May, 2026; Salcedo-La Viña et al., 2025; Barber et al., 2025). In this perspective, the TFFF functions as a redistributive complement to regulatory instruments such as the EU Deforestation Regulation and as a core element of an emerging hybrid governance regime (Muradian & May, 2026). Climate justice, therefore, operates not as a corrective add-on but as a constitutive design principle.

This chapter adopts a multidimensional understanding of climate justice that integrates distributive, procedural, recognition-based, and intergenerational dimensions (Schlosberg, 2007; Fraser, 2009; Okereke & Coventry, 2016). The institutional architecture of the TFFF determines whether these dimensions are translated into durable governance arrangements or remain rhetorical commitments.

From a distributive perspective, justice concerns arise primarily from the allocation of costs, benefits, and risks among sponsor states, recipient countries, and local actors. By mobilizing capital from high-emission economies and channeling investment returns into forest protection payments, the TFFF institutionalizes a form of global burden-sharing. However, where capital market volatility is insufficiently buffered, recipient countries may indirectly bear financial risks they did not create (Andrae, 2026; Barber et al., 2025). Distributive justice, therefore, requires legally secured stabilization mechanisms and differentiated payment structures reflecting heterogeneous opportunity costs.

Procedural justice concerns the conditions under which affected actors participate in decision-making and contest institutional outcomes. Although the TFFF formally provides participatory rights for recipient countries and indigenous peoples, their effectiveness depends on voting arrangements, access to information, and legal remedies. Comparative studies of climate finance governance show that weak procedural safeguards systematically undermine accountability and implementation capacity (Salcedo-La Viña et al., 2025).

Recognition-based justice addresses the legal and symbolic status of indigenous peoples and local communities within the governance regime. Secure land tenure, legal personality, and direct access to financial mechanisms are central preconditions for meaningful participation. Extensive empirical evidence demonstrates that where community land and resource rights are legally protected, deforestation rates decline and governance performance improves (Larson et al., 2013; Rights and Resources Initiative, 2023). The TFFF's design must therefore embed recognition not merely rhetorically but through enforceable legal entitlements and third-party beneficiary structures.

Intergenerational justice is reflected in the facility's "Forever" logic. The endowment-based financing model may institutionalize long-term obligations to future generations. Yet this promise depends on strict asset-preservation rules, insulation from political budget cycles, and robust fiduciary governance. Without such safeguards, long-term commitments remain vulnerable to short-term political incentives (Andrae, 2026; Barber et al., 2025).

These justice dimensions are mediated through concrete institutional choices. Risk-layering arrangements shape distributive outcomes; fiduciary duties structure procedural fairness; MRV systems influence epistemic authority; and third-party beneficiary rules determine recognition (Salcedo-La Viña et al., 2025; Rights and Resources Initiative, 2023). Climate justice within the TFFF is therefore not produced by ethical declarations but by legally structured governance mechanisms.

Persistent tensions nevertheless remain. The integration of capital markets may privilege return-oriented rationalities over ecological priorities. Global standardization facilitates scalability but risks marginalizing local contexts. Strict conditionality strengthens accountability but may conflict with national sovereignty. Empirical analyses of emerging forest finance instruments suggest that unresolved distributional and participation conflicts constitute major sources of institutional fragility (Barber et al., 2025; Andrae, 2026).

Overall, the analysis demonstrates that climate justice within the TFFF emerges from institutional design rather than moral intent. Distributive, procedural, recognition-based, and intergenerational justice are embedded in payment rules, participation rights, fiduciary structures, and enforcement mechanisms. The TFFF thus functions as a laboratory for institutionalizing climate justice. Its success will depend less on political rhetoric than on the consistent legal implementation of its normative commitments.

MONITORING, VERIFICATION, AND INTEGRITY RISKS

The effectiveness of the Tropical Forest Forever Facility depends to a large extent on the credibility of its monitoring, reporting, and verification (MRV) system. Unlike project-based climate instruments, the TFFF is entirely linked to objective performance indicators. Forest condition, deforestation rates, and degradation processes form the immediate basis for payment entitlements and sanction mechanisms. The MRV system is therefore not a technical ancillary instrument, but rather the epistemic foundation of the entire regime. Methodological weaknesses, distortions, or political manipulations directly affect the distribution of global financial flows (Herold & Skutsch, 2011; Grassi et al., 2018; Neumann et al., 2025; GOFC-GOLD, 2013).

The architecture of the TFFF is primarily based on satellite-based remote sensing, supplemented by national monitoring programs and external scientific validation. On paper, this creates a multi-level security system consisting of data collection, algorithmic classification, and independent verification (Moura Costa, 2024). In practice, however, there are structural uncertainties at all interfaces. Limited spatial and temporal resolution makes it difficult to detect small-scale clearing, selective logging, and gradual degradation processes. Classification models are also based on thresholds and training data, the determination of which is not politically neutral. Even the definition of forest areas—for example, a minimum tree cover of 20%—influences which areas are considered worthy of protection and which forms of degradation remain invisible (Sutherland et al., 2026).

In operational terms, MRV procedures are expected to follow a multi-stage process combining satellite-based detection, national reporting, independent peer review, and final verification by expert panels. Initial deforestation signals generated through remote sensing are first processed and classified by centralized data platforms. These preliminary assessments are then transmitted to national authorities for validation and contextualization. Subsequently, independent scientific reviewers examine methodological consistency and error margins. Final payment and sanction decisions are adopted by designated verification bodies following formal objection periods and technical reassessments. Comparable procedures are currently applied in REDD+ frameworks and World Bank forest monitoring programs, where institutionalized dispute resolution mechanisms play a central role in maintaining system credibility.

In addition, the technical definitions of forest currently under discussion set comparatively low thresholds, such as a crown density of only 20 to 30% at a minimum height of 5 meters. Although plantations are formally excluded, this creates gray areas in which ecologically degraded areas can still be classified as forests worthy of protection. Furthermore, monitoring has so far focused heavily on fire-related damage, while selective logging, infrastructure fragmentation, and creeping degradation are insufficiently recorded. These systematic blind spots increase the risk of strategic adjustments and undermine the ecological integrity of the payment logic (Barber et al., 2025).

The strong focus on formal land loss is particularly problematic. Many ecological functions decline gradually through biomass decline, fragmentation, or repeated disturbances. If these processes are not taken into account, there is an incentive for 'legalized erosion,' in which nominal forest areas are preserved while their ecological integrity declines. Comparable distortions have already been documented in the context of REDD+ programs (Angelsen et al., 2018; Corbera & Schroeder, 2011).

The increasing use of AI-supported classification models exacerbates this problem. The more payment decisions are based on algorithmically generated data, the greater the power imbalance between data producers, model operators, and recipient countries. Without transparent disclosure of training data, error rates, and weighting mechanisms, forms of black-box governance emerge whose decisions can only be verified to a limited extent (Kitchin, 2014). This raises fundamental constitutional questions about traceability, evidentiary value, and the right of appeal.

In addition to methodological uncertainties, there are considerable political integrity risks. National authorities are often involved in data collection and validation. In politically sensitive contexts, definitions can be adjusted, publications delayed, or classifications selectively interpreted. Even without overt manipulation, strategic information control can effectively undermine sanctions. At the same time, there is a risk of leakage effects, whereby deforestation is shifted to other regions or countries with weaker standards (Angelsen et al., 2018).

These integrity issues are highly relevant from a legal perspective. Payment decisions must be justified robustly in the event of a dispute. However, it often remains unclear who is liable for methodological errors: national data providers, model operators, verification bodies, or the TFFF itself. Without a clear allocation of responsibility, diffuse liability and accountability structures arise, as is typical for transnational regimes (Abbott et al., 2017; Young, 2017).

The institutional stability of the TFFF, therefore, requires robust legal safeguards for its MRV system. Multi-source verification, open data standards, transparent algorithms, independent audits, and the systematic integration of local knowledge are not optional additions, but constitutive elements of the governance architecture. MRV does not function as a downstream control mechanism, but as a constitutive component of normative legitimacy.

FINANCIAL AND POLITICAL RISK PROFILE

Even a technically robust MRV system does not guarantee institutional stability. The Tropical Forest Forever Facility is designed as a capital market-based financing instrument and is therefore inherently exposed to cyclical market movements and political power constellations. Its distributions are not based on guaranteed budgetary resources, but on returns from a diversified investment portfolio. This shifts a significant portion of the financial risk to recipient countries, whose payment entitlements depend on global market conditions (de Nevers et al., 2018).

These structural risks are exacerbated by the limited success of the public tranche in mobilizing funds to date. Of the targeted US$25 billion in government start-up capital, only around US$6.7 billion has been committed so far. This financing gap weakens the signal effect on private investors, increases the perceived refinancing risks, and makes it more difficult to build the intended leverage effect. Without a substantial increase in public anchor financing, the TFFF risks entering a fragile pre-financing phase (Barber et al., 2025).

Capital market volatility represents the first systemic vulnerability. Fluctuating interest rates, changes in the valuation of bond and equity portfolios, and ESG-specific market dynamics directly influence the fund's profitability. In periods of low interest rates, there is a risk of structural yield bottlenecks, while financial crises can lead to abrupt losses. Comparable risks have been described many times in the literature on development policy and financial market integration (Mosley, 2010). Without legally binding minimum payment guarantees, there will be years with reduced or no distributions – a structural pattern of voluntary climate finance instruments (Newell & Mulvaney, 2013; World Bank, 2017).

Exchange rate movements further exacerbate these uncertainties. Revenues, investments, and disbursements take place in different currency areas. Devaluations in recipient countries can reduce real payment effects, while appreciations make commitments more expensive. Without systematic hedging, unintended distribution shifts occur. In addition to financial risks, political cycles are a second dimension of vulnerability. The TFFF is designed for long-term stability, but operates in political systems with short-term decision-making logics. Changes in government can call sponsor contributions into question, renegotiate conditionalities, or block governance bodies. In the absence of a binding multilateral legal basis, commitments remain politically disposable (Keohane & Victor, 2011).

Stress scenarios illustrate these vulnerabilities. A prolonged global recession could reduce investment returns below projected thresholds, thereby destabilizing the expected flow of performance-based payments. In such a scenario, financial volatility would be transmitted to forest countries through declining or delayed transfers. A second scenario concerns political reversals in major donor states. Changes in fiscal priorities, parliamentary resistance, or climate policy rollbacks may delay capital injections or weaken guarantee commitments. Given the hybrid public–private structure of the TFFF, even moderate political uncertainty can amplify market risk premia and erode investor confidence. A third scenario involves abrupt regulatory or monetary shifts, including changes in financial market regulation, capital controls, or sovereign risk ratings. Such disruptions may trigger capital reallocation, currency instability, or liquidity constraints, undermining the long-term investment strategy of the facility.

Geopolitical tensions exacerbate this dynamic. Climate financing is increasingly embedded in strategic power constellations. Rivalries between major powers, fiscal constraints, or economic recessions can politicize commitments and create instruments of external influence ('weaponized interdependence') (Farrell & Newman, 2019). Another structural risk lies in governance capture. Close ties between fund management, political decision-makers, and investors can foster conflicts of interest. Susceptibility to corruption, opaque fund transfers, or shortcomings in implementing the intended participation quotas for indigenous and local communities undermine the legitimacy of the instrument (Young, 2017; Muradian & May, 2026).

The interdependence of these risk dimensions is particularly problematic. Financial losses can create political pressure to relax MRV standards to maintain cash flows. Data conflicts can unsettle investors and increase capital outflows. Sovereignty conflicts can encourage selective sanctions waivers. This creates self-reinforcing spirals of erosion, as is typical for complex regime complexes (Ostrom, 2010; Keohane & Victor, 2011).

The long-term stability of the TFFF, therefore, requires an integrated resilience architecture. This includes countercyclical reserves, legally secured minimum payments, multi-year commitment mechanisms, strict compliance rules, independent ethics and supervisory bodies, and transparent accountability structures. Only when financial security, legal self-commitment, and political legitimacy are systematically interlinked can the TFFF evolve from an experimental hybrid instrument into a permanently stable institution of global climate governance.

LEGAL STRESS TEST

The preceding sections have shown that the Tropical Forest Forever Facility, as an institutional project, is intended to perform financial, legal, epistemic, and political regulatory functions. It is intended to mobilize capital markets, ensure ecological integrity, institutionalize social justice, and guarantee political stability. This multifunctionality makes the TFFF innovative, but it also increases its structural fragility. A legal stress test systematically reveals this fragility.

Such a legal stress test does not ask whether the TFFF functions under ideal conditions, but whether it remains capable of acting under realistic crisis scenarios: market turbulence, political power shifts, data conflicts, and governance disruptions. Only under these conditions will it become clear whether a political experiment can become a resilient institution. Comparable systemic vulnerabilities have also been documented in other complex transnational regulatory regimes (Abbott et al., 2017; Young, 2017).

The central question here is whether the TFFF's legal binding mechanisms are strong enough to absorb economic and political pressure without sacrificing its fundamental normative principles. The first dimension of stress concerns financial shocks. The TFFF's capital market-based returns are subject to cyclical fluctuations. In times of global financial crises, rising interest rates, or massive capital outflows, returns can plummet. In such a scenario, distributions to recipient countries come under pressure. Without legally guaranteed minimum payments, a conflict arises between investors' interests and climate protection obligations. The lack of legally guaranteed minimum payments corresponds to a structural pattern in which voluntary climate finance instruments suffer from a permanent credibility deficit (Newell & Mulvaney, 2013; World Bank, 2017). Legally, the question arises whether payments are intended as non-binding discretionary payments or as enforceable claims. If the latter is not the case, the risk shifts structurally to the detriment of forest countries. The stress test reveals a key weakness: as long as the TFFF does not include hard payment guarantees, it remains vulnerable to crises and politically susceptible to blackmail.

A second dimension of stress arises from political upheavals. Changes of government in donor or recipient countries, national sovereignty discourses, or geopolitical conflicts can destabilize the basis for cooperation under the TFFF. In such situations, pressure mounts to relax conditionalities, suspend sanctions, or politicize governance decisions. The legal relevance here is whether obligations have multilateral treaty status or are based solely on political declarations of intent. In the absence of a binding legal basis, deviations from the rules will in fact have no consequences. In this case, the TFFF would regress from a rule-bound institution to a negotiable political arrangement. The stress test thus reveals a legitimacy risk: the weaker the formal legal framework, the higher the probability of selective application.

The third dimension of stress concerns the MRV system, which serves as the epistemic foundation of the TFFF. Data conflicts, methodological uncertainties, or allegations of manipulation can call into question the basis for payments and sanctions. In such a scenario, conflicts shift from the factual level to the knowledge level. From a legal perspective, the system must have clear rules of evidence, appeal procedures, and independent review bodies. Without these safeguards, there is a risk of protracted disputes over data validity that block payment processes. The stress test reveals a structural vulnerability: as long as algorithmic classification systems are not fully transparent and verifiable, the TFFF remains vulnerable to epistemic conflicts of authority.

A fourth dimension of stress arises from governance disruptions and institutional erosion. Conflicts of interest, political influence, or bureaucratization can undermine the fiduciary logic of TFFF. When fund management, political decision-makers, and investors are functionally intertwined, the institution loses its independence. Legally, this raises the question of whether compliance rules, liability standards, and sanction mechanisms are effective. In many international funds, governance misconduct has virtually no consequences because responsibilities are diffuse. The stress test reveals that TFFF is only resilient if individual and institutional liability is clearly assigned. Without personal accountability, integrity remains programmatic.

A fifth dimension of stress concerns the involvement of indigenous and local communities. In crises, participatory structures are often treated as dispensable. Budget cuts, administrative simplification, or political centralization can effectively undermine direct access and co-decision rights. It is therefore legally crucial whether IPLC rights are designed as enforceable third-party beneficiary positions or merely as programmatic targets. The stress test shows that without justiciable rights, social inclusion risks becoming symbolic rhetoric.

A sixth dimension of stress arises from cumulative interactions. Financial crises, political conflicts, data problems, and governance scandals rarely occur in isolation. They reinforce each other. For example, a slump in returns can create political pressure to relax MRV standards to maintain payments. This, in turn, undermines integrity and reputation, unsettling investors and increasing capital outflows. Self-reinforcing spirals of erosion emerge.

The legal stress test clearly shows that the TFFF's stability is not additive but systemic. Weaknesses in one area can destabilize the entire regime.

From a legal doctrine perspective, four minimum structural requirements can be derived from this stress test.

1. First, the TFFF requires the strict legal codification of key obligations. Payment claims, sanction mechanisms, and governance structures must be set out in contracts or international law to remove them from political opportunism.
2. Second, institutionalized risk buffering is necessary. Countercyclical reserves, minimum payment guarantees, and endowment structures are not merely financial instruments, but legal anchors of stability.
3. Third, epistemic authority must be legally controllable. Transparent algorithms, open data, and independent review bodies are constitutional elements of the TFFF.
4. Fourth, climate justice requires enforceable subject positions. Without legally protected participation and entitlement rights, justice remains politically contingent.

Taken together, the legal stress test reveals that financial fragility, governance capture, and weak enforceability constitute mutually reinforcing risk clusters rather than isolated vulnerabilities. Market volatility amplifies political discretion; weak accountability enables selective enforcement; and institutional opacity erodes fiduciary discipline. Institutional resilience, therefore, depends on the simultaneous mitigation of these risks through binding fiduciary duties, transparent and rule-based conditionality, and credible, independent dispute-resolution mechanisms. Only where financial architecture, legal safeguards, and procedural guarantees are systematically aligned can the TFFF withstand prolonged political and economic stress.

CASE VIGNETTES AND REFORM PATHWAYS

The previous Legal Stress Test revealed the structural vulnerabilities of the Tropical Forest Forever Facility. However, these risks are not abstract; they manifest in specific political, legal, and socioeconomic contexts. To illustrate the institutional challenges facing the TFFF, three regional case vignettes are analyzed below. These case vignettes illustrate how the abstract design principles analyzed in Sections 2 to 4 translate into concrete governance outcomes under heterogeneous political, legal, and socio-economic conditions. They demonstrate that institutional coherence, rather than financial volume alone, determines long-term effectiveness. In particular, the interaction between capital structure, fiduciary governance, participation rights, and monitoring systems shapes whether redistributive mechanisms generate durable behavioral change or merely short-term compliance effects. The cases therefore function as institutional stress tests that reveal how legal design, justice arrangements, and financial architecture jointly condition implementation capacity.

Brazil: Political Volatility and Institutional Resilience

As the initiator of the TFFF and the country with the largest share of tropical rainforests, Brazil is a key reference case. At the same time, it is characterized by extreme political cycles that fluctuate between environmental reform and extractivist deregulation.

In phases of strong environmental protection policy, Brazil has comparatively efficient MRV systems and specialized environmental agencies. In politically regressive phases, however, control bodies are weakened, budgets are cut, and enforcement capacities are reduced. The stress test becomes empirically tangible here: institutions that are not sufficiently secured by law are subject to political reversibility. The Brazilian experience reflects well-known tensions between institutional enforcement capacity and political cyclicality, as described many times in the REDD+ literature (Angelsen et al., 2018; Gupta et al., 2012).

At the same time, there are significant problems of illegal land grabbing, organized environmental crime, and regional power structures that undermine national control instruments. In this context, TFFF payments could either act as a stabilizer or, if conditionality is weak, reproduce existing power asymmetries.

Reform path: For Brazil, it is crucial to link payments to enforceable national protection instruments. These include:

- legally established forest protection funds,
- independent environmental prosecutors,
- legally secured budget commitments,
- direct financial channels for indigenous territories.

The TFFF should act as a reinforcement of the institutional rule of law, not as a substitute. On this basis, reform paths will then be developed that enable the transition from an experimental financial instrument to an adaptive governance regime.

Indonesia: Financial Market Integration and Sectoral Leakage

A strong export orientation, intensive palm oil and timber production, and close links between politics and corporate structures characterize Indonesia. Deforestation here is caused less by smallholder expansion than by industrial land use. Similar dynamics can also be observed in export-oriented forest economies, where industrial land use and political interdependence create structural governance risks (Corbera & Schroeder, 2011; Streck, 2012).

The MRV system is comparatively effective, but sectoral control remains fragmented. While formal forest areas are protected, activities are shifting to peatlands, secondary forests, or agroforestry zones. The leakage effect identified in the stress test is particularly evident here.

In addition, there is strong integration into global capital markets. Sustainability standards compete with short-term return interests. Without being linked to financial market regulation, the TFFF risks having an isolated effect.

Reform path: For Indonesia, the focus is on regulatory integration:

- linking TFFF payments to lending standards,
- involving banking and stock exchange supervision,
- cross-sector land use databases,
- Commitment to corporate disclosure.

The TFFF should serve as a lever for deforestation-free value chains.

Democratic Republic of Congo and Ghana: Tenure, Capacity, and Fiscal Transmission

The DR Congo and Ghana represent contexts of weak statehood, fragmented land rights, and limited administrative capacities. Here, the main problem is not political volatility or financial market integration, but fundamental institutional weaknesses.

Land use rights are often informal, overlapping, or disputed. Payments to central governments often do not reach local actors adequately. The disconnect between formal conditionality and actual impact identified in the stress test is particularly evident here.

At the same time, there is great potential for community-based protection structures, which have so far been insufficiently integrated into the institutional framework.

Reform path: In these contexts, institutional preconditions are central:

- Support for land rights mapping,
- legal recognition of collective forms of ownership,
- establishment of local trust funds,
- decentralized payment systems,
- long-term capacity programs.

The TFFF must play a stronger role here as an institutional development partner.

Systemic Reform Paths: From Case Contexts to Governance Architecture

Overarching reform modules can be derived from the case vignettes, which must be combined in a context-sensitive manner.

1. Differentiated payment models. Uniform per-hectare rates should be supplemented with context-specific surcharges that account for opportunity costs and governance performance.
2. Legally secure direct channels. Payments to IPLCs must be made through fiduciary-protected structures that are immune to political diversion.
3. Hybrid financing instruments. Combinations of grants, guarantees, and reserve funds increase stability and adaptability.
4. Regulatory coupling. Integration with supply chain law, banking supervision, and sustainability reporting enhances impact.

5. MRV architectures. Regional add-on modules and community monitoring reduce blind spots.
6. Institutionalized learning mechanisms. Regular design reviews, sunset clauses, and pilot programs ensure reformability.

From the Experimental Phase to Institutional Maturity

The case vignettes show that the TFFF cannot function as a standardized instrument. Its effectiveness depends on the ability to translate global principles into local institutions.

Legally, this means a transition

- from blanket conditionality to contextualized commitment,
- from technocratic control to rights-based co-production,
- from project-based intervention to institutional embedding.

Only if this transformation succeeds can the TFFF break out of the logic of temporary climate financing.

Taken together, the three cases confirm that variation in outcomes is driven less by funding levels than by differences in legal embedding, governance capacity, and participation structures. This reinforces the central argument that institutional design conditions distributive and ecological performance.

Reform is therefore not a one-off act, but a permanent institutional process. The TFFF must be constructed as a learning organization whose legal and financial architecture is systematically adapted to new risks. Only under this condition can it withstand the stresses identified in the legal stress test and unfold its potential as an instrument of global climate justice.

CONCLUSION

This chapter advances three central conclusions.

First, durable tropical forest protection requires legally anchored trust and endowment structures rather than fragmented, project-based financing. Only institutionally insulated and fiduciary-bound arrangements can stabilize long-term conservation incentives and protect environmental commitments against political and economic volatility.

Second, climate justice within global forest governance depends on enforceable participation, tenure security, and third-party beneficiary rights. Without legally embedded access, complaint mechanisms, and recognition of indigenous and local communities, redistributive finance risks reproducing existing asymmetries.

Third, financial scale alone cannot substitute for institutional coherence. High capital volumes do not compensate for weak governance, opaque monitoring systems, or discretionary enforcement. Sustainable effectiveness emerges from the alignment of legal safeguards, accountability mechanisms, and financial architecture.

Based on these findings, three policy priorities emerge.

First, fiduciary governance must be strengthened through independent oversight, binding asset-protection rules, and transparent accountability structures. This includes clearly defined duties of care, loyalty, and prudence for all managing entities.

Second, IPLC participation should be institutionalized through legally enforceable rights, direct access mechanisms, and robust grievance procedures. Participation must function as a governance instrument rather than a symbolic consultation format.

Third, capital market integration must be systematically aligned with ecological risk management. Stabilization buffers, countercyclical reserves, and precautionary investment rules are necessary to prevent financial volatility from undermining conservation outcomes.

Taken together, these priorities underline that the TFFF's long-term success depends less on political symbolism or headline funding volumes than on the robustness of its legal and institutional foundations. The facility constitutes a critical experiment in institutionalizing global environmental responsibility. Whether this experiment succeeds will be determined by its capacity to translate normative ambitions into enforceable governance arrangements.

REFERENCES

Abbott, K. W., Levi-Faur, D., & Snidal, D. (2017). Theorizing regulatory intermediaries: The RIT model. *The Annals of the American Academy of Political and Social Science, 670*(1), 14–35. DOI: 10.1177/0002716216688272

Andrae, S. (2025). Financing mechanisms for the conservation of biodiversity: A multi-country comparison. In Walker, T., Naffa, H., Tharumar, R. K., & Donders, S. (Eds.), *Biodiversity Finance: The Economic, Operational, and Societal Impacts of Biodiversity Loss* (pp. 269–299). Springer., DOI: 10.1007/978-3-032-02029-1_11

Andrae, S. (2026). Justice and climate change: An interdisciplinary perspective on global environmental limits. In Meçik, O. (Ed.), *Economic Growth Amidst Social and Environmental Uncertainty* (pp. 1–44). IGI Global Publishing., DOI: 10.4018/979-8-3373-1529-4.ch001

Angelsen, A., Hermansen, E. A. T., Rajão, R., & van der Hoff, R. (2018). Results-based payment: Who should be paid, and for what? In Angelsen, A., Martius, C., De Sy, V., Duchelle, A. E., Larson, A. M. & Pham, T. T. (eds.). *Transforming REDD+: Lessons and new directions*, CIFOR, 41–53. https://www.cifor-icraf.org/publications/pdf_files/Books/BAngelsen180104.pdf

Barber, C., Davey, D., Zalles, V., & Oliveira, M. (2025). The Tropical Forests Forever Facility could finally finance nature conservation. Will funders back it? WRI, November 26, 2025. https://www.wri.org/insights/financing-nature-conservation-tropical-forest-forever-facility

Corbera, E., & Schroeder, H. (2011). Governing and implementing REDD+. *Environmental Science & Policy, 14*(2), 89–99. DOI: 10.1016/j.envsci.2010.11.002

de Nevers, M., Lay, K., Wolosin, M., & Bliss-Guest, P. (2018). Creating a Multilateral Wealth Fund for a global public good: A proposal for a Tropical Forest Finance Facility. CGD Policy Paper 125. https://doi.org/DOI: 10.13140/RG.2.2.35134.09281

DeMott, D. A. (2019). Fiduciary obligation. In Criddle, E. J., Miller, P. B., & Sitkoff, R. H. (Eds.), *The Oxford Handbook of Fiduciary Law* (pp. 45–68). Oxford University Press., DOI: 10.1093/oxfordhb/9780190634100.013.2

Farrell, H., & Newman, A. L. (2019). Weaponized Interdependence: How Global Economic Networks Shape State Coercion. *International Security, 44*(1), 42–79. DOI: 10.1162/isec_a_00351

Franco, M. A., Rizzo, L. V., Teixeira, M. J., Artaxo, P., Azevedo, T., Lelieveld, J., Nobre, C. A., Pöhlker, C., Pöschl, U., Shimbo, J., Xu, X., & Machado, L. A. T. (2025). How climate change and deforestation interact in the transformation of the Amazon rainforest. *Nature Communications*, *16*(1), 7944. DOI: 10.1038/s41467-025-63156-0

Fraser, N. (2009). *Scales of Justice: Reimagining Political Space in a Globalizing World*. Columbia University Press.

GOFC-GOLD. (2013). *A sourcebook of methods and procedures for monitoring forests*. FAO.

Grassi, G., Camia, A., Fiorese, G., House, J., Jonsson, R., Kurz, W. A., Matthews, R., Pilli, R., Robert, N., & Vizzarri, M. (2018). Wrong premises mislead the conclusions by Kallio et al. on forest reference levels in the EU. *Forest Policy and Economics*, *95*, 10–12. DOI: 10.1016/j.forpol.2018.07.002

Gupta, A., Lövbrand, E., Turnhout, E., & Vijge, M. J. (2012). In pursuit of carbon accountability: The politics of REDD+ measuring, reporting and verification systems. *Current Opinion in Environmental Sustainability*, *4*(6), 726–731. DOI: 10.1016/j.cosust.2012.10.004

Hansmann, H., & Kraakman, R. (2000). Organizational law as asset partitioning. *European Economic Review*, *44*(4–6), 807–817. DOI: 10.1016/S0014-2921(00)00046-5

Herold, M., & Skutsch, M. (2011). Monitoring, reporting and verification for national REDD + programmes: Two proposals. *Environmental Research Letters*, *6*(1), 014002. DOI: 10.1088/1748-9326/6/1/014002

Keohane, R. O., & Victor, D. G. (2011). The regime complex for climate change. *Perspectives on Politics*, *9*(1), 7–23. DOI: 10.1017/S1537592710004068

Kitchin, R. (2014). Big data, new epistemologies. *Big Data & Society*, *1*(1), 2053951714528481. Advance online publication. DOI: 10.1177/2053951714528481

Langbein, J. H. (1995). The contractarian basis of the law of trusts. *The Yale Law Journal*, *105*(3), 625–675. DOI: 10.2307/797196

Larson, A. M., Brockhaus, M., Sunderlin, W. D., Duchelle, A., Babon, A., Dokken, T., Pham, T. T., Resosudarmo, I. A. P., Selaya, G., Awono, A., & Huynh, T.-B. (2013). Land tenure and REDD+: The good, the bad and the ugly. *Global Environmental Change*, *23*(3), 678–689. DOI: 10.1016/j.gloenvcha.2013.02.014

Mosley, L. (2010). *Global Capital and National Governments*. Cambridge University Press.

Moura Costa, P. (2024). Tropical Forests Mechanism: A new approach for funding forest conservation http://dx.doi.org/DOI: 10.2139/ssrn.4994337

Muradian, R., & May, P. (2026). Innovations and dilemmas in Global Forest Governance - The tortuous pathway toward a deforestation-free world: Introduction to the Special Issue. *Ecological Economics*, *241*, 108840. DOI: 10.1016/j.ecolecon.2025.108840

Neumann, M., Raichuk, A., Jiang, Y., Rey, M., Stanimirova, R., Sims, M. J., Carter, S., Goldman, E., Anderson, K., Poklukar, P., Tarrio, K., Lesiv, M., Fritz, S., Clinton, N., Stanton, C., Morris, D., & Purves, D. (2025). Natural forests of the world – a 2020 baseline for deforestation and degradation monitoring. *Scientific Data*, *12*(1), 1715. DOI: 10.1038/s41597-025-06097-z

Newell, P., & Mulvaney, D. (2013). The political economy of the 'just transition'. *The Geographical Journal*, *179*(2), 132–140. DOI: 10.1111/geoj.12008

Okereke, C., & Coventry, P. (2016). Climate justice and the international regime: Before, during, and after Paris. *Wiley Interdisciplinary Reviews: Climate Change*, *7*(6), 834–851. DOI: 10.1002/wcc.419

Ostrom, E. (2010). Polycentric systems for climate governance. *Global Environmental Change*, *20*(4), 550–557. DOI: 10.1016/j.gloenvcha.2010.07.004

Rights and Resources Initiative. (2023). Who owns the world's land? 2th. Ed. RRI. https://rightsandresources.org/wp-content/uploads/Who-Owns-the-Worlds-Land_Final-EN.pdf

Salcedo-La Viña, C., Degawan, M., Currey, K., & Davey, E. (2025). For new global forest pledges to succeed, they must center forest communities. WRI, November 7, 2025. https://www.wri.org/insights/global-forest-pledges-must-center-forest-communities

Schlosberg, D. (2007). *Defining Environmental Justice: Theories, Movements, and Nature*. Oxford University Press. DOI: 10.1093/acprof:oso/9780199286294.001.0001

Sitkoff, R. H., & Dukeminier, J. (2021). *Wills, Trusts, and Estates* (11th ed.). Wolters Kluwer.

Streck, C. (2012). Financing REDD+: Matching needs and ends. *Current Opinion in Environmental Sustainability*, *4*(6), 628–637. DOI: 10.1016/j.cosust.2012.10.001

Sutherland, W. J., Butchart, S. H. M., Clarke, S. J., Doar, N. R., Doran, H., Douglas, I. C., Field, D. J., Fleishman, E., Gaston, K. J., Herbert-Read, J. E., Hughes, A. C., Kaartokallio, H., Maggs, L., Palardy, J. E., Pearce-Higgins, J. W., Peck, L. S., Pettorelli, N., Schloss, I. R., Spalding, M. D., & Thornton, A. (2026). A horizon scan of biological conservation issues for 2026. *Trends in Ecology & Evolution, 41*(1), 91–101. DOI: 10.1016/j.tree.2025.10.016

World Bank. (2017). Results-based climate finance in practice: Delivering climate finance for low-carbon development. World Bank. https://documents1.worldbank.org/curated/en/410371494873772578/pdf/115053-WP-PUBLIC-111p-RBCFinPracticeFinalMay.pdf

Young, O. R. (2017). *Governing Complex Systems: Social Capital for the Anthropocene Get access Arrow*. MIT Press. DOI: 10.7551/mitpress/9780262035934.001.0001

About the Contributors

Enas Mohammed AlQodsi is an Associate Professor of Private Law at the College of Law, United Arab Emirates University, and Coordinator of the Master's Programs in Public and Private Law. She holds a Ph.D. in Civil Law and brings over two decades of distinguished academic, judicial, and legal experience. Dr. AlQodsi has an outstanding international publication record, with numerous peer-reviewed articles published in leading journals indexed in Scopus and Web of Science, including journals ranked among the Top 1%, Top 5%, and Top 10% worldwide. Her scholarly contributions span civil law, contract law, civil liability, intellectual property, data protection, and the legal implications of emerging technologies and sustainability. She has served as editor and contributor to internationally indexed books, and as a reviewer and editorial board member for multiple high-impact journals. Her academic excellence is further demonstrated by securing competitive internal and external research grants and supervising graduate research at both Master's and Ph.D. levels. Dr. AlQodsi is a recipient of multiple excellence awards from both the United Arab Emirates University and Abu Dhabi University, recognizing her outstanding achievements in teaching, research, and community service. Her work continues to contribute significantly to advancing legal scholarship and shaping contemporary legal discourse at both regional and international levels.

Iyad Jadalhaq holds a doctorate degree in Civil Law from University of Athens in Greece. His teaching experience in many universities is over 20 years. He have many publications in international journals indexed in SCOPUS, so he has been a recipient of many awards during his previous academic work, such as, the University Research Award, University Research Fellow Award, College of Law Teaching Award, and twice received the College of Law Research Award. In addition, he has 8 authorized books published in Palestine and UAE.

Mohammed El Hadi El Maknouzi is an Associate Professor of Commercial Law at the University of Sharjah, College of Law. Dr. El Maknouzi holds a PhD in

Commercial and Business Law and a _Diplôme d'Études Supérieures Approfondies_ (DESA) from Mohammed V University in Rabat. He was also awarded an LL.M. in Enterprise Law jointly from Toulouse University in France and the _Institut Supérieur de Commerce et d'Administration des Entreprises _(ISCAE) in Morocco. Furthermore, he has published numerous articles in international journals indexed in Scopus.

***

Lawal Abdulwahab Olamilekan is a dedicated academic with a PhD in Economics from the Universiti Brunei Darussalam, School of Business and Economics (AACSB-accredited). He holds a master's degree in health economics and a bachelor's in economics education from Taraba State University, Nigeria. His research interests span natural resource economics, health economics, development, and macroeconomics, particularly in Africa and emerging economies. With over six years of academic experience, he has taught and supervised students across Nigeria, Somalia, Somaliland, and Brunei. His teaching expertise covers economics, management, and public health, while his research contributions include conference papers, journal articles, and book chapters. Lawal's proficiency in quantitative and qualitative methodologies, coupled with his advanced skills in econometric modelling and using software such as Stata, SPSS, and EViews, instills confidence in his technical abilities. He is a member of the World Economics Association and the Institute of Chartered Economists of Nigeria. His active engagement in collaborative research that bridges health policy, governance, and economic development to provide evidence-based insights for sustainable growth is truly inspiring.

Riad Al Ajlani is an Academician and Associate Professor of International Law in United Arab Emirates University, earned his PhD from Glasgow University, served as a Secretary-General of Higher Education Council, Ministry of Higher Education 2003-6, Secretary-General of Damascus University, 2001-2003, & Vice-Dean for Administrative Affairs, Faculty of Law, Damascus University, 1998-2001. Scopus ID: 58664848500 ORCID ID: 0000-0002-2587-6733

Ankit Anand is a Ph.D Research Scholar at the School of Law, SRM Institute of Science and Technology (KTR). He holds a B.A. LL.B. (Hons.) degree from Chanakya National Law University, Patna and an LL.M. in International Commercial Laws from Pondicherry University. His research interests include alternative dispute resolution, forensic law, public health law, maritime security, forensic science and human rights. He has published extensively in Scopus-indexed, Web of Science, and UGC-CARE journals, as well as in edited book chapters. He has also actively contributed to academic administration, moot court activities, mediation training, and interdisciplinary legal research integrating law, forensic science, and conflict resolution. His Scopus Author ID is 59422199900, and his Web of Science Researcher ID is ONJ-4100-2025.

Deepak Kumar Chauhan is Professor of Law at the School of Legal Studies, Central University of Punjab, Bathinda. With over 17 years of teaching and 19 years of research experience, he specializes in Environmental Law, Constitutional Law, and Drug Abuse Policy. A recipient of the 'Young Scientist Award' by the Ministry of Petroleum & Natural Gas, he has led multiple government-funded research projects and guided numerous Ph.D. and LL.M. scholars. As a founder member of the Law Department, he has contributed significantly to curriculum development, legal research, institutional building, and public engagement on pressing socio-legal issues.

Driss Ed. Daran is an Academician and Assistant Professor of International Law, with over eleven years of teaching experience across diverse universities. He earned his PhD from the College of Law, Wuhan University, and has served as a Legal Counsellor on legal and judicial matters with a local government in Morocco. Dr. Ed.daran was awarded a research grant from the United Arab Emirates University (UAEU) for his project on the legal and ethical justification of human organ donation. His research focuses on International Law and Sustainable Development, with expertise in comparative legal systems and the ethical dimensions of global governance. Scopus ID: 55863701400 Orcid ID: 0000-0001-9954-7249

Oindrilla Ghosh is an Assistant Professor of Law hailing from the School of Law at Christ (Deemed to be University) Bangalore, India, focuses her research on health of tribal communities. Her work includes total 42 publications on Neuroscience on Organizational Behavior at IGI Global and also Environmental Disasters that are indexed in SCOPUS and Web of Sciences conference proceedings. She completed her early education at Loreto Day School, Bowbazar and went on to achieve distinction in her undergraduate (B.A/LL.B) and postgraduate (LL.M) studies at Calcutta University. Having served as an Assistant Professor of Law at Kingston Law College,Barasat, Kolkata, Ghosh has actively participated in numerous National Conferences, International Conferences, UN Global Summits, webinars, and seminars organized by prestigious academic institutions. Her expertise lies in health perspectives, environmental studies, human behavior, and related areas. Additionally, she has conducted various workshops, lectures in educational institutions, and is engaged with NGO’s to help the needy.

Abby Balkis Gorayah is a Chartered Quantity Surveyor (MRICS) and Senior Lecturer and Program Manager at National University of Science and Technology, Oman with over 30 years of industry and academic experience. Specialising in contract law, cost management, and project management, he has led major international projects, supervised research, and contributed to accreditation, consultancy, and professional development in construction education.

Jaishri Gothania is an academician, author, researcher, and data scientist with over 9 years of experience. She has published 01 book, 17 book chapters, 03 paper presentations, 09 paper publications, and 3 patents. Her work bridges data science and innovation, contributing significantly to academia, research, and technological advancement.

Gurminder Kaur is an Assistant Professor of Law at the School of Law, Pondicherry University, India. She holds a PhD in Law from Panjab University, Chandigarh, and specialises in Criminal Law, Human Rights, Feminist Jurisprudence, Forensic and Scientific Evidence, and Technology Law. Her research focuses on criminal justice, evidentiary innovation, cyber harms, and interdisciplinary legal studies. She has published extensively in national and international journals, including those indexed in Scopus and Web of Science, and has presented papers at various academic forums in India and abroad. She also supervises research scholars and actively contributes to interdisciplinary and comparative legal research.

Ashok Kumar is a Professor, Dean School of Law at Central university of South Bihar, Gaya, and has got teaching experience of more than a decades in U.G.,P.G. and Research Level. He is Chairman, FST, CUSB, Coordinator, Admission at School of Law and Governance, CUSB. Prior to joining Central University of South Bihar he has taught at P.G.Department of Law, School for Legal Studies, M.J.P.Rohilkhand University (A State Government University), Bareilly. Prof. Kumar also Nominated for Educator of the Year-2012 Award, International Biographical Centre, Cambridge, UK. Professor Kumar has professional experience as an author, researcher, teacher, policy maker and administrator and has made significant contributions to the growth and expansion of multiple horizons of academic in Universities. Professor Kumar has been instrumental in institutionalizing legal research publication in Universities. He has successfully guided 105 Ph.Ds./M.Phil./P.G. research scholars. Prof. Kumar has authored fifteen books and more than fifty research /Policy papers. He is the Editor of two prestigious international journals: Journal of Law and Public Policy and CUSB Journal of Law and Governance. He is also worked as Chairman/ Convener, Board of Studies, School for Legal Studies at of M.J.P.Rohilkhand University, Bareilly. Prof. Kumar is also appointed as Judicial Magistrate First Class by Government of Uttar Pradesh. Prof. Kumar is Member of Executive Council and Academic Council of M.J.P.Rohilkhand University, Bareilly. He is also holding the position of Legal Advisor /Nodal Officer at M.J.P.Rohilkhand University, Bareilly advice the University on various issues related to Law, Governance and University Administration. Prof. Kumar has vast experience of being a resource person and has conducted training programmes/ workshops on Gender Sensitization and 'Sexual Harassment of Women at Workplace' ands, PCS Officers, Doordarshan & All India Radio (AIR) etc. He has participated and presented papers more than 100 National and International Conference, Seminars and workshop. He has various International academic collaboration with many countries like USA South Korea,South America, Nepal, Sri Lanka,Bhutan, Ethiopia etc. His areas of interests include Constitutional Law, Jurisprudence, Human Rights, Media Law, RTI, Child Rights, Women Rights and Copyright Law. He is a member of the Various Advisory and Editorial Board in National and International Journals and an external expert in the various State and Central Universities. He is a member of Executive Committee of SRNCT, Lucknow. He has been a member of various committees/selection committees of Universities, Public Service Commissions,NAAC. As a distinguished academician and policy maker, Prof. Kumar has played a key role as a member-designate for several policies making at M.J.P.Rohilkhand University. He is also serves as Nodal Officer and Master Trainer, IPR-NIPAM, Ministry of Commerce and Industry, Government of India. Prof. Kumar is the recipient of several prestigious awards; Best Law Teacher for his outstanding contribution to Legal Education, Legal Research and Social Justice. 'Excellence in Higher Education" Govt. of UP, 2021.

Binod Kumar is a graduate student known for his academic excellence, with his total 31 scholarly works indexed in Scopus database on Neuroscience in Organizational Behavior, Neuromarketing and AI Applications on Wildlife Conservation. Hailing from Ranchi, Jharkhand in India, Binod is currently pursuing a Master of Laws (LL.M) at the prestigious Radha Govind University. His passion for technology and innovation along with climate changes is evident through his high academic performance and numerous projects. Binod has a keen interest in artificial intelligence, machine learning, neuropsychology, climate changes and disaster management, fields in which he has undertaken several research projects. He is particularly fascinated by the potential of AI to solve real-world problems and is dedicated to learning and applying new techniques in these areas. He volunteers at local NGOs, teaching underprivileged children basic computer skills. With a strong foundation in law and a commitment to making a positive impact, Binod Kumar is poised to become a future leader in the legal industry.

Rajneesh Kumar is a Ph.D Research Scholar at School of Law, Pondicherry University, India. He holds a B.B.A. LL.B. (Hons.) degree from Chanakya National Law University, Patna and an LL.M. in Criminal Law from the Central University of South Bihar. His research interests lie in Criminal Law, Forensic and Scientific Evidence, Human Rights, and the Criminal Justice System. His doctoral research focuses on forensic evidence and fair trial rights in Indian judicial proceedings. He has published extensively in national and international journals, including those indexed in Scopus and Web of Science, and has presented papers at various academic conferences and workshops. His Scopus Author ID is 59736256500, and his Web of Science Researcher ID is OZF-8381-2025.

Hrishikesh Manu is an Assistant Professor of Law at Chanakya National Law University, Patna. He holds an LL.B. degree from Banaras Hindu University and an LL.M. from the University of Delhi. He has completed his Ph.D. from Chanakya National Law University in Climate Change Law. He teaches Environmental Law, Alternative Dispute Resolution and Criminal Justice System at both undergraduate and postgraduate levels. He has presented papers at national and international conferences and has national and international publications to his name.

Froilan D. Mobo, DPA, Ph.D. is a distinguished Filipino educator, researcher, and academic leader known for his outstanding contributions to education, public administration, and development studies. He currently serves as a Full Professor II at the Philippine Merchant Marine Academy (PMMA), where he has devoted over a decade of service in shaping future maritime professionals. In addition to his academic role, he is the Editor-in-Chief of the International Journal of Multidisciplinary: Applied Business and Education Research (IJMABER), a reputable international refereed journal. Prof. Mobo is a strong advocate of lifelong learning, having earned dual doctorate degrees—Doctor of Public Administration from Urdaneta City University and Doctor of Philosophy in Development Education from Central Luzon State University—as well as triple master's degrees, including a Master of Science in Computer Science, Master of Business Administration, and Master of Arts in Social Studies Education from Bicol University Open University. He also holds a Diploma in Social Studies Education from UP Open University and is a Licensed Professional Teacher with a specialization in Social Studies.

Pallavi is research scholar at Central University of South Bihar, Gaya. She has completed her LLM with specialization in cyber law from University of Petroleum and Energy, Dehradun. She has qualified for the prestigious UGC-NET. She has completed BA.LLB (Criminal Law hons.) from School of Law, UPES, Dehradun.

M. Shamima Parveen is an Assistant Professor at Saveetha School of Law, Saveetha Institute of Medical and Technical Sciences, with academic and professional expertise in constitutional law, human rights, and international law. She holds a PhD in Law from Bhagwant University, Ajmer, Rajasthan. She has teaching and litigation experience across reputed institutions and courts, including practice before the Madras High Court. Her research interests include human rights, forensic science, euthanasia law, gender justice, and socio-legal studies. She has published extensively in national and international journals, including those indexed in Scopus and Web of Science, and has presented papers at various academic forums. She has actively contributed to moot court activities, legal education, mediation initiatives, and interdisciplinary legal research through seminars, workshops, and academic mentoring.

Naresh Prajapati is an Assistant Professor of Law at the School of Law, Bennett University, Greater Noida, India. A recipient of the Chancellor's Award for Best Ph.D. Scholar, his research focuses on international climate change law, human rights, gender justice, and comparative environmental governance. He completed his Ph.D. in Law from the Central University of Punjab, where his doctoral research examines judicial approaches to human rights protection amid climate change in South Asia. Dr. Prajapati has published extensively in Scopus-indexed journals and edited volumes on climate justice, environmental law, and gender studies, including contributions and book reviews in leading journals such as International & Comparative Law Quarterly (Cambridge University Press, Q1) and Global Knowledge, Memory and Communication (Q1). His research has been presented internationally, and he has undertaken academic visits and collaborations in Kazakhstan, Türkiye, Nepal, Bangladesh, Sri Lanka, and the Maldives. He also serves as reviewer, editor, and session chair for several international journals and conferences, contributing to interdisciplinary scholarship on law, sustainability, and global environmental governance.

Sitesh Kumar Singh is an accomplished academic and Chartered Engineer with a Ph.D. in Civil Engineering and about ten years of diverse experience in education, research, consultancy, and educational administration. Currently serving as an Assistant Professor and Program Manager in the Department of Civil and Environmental Engineering at the National University of Science and Technology, Muscat, Sultanate of Oman, he specializes in Civil and Transportation Engineering, Project Management, and Data Analytics. Dr. Singh holds multiple advanced degrees including M.Tech, B.Tech, PGDIPR, PGDUPDL. His career encompasses curriculum development, research supervision, quality assurance, and strategic academic planning. He has supervised over 40 postgraduate and 2 doctoral students, contributing to academic leadership through workshops, conferences, and editorial roles, and has published over 5 books, 27 book chapters, 16 paper presentations and publications in conference proceedings, 43 paper publications in journals, and 7 published patents.

Malik Zia-ud-Din is a legal academic and researcher, currently serving as an Assistant Professor in the Department of Law at The Islamia University of Bahawalpur, where he also holds additional responsibilities as Head of the Department of Criminology. A PhD in Economic Law from Wuhan University, China, Dr. Zia-ud-Din brings a diverse academic background with qualifications in law, political science, rural development, and business administration. His prolific research portfolio spans over 30 international publications, many in SCOPUS and SSCI-indexed journals, with a focus on environmental law, human rights, marine biodiversity, and legal education. As an HEC-approved PhD supervisor, he actively contributes to the development of legal scholarship in Pakistan and beyond. Scopus ID: 57203068017 Orcid ID: 0000-0001-6015-0914

Index

A

C

D

E

H

I

J

L

M

N

P

R

S

T

U

V

www.ingramcontent.com/pod-product-compliance
Lightning Source LLC
LaVergne TN
LVHW061527290926
849162LV00004B/699
9798337366357